T0211448

STEAM PLANT
CALCULATIONS
MANUAL

MECHANICAL ENGINEERING

A Series of Textbooks and Reference Books

Editor

L. L. Faulkner

*Columbus Division, Battelle Memorial Institute
and Department of Mechanical Engineering
The Ohio State University
Columbus, Ohio*

Additional Volumes in Preparation

Mechanical Engineering Software

Spring Design with an IBM PC, Al Dietrich

Mechanical Design Failure Analysis: With Failure Analysis System Software for the IBM PC, David G. Ullman

STEAM PLANT CALCULATIONS MANUAL

SECOND EDITION, REVISED AND EXPANDED

V. GANAPATHY

ABCO Industries
Abilene, Texas

CRC Press
Taylor & Francis Group
Boca Raton London New York

CRC Press is an imprint of the
Taylor & Francis Group, an Informa business

CRC Press
Taylor & Francis Group
6000 Broken Sound Parkway NW, Suite 300
Boca Raton, FL 33487-2742

First issued in paperback 2019

© 1994 by Taylor Francis Group, LLC
CRC Press is an imprint of Taylor & Francis Group, an Informa business

No claim to original U.S. Government works

ISBN-13: 978-0-8247-9147-6 (hbk)
ISBN-13: 978-0-367-40229-7 (pbk)

This book contains information obtained from authentic and highly regarded sources. Reasonable efforts have been made to publish reliable data and information, but the author and publisher cannot assume responsibility for the validity of all materials or the consequences of their use. The authors and publishers have attempted to trace the copyright holders of all material reproduced in this publication and apologize to copyright holders if permission to publish in this form has not been obtained. If any copyright material has not been acknowledged please write and let us know so we may rectify in any future reprint.

Except as permitted under U.S. Copyright Law, no part of this book may be reprinted, reproduced, transmitted, or utilized in any form by any electronic, mechanical, or other means, now known or hereafter invented, including photocopying, microfilming, and recording, or in any information storage or retrieval system, without written permission from the publishers.

For permission to photocopy or use material electronically from this work, please access www.copyright. com (http://www.copyright.com/) or contact the Copyright Clearance Center, Inc. (CCC), 222 Rosewood Drive, Danvers, MA 01923, 978-750-8400. CCC is a not-for-profit organization that provides licenses and registration for a variety of users. For organizations that have been granted a photocopy license by the CCC, a separate system of payment has been arranged.

Trademark Notice: Product or corporate names may be trademarks or registered trademarks, and are used only for identification and explanation without intent to infringe.

Library of Congress Cataloging-in-Publication Data

Ganapathy, V.
    Steam plant calculations manual / V. Ganapathy. – 2nd ed., rev.
and expanded.
        p.   cm. – (Mechanical engineering ; 87)
    Includes bibliographical references and index.
    ISBN 0-8247-9147-9
    1. Steam power-plants–Handbooks, manuals, etc.   I. Title.
II. Series: Mechanical engineering (Marcel Dekker, Inc.) ; 87
TJ395.G35 1993
621.1'8–dc20   93-9048
                                                                      CIP

Visit the Taylor & Francis Web site at
http://www.taylorandfrancis.com

and the CRC Press Web site at
http://www.crcpress.com

To my family–G. Shantha, G. Padma, G. Sivapriya

Preface to the Second Edition

The second edition of *Steam Plant Calculations Manual* is completely revised, with 70 additional problems covering emissions, boiler efficiency, heat transfer equipment design and performance, circulation, and various other aspects of steam plants.

The first chapter contains a few new problems related to estimating deaeration steam quantity based on system water chemistry per ASME and ABMA boiler water guidelines. Examples illustrate the computation of steam purity and quality and their interrelationship. Conversion between boiler horse power and steam production is also explained.

Because of regulations concerning NO_x and CO, plant engineers frequently have to compute emission of these pollutants or relate them from mass units such as lb/h to ppm or vice versa. In Chapter 2 this conversion is explained for both gas turbine exhaust and conventional fired boilers. Cogeneration and combined cycle plants use heat recovery steam generators (HRSGs), which are often fired with auxiliary fuel using the oxygen in the exhaust gases. The relationship between oxygen consumption and amount of fuel fired is derived and explained with an example. Correlations for dew points of various acid gases, such as hydrochloric, sulfuric, and hydrobromic acid, are cited and the effect of gas temperature on tube metal temperature is explained to illustrate the possibility of corrosion. The effect of excess air on efficiency of different fuels is explained and simple equations are developed for computing boiler efficiency.

Boiler circulation is explained in Chapter 3 and examples illustrate its computation for both fire tube and water tube boilers. Importance of steam quality and factors affecting departure from nucleate boiling (DNB) conditions are explained. Determination of steam flow in blow-off lines and flow in blow-down lines are illustrated with examples.

Over 45 additional problems, covering various aspects of heat transfer equipment design, are included in Chapter 4. These include:

Effect of fouling on tube wall temperature and duty in fire tube and water tube boilers

Computation of natural convection heat transfer

Design and off-design performance and simplified design procedures for fire tube and water tube boilers and air heaters

Simulation of HRSG design and off-design performance and understanding pinch and approach points

Prediction of furnace heat transfer in both fired and unfired (waste heat) boilers and distribution of radiation to tube banks

Correlations for critical heat flux in both fire tubes and water tubes and several more on equipment design

Several examples related to HRSGs explain the method of evaluating alternative designs considering initial and operation costs. The effect of gas analysis and pressure on heat transfer is also explained. A few examples illustrate the effect of wind velocity, casing emissivity on heat losses, and casing temperature. The principle behind the use of hot casing is explained. Consideration of variables affecting HRSG performance testing is explained with an example.

A few problems have been added in Chapter 5 to explain the importance of ambient temperature on boiler fan sizing and the effect of feedwater temperature on pump performance. Since gas turbines are becoming the workhorses of the future, examples show the effect of variables on the efficiency of a simple Bryaton cycle, with and without regeneration.

Steam plant engineers who found the first edition useful will find the new edition packed solid with information and with several completely solved problems of practical significance.

V. Ganapathy

Preface to the First Edition

Engineers connected with the performance, operation, and maintenance of steam power and process plants often have to perform simple and sometimes involved calculations related to boilers, pumps, fans, fuels, combustion, fluid flow, valve selection, heat transfer, and energy utilization. These are not the routine, lengthy design calculations a design office would perform, but rather simple calculations done to realize the following objectives:

To understand the performance characteristics of the equipment
To check if the equipment is performing within predicted range of
   parameters
To evaluate cost effectiveness of proposals
To utilize energy in an optimal manner
To specify equipment for different service conditions

Plant engineers of today have to be more energy and cost conscious than they were a few years ago, when knowledge of equipment and its performance alone was adequate. With the increasing cost of energy, all aspects of energy utilization and economics of operation must be considered by plant engineers in their day-to-day work. Examples have been dispersed throughout this text to illustrate the above-mentioned subjects.

The book is divided into five chapters and is written in a question-and-answer style. This approach, it is felt, will be appealing to plant engineers, who have little time to go into theory.

Chapter 1 deals with the general category of calculations such as conversion of mass-to-volume flow rates, energy utilization from boiler blow-down and exhaust gases, ASME code calculations to figure pipe sizes for external and internal pressure applications, life cycle costing methods, and estimation of noise levels. A few examples illustrate how gas leakage across dampers, its cost, and leakage rates of steam through openings can be found. Importance of moisture in air and water dew point is also explained. Application of life cycle costing to equipment selection is explained. Purchase of equipment based on initial cost may not be generally a good proposition.

Chapter 2 deals with fuels, combustion, and boiler or heater efficiency. Often, fuel analysis will not be available and plant engineers may be required to estimate the combustion air requirements, the excess air, or the boiler efficiency. A few examples illustrate how these can be done. The dollar savings that can be realized by reducing the oxygen levels in the flue gas can also be estimated. Engineers are often confused between the efficiencies based on higher and lower heating value of the fuel. Often furnaces are designed to burn a particular fuel and may be required to burn a different one later. The factors that are to be considered in burner are also discussed.

Chapter 3 deals with fluid flow, sizing of flow meters, and selection of control and safety valves. Importance of permanent pressure drop in flow meters and its cost is discussed, followed by examples on selection of safety and control valves. Relieving capacity of a given safety valve when used on a different fluid and different pressure conditions is also discussed. Correction of orifice meter readings for different steam parameters is discussed, followed by pressure drop calculations for fluids inside pipes, and flow over plain and finned tube bundles. Knowledge of pressure drop also helps the plant engineer to check whether or not fouling has occurred. A large increase in gas pressure drop across an economizer tube bundle, for example, means that fouling could have taken place and cleaning cycles may have to be initiated.

In Chapter 4 on heat transfer, several problems covering estimation of heat transfer coefficients for flow over plain and finned tubes, flow inside tubes, nonluminous radiation and prediction of the performance of heat transfer devices are illustrated. A simple approach has been used, and lengthy, routine methods are avoided. Estimation of performance of a given thickness of insulation and determining the optimum thickness of insulation using life cycle costing are also discussed.

The last chapter deals with pumps, fans, and turbines. Often plant engineers switch fans from one site to another without considering aspects such as the ambient temperature or elevation on the fan performance. Based on motor current readings, one can double-check whether a fan or a pump is working well. Examples have been provided to illustrate these points. Net positive suction head (NPSH) and power requirements for centrifugal and reciprocating pumps are also covered. Cogeneration is an important topic and several plants are using steam turbines for power generation and process steam applications. A few examples illustrate how the steam requirements and power generation can be evaluated.

In all, over 125 examples covering practical aspects of equipment utilization and energy management are worked out, which should make this book a good companion for plant engineers, operators, managers, and design engineers. It would also be of interest to engineers perparing for professional license examinations.

During the past several years, I have had the privilege of authoring several articles in various magazines such as *Plant Engineering, Power Engineering, Chemical Engineering, Hydrocarbon Processing*, and *Oil and Gas Journal*. Based on the interaction I have had with several readers, I felt that a book of this nature would be helpful to a large cross section of steam plant engineers and managers.

This being the first edition, there are likely to be a few errors and topics that have been missed. I shall be glad if the readers could bring these to my attention.

Finally I would like to thank the various journals and organizations that gave me permission to reproduce material from their publications.

V. Ganapathy

Contents

Basic Steam Plant Calculations

1.01: Converting liquid flow in lb/hr to gpm, and vice versa; relating density, specific gravity, and specific volume

1.02: Relating head of liquid or gas column to pressure; converting feet of liquid to psi; relating inches of water column of gas to psi and feet of gas column

1.03: Estimating density of gases; relating molecular weight and density; effect of elevation on gas density; simplified formula for density of air and flue gases at sea level

1.04: Relating actual and standard cubic feet of gas per minute to lb/hr

1.05: Computing density of gas mixture; relating mass to volumetric flow; computing velocity of gas in duct or pipe

1.06: Relating mass and linear velocities

1.07:   Calculating velocity of wet and superheated steam in pipes; computing specific volume of wet steam; use of steam tables

1.08:   Relating boiler horsepower to steam output

1.09:   Calculating amount of moisture in air; relative humidity and saturation vapor pressure

1.10:   Water dew point of air and flue gases; partial pressure of water vapor

1.11:   Energy absorbed by wet and superheated steam in boilers; enthalpy of wet and dry steam; use of steam tables; converting MM Btu/hr (million Btu/hr) to kilowatts

1.12:   Relating steam by volume, steam by weight, and steam quality; relating circulation ratio and quality

1.13a: Determining steam quality using throttling calorimeter

1.13b: Relating steam quality to steam purity

1.14:   Water required for desuperheating steam; energy balance in attemperators, desuperheaters

1.15:   Water required for cooling gas streams

1.16:   Calculating steam volume after throttling process; use of steam tables

1.17:   Determining blowdown and steam for deaeration

1.18:   Calculating flash steam from boiler blowdown; economics of flash steam recovery

1.19a: Estimating leakage of steam through openings; effect of wetness of steam on leakage

1.01

Q: Convert 50,000 lb/hr of hot water at a pressure of 1000 psia and 390°F to gpm.

A: To convert from lb/hr to gpm, or vice versa, for any liquid, we
    can use the following expressions:

$$W = 8 \frac{q}{v} \tag{1}$$

$$\rho = 62.4s = \frac{1}{v} \tag{2}$$

where

> W = flow, lb/hr
> q = flow, gpm (gallons per minute)
> ρ = density of liquid, lb/cu ft
> s = specific gravity of liquid
> v = specific volume of liquid, cu ft/lb

For hot water we can obtain the specific volume from the steam
tables (see the Appendix). v at 1000 psia and 390°F is 0.0185 cu
ft/lb. Then, from Eq. (1),

$$q = 50{,}000 \times \frac{0.0185}{8} = 115.6 \text{ gpm}$$

For water at temperatures of 40 to 100°F, for quick estimates we
divide lb/hr by 500 to obtain gpm. For example, 50,000 lb/hr of
water at 70°F would be 100 gpm.

1.02a

Q: Estimate the head in feet developed by a pump when it is
    pumping oil with a specific gravity of 0.8 through a differential
    pressure of 150 psi.

A: Conversion from feet of liquid to psi, or vice versa, is needed in
    pump calculations. The expression relating the variables is

$$H_l = 144 \, \Delta P \, v = 2.3 \frac{\Delta P}{s} \tag{3}$$

where

> ΔP = differential pressure, psi
> H_l = head, feet of liquid

Substituting for ΔP and s, we have

$$H_l = 2.3 \times \frac{150}{0.8} = 431.2 \text{ ft}$$

1.02b

Q: If a fan develops 8 in. WC (inches of water column) with a flue gas density of 0.05 lb/cu ft, what is the head in feet of gas and in psi?

A: Use the expressions

$$H_g = 144 \frac{\Delta P}{\rho_g} \tag{4}$$

$$H_w = 27.7 \, \Delta P \tag{5}$$

where

H_g = head, feet of gas
H_w = head, in. WC
ρ_g = gas density, lb/cu ft

Combining Eqs. (4) and (5), we have

$$H_g = 144 \times \frac{8}{27.7 \times 0.05} = 835 \text{ ft}$$

$$\Delta P = \frac{8}{27.7} = 0.29 \text{ psi}$$

1.03

Q: Estimate the density of air at 5000 ft elevation and 200°F.

A: The density of any gas can be estimated from

$$\rho_g = 492 \times MW \times \frac{P}{359 \times (460 + t) \times 14.7} \tag{6}$$

where

P = gas pressure, psia
MW = gas molecular weight (Table 1.1)

Table 1.1 Gas Molecular Weights

Gas	MW
Hydrogen	2.016
Oxygen	32.0
Nitrogen	28.016
Air	29.2
Methane	16.04
Ethane	30.07
Propane	44.09
n-Butane	58.12
Ammonia	17.03
Carbon dioxide	44.01
Carbon monoxide	28.01
Nitrous oxide	44.02
Nitric oxide	30.01
Nitrogen dioxide	46.01
Sulfur dioxide	64.06
Sulfur trioxide	80.06
Water	18.02

t = gas temperature, °F

ρ_g = gas density, lb/cu ft

The pressure of air decreases as the elevation increases, as shown in Table 1.2, which gives the term $(P/14.7) \times$ MW of air = 29. Substituting the various terms, we have

$$\rho_g = 29 \times 492 \times \frac{0.832}{359 \times 660} = 0.05 \text{ lb/cu ft}$$

A simplified expression for air at atmospheric pressure and at temperature t at sea level is

$$\rho_g = \frac{40}{460 + t} \tag{7}$$

For a gas mixture such as flue gas, the molecular weight (MW) can be obtained as discussed in Q1.05. In the absence of

Table 1.2 Density Correction for Altitude

Altitude (ft)	Factor
0	1.0
1000	0.964
2000	0.930
3000	0.896
4000	0.864
5000	0.832
6000	0.801
7000	0.772
8000	0.743

data on flue gas analysis, Eq. (7) also gives a good estimate of density.

When sizing fans, it is the usual practice to refer to 70°F and sea level as standard conditions for air or flue gas density calculations.

1.04a

Q: What is acfm (actual cubic feet per minute), and how does it differ from scfm (standard cubic feet per minute)?

A: acfm is computed using the density of the gas at given conditions of pressure and temperature, and scfm is computed using the gas density at 70°F and at sea level (standard conditions).

$$q = \frac{W}{60\,\rho_g} \tag{8}$$

where

q = gas flow in acfm (at 70°F and sea level, scfm and acfm are equal; then $q = W/4.5$)

ρ_g = gas density in lb/cu ft (at standard conditions ρ_g = 0.075 lb/cu ft)

W = gas flow in lb/hr = $4.5q$ at standard conditions

1.04b

Q: Convert 10,000 lb/hr of air to scfm.

A: Using Eq. (6), it can be shown that at $P = 14.7$ and $t = 70$, for air $\rho_g = 0.75$ lb/cu ft. Hence, from Eq. (8),

$$q = \frac{10,000}{60 \times 0.075} = 2222 \text{ scfm}$$

1.04c

Q: Convert 3000 scfm to acfm at 35 psia and 275°F. What is the flow in lb/hr? The fluid is air.

A: Calculate the density at the actual conditions.

$$\rho_g = 29 \times 492 \times \frac{35}{359 \times 735 \times 14.7} = 0.129 \text{ lb/cu ft}$$

From the above,

$$W = 4.5 \times 3000 = 13,500 \text{ lb/hr}$$

Hence

$$\text{acfm} = \frac{13,500}{60 \times 0.129} = 1744 \text{ cfm}$$

1.05

Q: In a process plant, 35,000 lb/hr of flue gas having a composition $N_2 = 75\%$, $O_2 = 2\%$, $CO_2 = 15\%$, and $H_2O = 8\%$, all by volume, flows through a duct of cross section 3 ft^2 at a temperature of 350°F. Estimate the gas density and velocity. Since the gas pressure is only a few inches of water column, for quick estimates the gas pressure may be taken as atmospheric.

A: To compute the density of gas, we need the molecular weight. For a gas mixture, molecular weight is calculated as follows:

$$MW = \Sigma(MW_i \times y_i)$$

where

y_i = volume fraction of gas i
MW_i = molecular weight of gas i

Hence

$$MW = 0.75 \times 28 + 0.02 \times 32 + 0.15 \times 44 + 0.08 \times 18 = 29.68$$

From Eq. (6),

$$\rho_g = 29.68 \times \frac{492}{359 \times 810} = 0.05 \text{ lb/cu ft}$$

The gas velocity V_g can be obtained as follows:

$$V_g = \frac{W}{60 \, \rho_g A} \tag{9}$$

where

V_g = velocity, fpm
A = cross section, ft^2

Hence

$$V_g = \frac{35,000}{60 \times 0.05 \times 3} = 3888 \text{ fpm}$$

The normal range of air or flue gas velocities in ducts is 2000 to 4000 fpm. Equation (9) can also be used in estimating the duct size.

In the absence of flue gas analysis, we could have used Eq. (7) to estimate the gas density.

1.06

Q: A term that is frequently used by engineers to describe the gas flow rate across heating surfaces is *gas mass velocity*. How do we convert this to linear velocity? Convert 5000 lb/ft^2 hr of hot air flow at 130°F and atmospheric pressure to fpm.

A: Use the expression

$$V_g = \frac{G}{60\,\rho_g} \tag{10}$$

where G is the gas mass velocity in lb/ft^2 hr. Use Eq. (7) to calculate ρ_g.

$$\rho_g = \frac{40}{460 + 130} = 0.0678 \text{ lb/cu ft}$$

Hence

$$V_g = \frac{5000}{60 \times 0.0678} = 1230 \text{ fpm}$$

1.07a

Q: What is the velocity when 25,000 lb/hr of superheated steam at 800 psia and 900°F flows through a pipe of inner diameter 2.9 in.?

A: Use expression (11) to determine the velocity of any fluid inside tubes, pipes, or cylindrical ducts.

$$V = 0.05 \times W \times \frac{v}{d_i^2} \tag{11}$$

where

   V = velocity, fps
   v = specific volume of the fluid, cu ft/lb
   d_i = inner diameter of pipe, in.

For steam, v can be obtained from the steam tables in the Appendix.

   $v = 0.9633$ cu ft/lb

Hence

$$V = 0.05 \times 25,000 \times \frac{0.9633}{2.9^2} = 143 \text{ fps}$$

The normal ranges of fluid velocities are

Water: 3 to 12 fps
Steam: 100 to 200 fps

1.07b

Q: Estimate the velocity of 70% quality steam in a 3-in. schedule 80 pipe when the flow is 45,000 lb/hr and steam pressure is 1000 psia.

A: We need to estimate the specific volume of wet steam.

$$v = xv_g + (1 - x)v_f$$

where v_g and v_f are specific volumes of saturated vapor and liquid at the pressure in question, obtained from the steam tables, and x is the steam quality (see Q1.12 for a discussion of x). From the steam tables, at 1000 psia, $v_g = 0.4456$ and $v_f = 0.0216$ cu ft/lb. Hence the specific volume of wet steam is

$$v = 0.7 \times 0.4456 + 0.3 \times 0.0216 = 0.318 \text{ cu ft/lb}$$

The pipe inner diameter d_i from Table 1.3 is 2.9 in. Hence, from Eq. (11),

$$V = 0.05 \times 45,000 \times \frac{0.318}{2.9^2} = 85 \text{ fps}$$

1.08

Q: What is meant by *boiler horsepower*? How is it related to steam generation at different steam parameters?

A: Packaged fire tube boilers are traditionally rated and purchased in terms of boiler horsepower (BHP). BHP refers to a steam capacity of 34.5 lb/hr of steam at atmospheric pressure with feedwater at 212°F. However, a boiler plant operates at different pressures and with different feedwater temperatures. Hence conversion between BHP and steam generation becomes necessary.

$$W = \frac{33,475 \times \text{BHP}}{\Delta h} \tag{12}$$

Table 1.3 Dimensions of Steel Pipe (IPS)

Nominal pipe size, IPS (in.)	OD (in.)	Schedule No.	ID (in.)	Flow area per pipe (in.²)	Surface per linear ft (ft²/ft)		Weight per lin ft (lb steel)
					Outside	Inside	
1/3	0.405	40[a]	0.269	0.058	0.106	0.070	0.25
		80[b]	0.215	0.036		0.056	0.32
1/4	0.540	40[a]	0.364	0.104	0.141	0.095	0.43
		80[b]	0.302	0.072		0.079	0.54
2/3	0.675	40[a]	0.493	0.192	0.177	0.129	0.57
		80[b]	0.423	0.141		0.111	0.74
1/2	0.840	40[a]	0.622	0.304	0.220	0.163	0.85
		80[b]	0.546	0.235		0.143	1.09
3/4	1.05	40[a]	0.824	0.534	0.275	0.216	1.13
		80[b]	0.742	0.432		0.194	1.48
1	1.32	40[a]	1.049	0.864	0.344	0.274	1.68
		80[b]	0.957	0.718		0.250	2.17
1 1/4	1.66	40[a]	1.380	1.50	0.435	0.362	2.28
		80[b]	1.278	1.28		0.335	3.00
1 1/2	1.90	40[a]	1.610	2.04	0.498	0.422	2.72
		80[b]	1.500	1.76		0.393	3.64
2	2.38	40[a]	2.067	3.35	0.622	0.542	3.66
		80[b]	1.939	2.95		0.508	5.03

2½	2.88	40a	2.469	4.79	0.753	0.647	5.80
		80b	2.323	4.23		0.609	7.67
3	3.50	40a	3.068	7.38	0.917	0.804	7.58
		80b	2.900	6.61		0.760	10.3
4	4.50	40a	4.026	12.7	1.178	1.055	10.8
		80b	3.826	11.5		1.002	15.0
6	6.625	40a	6.065	28.9	1.734	1.590	19.0
		80b	5.761	26.1		1.510	28.6
8	8.625	40a	7.981	50.0	2.258	2.090	28.6
		80b	7.625	45.7		2.000	43.4
10	10.75	40a	10.02	78.8	2.814	2.62	40.5
		60	9.75	74.6		2.55	54.8
12	12.75	30	12.09	115	3.338	3.17	43.8
14	14.0	30	13.25	138	3.665	3.47	54.5
16	16.0	30	15.25	183	4.189	4.00	62.6
18	18.0	20c	17.25	234	4.712	4.52	72.7
20	20.0	20	19.25	291	5.236	5.05	78.6
22	22.0	20c	21.25	355	5.747	5.56	84.0
24	24.0	20	23.25	425	6.283	6.09	94.7

[a]Commonly known as standard.
[b]Commonly known as extra heavy.
[c]Approximately.

where

W = steam flow, lb/hr
Δh = enthalpy absorbed by steam/water = $(h_g - h_{fw})$
  + BD × $(h_f - h_{fw})$

where

h_g = enthalpy of saturated steam at operating steam pressure, Btu/lb
h_f = enthalpy of saturated liquid, Btu/lb
h_{fw} = enthalpy of feedwater, Btu/lb
BD = blowdown fraction

For example, if a 500-BHP boiler generates saturated steam at 125 psig with a 5% blowdown and with feedwater at 230°F, the steam generation at 125 psig will be

$$W = \frac{500 \times 33{,}475}{(1193 - 198) + 0.05 \times (325 - 198)} = 16{,}714 \text{ lb/hr}$$

where 1193, 198, and 325 are the enthalpies of saturated steam, feed water, and saturated liquid, respectively, obtained from steam tables. (See Appendix.)

1.09a

Q: Why do we need to know the amount of moisture in air?

A: In combustion calculations (Chapter 2) we estimate the quantity of dry air required to burn a given amount of fuel. In reality, the atmospheric air is never dry; it consists of some moisture, depending on the relative humidity and dry bulb temperature. To compute the partial pressure of water vapor in the flue gas, which is required for calculating nonluminous heat transfer, we need to know the total quantity of water vapor in flue gases, a part of which comes from combustion air.

Also, when atmospheric air is compressed, the saturated vapor pressure (SVP) of water increases, and if the air is cooled below the corresponding water dew point temperature, water can condense. The amount of moisture in air or gas fixes the water dew point, so it is important to know the amount of water vapor in air or flue gas.

1.09b

Q: Estimate the pounds of water vapor to pounds of dry air when the dry bulb temperature is 80°F and the relative humidity is 65%.

A: Use the equation

$$M = 0.622 \times \frac{p_w}{14.7 - p_w} \tag{13}$$

where

M = lb water vapor/lb dry air
p_w = partial pressure of water vapor in air, psia

This may be estimated as the vol % of water vapor × total air pressure or as the product of relative humidity and the saturated vapor pressure (SVP). From the steam tables we note that at 80°F, SVP = 0.5069 psia (at 212°F, SVP = 14.7 psia). Hence p_w = 0.65 × 0.5069.

$$M = 0.622 \times 0.65 \times \frac{0.5069}{14.7 - 0.65 \times 0.5069} = 0.0142$$

Hence, if we needed 1000 lb of dry air for combustion, we would size the fan to deliver 1000 × 1.0142 = 1014.2 lb of atmospheric air.

1.10a

Q: What is the water dew point of the flue gases discussed in Q1.05?

A: The partial pressure of water vapor when the vol % is 8 and total pressure is 14.7 psia will be

$$p_w = 0.08 \times 14.7 = 1.19 \text{ psia}$$

From the steam tables, we note that the saturation temperature corresponding to 1.19 psia is 107°F. This is also the water dew point. If the gases are cooled below this temperature, water can condense, causing problems.

1.10b

Q: What is the water dew point of compressed air when ambient air at 80°F, 14.7 psia, and a relative humidity of 65% is compressed to 35 psia?

A: Use the following expression to get the partial pressure of water vapor after compression:

$$p_{w2} = p_{w1} \times \frac{P_2}{P_1} \tag{14}$$

where

p_w = partial pressure, psia
P = total pressure, psia

The subscripts 1 and 2 stand for initial and final conditions. From Q1.09b, p_{w1} = 0.65 × 0.5069.

$$p_{w2} = 0.65 \times 0.5069 \times \frac{35}{14.7} = 0.784 \text{ psia}$$

From the steam tables, we note that corresponding to 0.784 psia, the saturation temperature is 93°F. This is also the dew point after compression. Cooling the air to below 93°F would result in its condensation.

1.11a

Q: Calculate the energy absorbed by steam in a boiler if 400,000 lb/hr of superheated steam at 1600 psia and 900°F is generated with feedwater at 250°F. What is the energy absorbed, in mega-watts?

A: The energy absorbed is given by

$$Q = W \times (h_2 - h_1) \quad \text{(neglecting blowdown)} \tag{15}$$

where

$$W = \text{steam flow, lb/hr}$$
$$h_2, h_1 = \text{steam enthalpy and water enthalpy, Btu/lb}$$
$$Q = \text{duty, Btu/hr}$$

From the steam tables, $h_2 = 1425.3$ Btu/lb and $h_1 = 224$ Btu/lb.

$$Q = 400,000 \times (1425.3 - 224)$$
$$= 480.5 \times 10^6 \text{ Btu/hr}$$
$$= 480.5 \text{ million Btu/hr (MM Btu/hr)}$$

Using the fact that 3413 Btu/hr = 1 kW, we have

$$Q = 480.5 \times \frac{10^6}{3413 \times 10^3} = 141 \text{ MW}$$

1.11b

Q: Estimate the energy absorbed by wet steam at 80% quality in a boiler at 1600 psia when the feedwater temperature is 250°F.

A: The enthalpy of wet steam can be computed as follows:

$$h = xh_g + (1 - x)h_f \tag{16}$$

where h is the enthalpy in Btu/lb. The subscripts g and f stand for saturated vapor and liquid at the referenced pressure, obtained from saturated steam properties. x is the steam quality fraction.

From the steam tables, $h_g = 1163$ Btu/lb and $h_f = 624$ Btu/lb at 1600 psia. The enthalpy of feedwater at 250°F is 226 Btu/lb.

$$h_2 = 0.8 \times 1163 + 0.2 \times 624 = 1054 \text{ Btu/lb,}$$
$$h_1 = 226 \text{ Btu/lb}$$
$$Q = 1054 - 226 = 828 \text{ Btu/lb}$$

If steam flow were 400,000 lb/hr, then

$$Q = 400,000 \times 828 = 331 \times 10^6 = 331 \text{ MM Btu/hr}$$

1.12

Q: How is the wetness in steam specified? How do we convert steam by volume (SBV) to steam by weight?

A: A steam–water mixture is described by the term *quality*, x, or dryness fraction. $x = 80\%$ means that in 1 lb of wet steam, 0.8 lb is steam and 0.2 lb is water. To relate these two terms, we use the expression

$$SBV = \frac{100}{1 + [(100 - x)/x] \times v_f/v_g} \tag{17}$$

where

v_f, v_g = specific volumes of saturated liquid and vapor, cu ft/lb

x = quality or dryness fraction

From the steam tables at 1000 psia, $v_f = 0.0216$ and $v_g = 0.4456$ cu ft/lb.

$$SBV = \frac{100}{1 + [(100 - 80)/80] \times 0.0216/0.4456} = 98.8\%$$

Circulation ratio (CR) is another term used by boiler engineers to describe the steam quality generated.

$$CR = \frac{1}{x} \tag{18}$$

A CR of 4 means that the steam quality is 0.25 or 25%; in other words, 1 lb of steam would have 0.75 lb of steam and the remainder water.

1.13a

Q: How is the quality of steam determined using a throttling calorimeter?

A: Throttling calorimeter (Fig. 1.1) is widely used in low-pressure steam boilers for determining the moisture or wetness (quality)

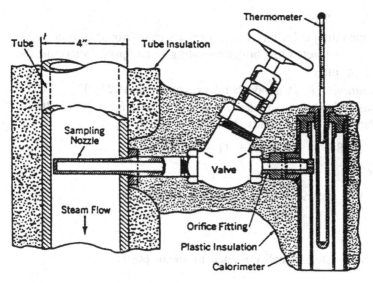

Figure 1.1   Throttling calorimeter.

of steam. A sampling nozzle is located preferably in the vertical section of the saturated steam line far from bends or fittings. Steam enters the calorimeter through a throttling orifice and into a well-insulated expansion chamber. Knowing that throttling is an isoenthalpic process, we can rewrite Eq. (16) for enthalpy balance as

$$h_s = h_m = xh_g + (1 - x)h_f$$

where

$$h_s, h_m, h_f, h_g = \text{enthalpies of steam, mixture, saturated}$$
$$\text{liquid, and saturated steam, respectively}$$
$$x = \text{steam quality fraction}$$

The steam temperature after throttling is measured at atmospheric pressure, and then the enthalpy is obtained with the help of steam tables. The steam is usually in superheated condition after throttling.

EXAMPLE

A throttling calorimeter measures a steam temperature of 250°F when connected to a boiler operating at 100 psia. Determine the steam quality.

Solution. h_s at atmospheric pressure and at 250°F = 1168.8 Btu/lb from steam tables; h_g = 1187.2 and h_f = 298.5 Btu/lb, also from steam tables. Hence

$$1168.8 = 1187.2x + (1 - x)298.5$$

or

$$x = 0.979 \text{ or } 97.9\% \text{ quality}$$

1.13b

Q: How is steam quality related to steam purity?

A: Steam purity refers to the impurities in wet steam in ppm. A typical value in low-pressure boilers would be 1 ppm of solids. However, quality refers to the moisture in steam.

   The boiler drum maintains a certain concentration of solids depending on ABMA or ASME recommendations as discussed in Q 1.17. If at 500 psig pressure the boiler water concentration is 2500 ppm, and if steam should have 0.5 ppm solids, then the quality can be estimated as follows:

$$\% \text{ Moisture in steam} = \frac{0.5}{2500} \times 100 = 0.02\%$$

or

$$\text{Steam quality} = 100 - 0.02 = 99.98\%$$

1.14

Q: How do we estimate the water required for desuperheating steam? Superheated steam at 700 psia and 800°F must be cooled to 700°F using a spray of water at 300°F. Estimate the quantity of water needed to do this.

A: From an energy balance across the desuperheater, we get

$$W_1 h_1 + W h_f = W_2 h_2 \qquad (19a)$$

where

   W_1, W_2 = steam flows before and after desuperheating
        W = water required
   h_1, h_2 = steam enthalpies before and after the process
        h_f = enthalpy of water

Also, from mass balance,

$$W_2 = W_1 + W$$

Hence we can show that

$$W = W_2 \times \frac{h_1 - h_2}{h_1 - h_f} \qquad (19b)$$

Neglecting the pressure drop across the desuperheater, we have from the steam tables: $h_1 = 1403$, $h_2 = 1346$, and $h_f = 271$, all in Btu/lb. Hence $W/W_2 = 0.05$. That is, 5% of the final steam flow is required for injection purposes.

1.15

Q: How is the water requirement for cooling a gas stream estimated? Estimate the water quantity required to cool 100,000 lb/hr of flue gas from 900°F to 400°F. What is the final volume of the gas?

A: From an energy balance it can be shown [1] that

$$q = 5.39 \times 10^{-4} \times (t_1 - t_2)$$
$$\times \frac{W}{1090 + 0.45 \times (t_2 - 150)} \qquad (20)$$

where

        q = water required, gpm
   t_1, t_2 = initial and final gas temperatures, °F
        W = gas flow entering the cooler, lb/hr

Substitution yields

$$q = 5.39 \times 10^{-4} \times (900 - 400)$$

$$\times \ \frac{100{,}000}{1090 + 0.45 \times (400 - 150)}$$

$$= 23 \text{ gpm}$$

The final gas volume is given by the expression

$$(460 + t_2) \times \left(\frac{W}{2361} + 0.341 \right)$$

The final volume is 43,000 acfm.

1.16

Q: In selecting silencers for vents or safety valves, we need to figure the volume of steam after the throttling process. Estimate the volume of steam when 60,000 lb/hr of superheated steam at 650 psia and 800°F is blown to the atmosphere in a safety valve.

A: We have to find the final temperature of steam after throttling, which may be considered an isoenthalpic process; that is, the steam enthalpy remains the same at 650 and 15 psia.
     From the steam tables, at 650 psia and 800°F, $h = 1402$ Btu/lb. At 15 psia (atmospheric conditions), the temperature corresponding to an enthalpy of 1402 Btu/lb is 745°F. Again from the steam tables, at a pressure of 15 psia and a temperature of 745°F, the specific volume of steam is 48 cu ft/lb. The total volume of steam is $60{,}000 \times 48 = 2{,}880{,}000$ cu ft/hr.

1.17

Q: How do we determine the steam required for deaeration and boiler blowdown water requirements?

A: Steam plant engineers have to frequently perform energy and mass balance calculations around the deaerator and boiler to obtain the values of makeup water, blowdown, or deaeration steam flows. Boiler blowdown quantity depends on the TDS (total dissolved solids) of boiler water and the incoming makeup water. Figure 1.2 shows the scheme around a simple deaerator. Note that there could be several condensate returns. This analysis does not consider venting of steam from the deaerator or the heating of makeup using the blowdown water. These refinements can be done later to fine tune the results.

ABMA (American Boiler Manufacturers Association) and ASME provide guidelines on the TDS of boiler water as a function of pressure (see Tables 1.4 and 1.5). The drum solids concentration can be at or less than the value shown in these tables. Plant water chemists usually set these values after reviewing the complete plant chemistry.

EXAMPLE

A boiler generates 50,000 lb/hr of saturated steam at 300 psia, out of which 10,000 lb/hr is taken for process and returns to the deaerator as condensate at 180°F. The rest is consumed. Makeup water enters the deaerator at 70°F, and steam is available at 300 psia for deaeration. The deaerator operates at a pressure of 25 psia. The blowdown has a total dissolved solids (TDS) of 1500 ppm, and the makeup has 100 ppm TDS.

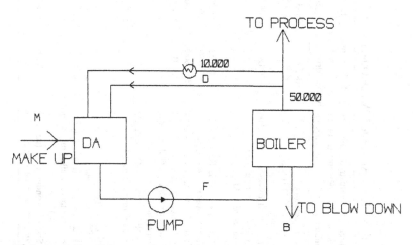

Figure 1.2   Scheme of deaeration system.

Table 1.4   Suggested Water Quality Limits[a]

Boiler type: industrial water tube, high duty, primary fuel fired, drum type
Makeup water percentage: up to 100% of feedwater
Conditions: includes superheater, turbine drives, or process restriction on steam purity

Drum operating Pressure,[b] MPa (psig)	0–2.07 (0–300)	2.08–3.10 (301–450)	3.11–4.14 (451–600)	4.15–5.17 (601–750)	5.18–6.21 (751–900)	6.22–6.89 (901–1000)	6.90–10.34 (1001–1500)	10.35–13.79 (1501–2000)
Feedwater[c]								
Dissolved oxygen (mg/L O_2) measured before oxygen scavenger addition[d]	<0.04	<0.04	<0.007	<0.007	<0.007	<0.007	<0.007	<0.007
Total iron (mg/L Fe)	≤0.100	≤0.050	≤0.030	≤0.025	≤0.020	≤0.020	≤0.010	≤0.010
Total copper (mg/L Cu)	≤0.050	≤0.025	≤0.020	≤0.020	≤0.015	≤0.015	≤0.010	≤0.010
Total hardness (mg/L $CaCO_3$)	≤0.300	≤0.300	≤0.200	≤0.200	≤0.100	≤0.050	n.d.	n.d.
pH range @ 25°C	7.5–10.0	7.5–10.0	7.5–10.0	7.5–10.0	7.5–10.0	8.5–9.5	9.0–9.6	9.0–9.6
Chemicals for preboiler system protection					Use only volatile alkaline materials			
Nonvolatile TOC (mg/L C)[e]	<1	<1	<0.5	<0.5	<0.5	—As low as possible, <0.2—		
Oily matter (mg/L)	<1	<1	<0.5	<0.5	<0.5	—As low as possible, <0.2—		
Boiler water								
Silica (mg/L SiO_2)	≤150	≤90	≤40	≤30	≤20	≤8	≤2	≤1
Total alkalinity (mg/L $CaCO_3$)	<350[f]	<300[f]	<250[f]	<200[f]	<150[f]	<100[f]	n.s.[g]	n.s.[g]
Free hydroxide alkalinity (mg/L $CaCO_3$)[b]	n.s.	n.s.	n.s.	n.s.	n.d.[g]	n.d.[g]	n.d.[g]	n.d.[g]
Specific conductance (μmho/cm) @ 25°C without neutralization	<3500[i]	<3000[i]	2500[i]	<2000[i]	<1500[i]	<1000[i]	≤150[i]	≤100[i]

[a]No values given for saturated steam purity target because steam purity achievable depends upon many variables, including boiler water total alkalinity and specific conductance as well as design of boiler, steam drum internals, and operating conditions (see footnote i). Since boilers in this category require a relatively high degree of steam purity, other operating parameters must be set as low as necessary to achieve this high purity for protection of the superheaters and turbines and/or to avoid process contamination.

[b]With local heat fluxes >473.2 kW/m^2 (>150,000 Btu/hr ft^2), use values for the next higher pressure range.

[c]Boilers below 6.21 MPa (900 psig) with large furnaces, large steam release space, and internal chelant, polymer, and/or antifoam treatment can sometimes tolerate higher levels of feedwater impurities than those in the table and still achieve adequate deposition control and steam purity. Removal of these impurities by external pretreatment is always a more positive solution. Alternatives must be evaluated as to practicality and economics in each case.

[d]Values in table assume the existence of a deaerator.

[e]Nonvolatile TOC is that organic carbon not intentionally added as part of the water treatment regime.

[f]Maximum total alkalinity consistent with acceptable steam purity. If necessary, should override conductance as blowdown control parameter. If makeup is demineralized water at 4.14 MPa (600 psig) to 6.89 MPa (1000 psig), boiler water alkalinity and conductance should be that in the table for 6.90 to 10.34 MPa (1001 to 1500 psig) range.

[g]"Not detectable" in these cases refers to free sodium or potassium hydroxide alkalinity. Some small variable amount of total alkalinity will be present and measurable with the assumed congruent or coordinated phosphate pH control or volatile treatment employed at these high pressure ranges.

[h]Minimum level of OH$^-$ alkalinity in boilers below 6.21 MPa (900 psig) must be individually specified with regard to silica solubility and other components of internal treatment.

[i]Maximum values often not achievable without exceeding suggested maximum total alkalinity values, especially in boilers below 6.21 MPa (900 psig) with >20% makeup of water whose total alkalinity is >20% of TDS naturally or after pretreatment with soda lime or sodium cycle ion-exchange softening. Actual permissible conductance values to achieve any desired steam purity must be established for each case by careful steam purity measurements. Relationship between conductance and steam purity is affected by too many variables to allow its reduction to a simple list of tabulated values.

Source: Adapted from ASME 1979 Consensus.

n.d. = not detectable; n.s. = not specified.

Table 1.5 Water Tube Boilers—Recommended Boiler Water Limits and Associated Steam Purity at Steady-State Full Load Operation—Drum-Type Boilers

Drum pressure (psig)	Range total dissolved solids,[a] boiler water (ppm) (max)	Range total alkalinity,[b] boiler water (ppm)	Suspended solids, boiler water (ppm)(max)	Range total dissolved solids,[b,c] steam (ppm) (max expected value)
0–300	700–3500	140–700	15	0.2–1.0
301–450	600–3000	120–600	10	0.2–1.0
451–600	500–2500	100–500	8	0.2–1.0
601–750	200–1000	40–200	3	0.1–0.5
751–900	150–750	30–150	2	0.1–0.5
901–1000	125–625	25–125	1	0.1–0.5
1001–1800	100	—[d]	1	0.1
1801–2350	50		n.a.	0.1
2351–2600	25		n.a.	0.05
2601–2900	15		n.a.	0.05
		once through boilers		
1400 and above	0.05	n.a.	n.a.	0.05

n.a. = not available.

[a] Actual values within the range reflect the TDS in the feedwater. Higher values are for high solids, lower values are for low solids in the feedwater.

[b] Actual values within the range are directly proportional to the actual value of TDS of boiler water. Higher values are for the high solids, lower values are for low solids in the boiler water.

[c] These values are exclusive of silica.

[d] Dictated by boiler water treatment.

Source: American Boiler Manufacturers Association, 1982.

Solution. This problem demonstrates the evaluation of two important variables: deaeration steam and blowdown water requirements.

From mass balance around the deaerator,

$$10,000 + D + M = F = 50,000 + B \qquad (21)$$

From energy balance around the deaerator,

$$10,000 \times 148 + 1202.8 \times D + M \times 38$$
$$= 209 \times F = 209 \times (50,000 + B) \qquad (22)$$

From balance of solids concentration,

$$100 \times M = 1500 \times B \qquad (23)$$

In Eq. (22), 1202.8 is the enthalpy of steam used for deaeration, 209 the enthalpy of boiler feedwater, 148 the enthalpy of condensate return, and 38 that of makeup, all in Btu/lb. The equation assumes that the amount of solids in returning condensate and steam is negligible, which is true. Steam usually has a TDS of 1 ppm or less, and so does the condensate. Hence, for practical purposes we can neglect it. The net solids enter the system in the form of makeup water and leave as blowdown. There are three unknowns, D, M, and B, and three equations. From Eq. (21),

$$D + M = 40,000 + B \qquad (24)$$

substituting (23) into (24),

$$D + 15B = 40,000 + B2 \qquad \text{or}$$
$$D + 14B = 40,000 \qquad (25)$$

From (22),

$$1,480,000 + 1202.8D + 38 \times 15B = 209 \times 50,000$$
$$+ 209B$$

Solving this equation, we have $B = 2375$ lb/hr, $D = 6750$ lb/hr, $M = 35,625$ lb/hr, and $F = 52,375$ lb/hr. Considering venting of steam from the deaerator to expel dissolved gases and the heat losses, we may consume 1 to 3% higher steam quantity.

1.18

Q: How can the boiler blowdown be utilized? A 600-psia boiler operates for 6000 hr annually and discharges 4000 lb/hr of blowdown. If this is flashed to steam at 100 psia, how much steam is generated? If the cost of the blowdown system is $8000, how long does it take to pay back? Assume that the cost of steam is $2 per 1000 lb.

A: To estimate the flash steam produced we may use the expression

$$h = xh_g + (1 - x)h_f \tag{26}$$

where

h = enthalpy of blowdown water at high pressure, Btu/lb

h_g, h_f = enthalpies of saturated steam and water at the flash pressure, Btu/lb

x = fraction of steam that is generated at the lower pressure

From the steam tables, at 600 psia, $h = 471.6$, and at 100 psia, $h_g = 1187$ and $h_f = 298$, all in Btu/lb. Using Eq. (26), we have

$$471.6 = 1187x + (1 - x)298$$

or

$$x = 0.195$$

About 20% of the initial blowdown is converted to flash steam, the quantity being $0.2 \times 4000 = 800$ lb/hr. This 800 lb/hr of 100-psia steam can be used for process. The resulting savings annually will be

$$800 \times 2 \times \frac{6000}{1000} = \$9600$$

Simple payback will be 8000/9600 = 0.8 year or about 10 months.

Tables are available that give the flash steam produced if the initial and flash pressures are known. Table 1.6 is one such table.

Table 1.6 Steam Flash and Heat Content at Differential Temperatures

Initial pressure (psig)	Temp. of liquid (°F)	Percent of flash at reduced pressures								
		Atm. pressure	5 lb	10 lb	15 lb	20 lb	25 lb	30 lb	35 lb	40 lb
100	338	13	11.5	10.3	9.3	8.4	7.6	6.9	6.3	5.5
125	353	14.5	13.3	11.8	10.9	10	9.2	8.5	7.9	7.2
150	366	16	14.6	13.2	12.3	11.4	10.6	9.9	9.3	8.5
175	377	17	15.8	14.4	13.4	12.5	11.6	11.1	10.4	9.7
200	388	18	16.9	15.5	14.6	13.7	12.9	12.2	11.6	10.9
225	397	19	17.8	16.5	15.5	14.7	13.9	13.2	12.6	11.9
250	406	20	18.8	17.4	16.5	15.6	14.9	14.2	13.6	12.9
300	421	21.5	20.3	19	18	17.2	16.5	15.8	15.2	14.5
350	435	23	21.8	20.5	19.5	18.7	18	17.3	16.7	16
400	448	24	23	21.8	21	20	19.3	18.7	18.1	17.5
450	459	25	24.3	23	22	21.3	20	19.9	19.3	18.7
500	470	26.5	25.4	24.1	23.2	22.4	21.7	21.1	20.5	19.9
550	480	27.5	26.5	25.2	24.3	23.5	22.8	22.2	21.6	20.9
600	488	28	27.3	26	25	24.3	23.6	23	22.4	21.8
Btu in flash per lb		1150	1155	1160	1164	1167	1169	1172	1174	1176
Temp. of liquid (°F)		212	225	240	250	259	267	274	280	287
Steam volume (cu ft/lb)		26.8	21	16.3	13.7	11.9	10.5	9.4	8.5	7.8

Source: Madden Corp. catalog.

1.19a

Q: Estimate the leakage of steam through a hole 1/8 in. in diameter
    in a pressure vessel at 100 psia, the steam being in a saturated
    condition.

A: The hourly loss of steam in lb/hr is given by [2]

$$W = 50 \, \frac{AP}{1 + 0.00065(t - t_{sat})} \tag{27}$$

where

W = steam leakage, lb/hr
A = hole area, in.^{2}
P = steam pressure, psia
t, t_{sat} = steam temperature and saturated steam
       temperature, °F

If the steam is saturated, $t = t_{sat}$. If the steam is wet with a steam
quality of x, fraction, then the leakage flow is obtained from Eq.
(27) divided by $\sqrt{x}$. Since the steam is saturated ($x = 1$),

$$W = 50 \times 3.14 \times \left(\frac{1}{8}\right)^2 \times \frac{1}{4} \times 100 = 61 \text{ lb/hr}$$

If the steam were superheated and at 900°F, then

$$W = \frac{61}{1 + 0.00065 \times (900 - 544)} = 50 \text{ lb/hr}$$

544°F is the saturation temperature at 1000 psia. If the steam
were wet with a quality of 80%, then

$$W = \frac{61}{\sqrt{0.8}} = 68 \text{ lb/hr}$$

1.19b

Q: How is the discharge flow of air from high pressure to atmo-
    spheric pressure determined?

A: Critical flow conditions for air are found in several industrial
    applications such as flow through soot blower nozzles, spray

guns, and safety valves and leakage through holes in pressure vessels. The expression that relates the variables is [8]

$$W = 356 \times AP \times \left(\frac{MW}{T}\right)^{0.5} \tag{28}$$

where

W = flow, lb/hr
A = area of opening, in.^{2}
MW = molecular weight of air, 28.9
T = absolute temperature, °R
P = relief or discharge pressure, psia

What is the leakage air flow from a pressure vessel at 40 psia if the hole is 0.25 in. in diameter? Air is at 60°F.

$$A = 3.14 \times 0.25 \times \frac{0.25}{4} = 0.049 \text{ in.}^2$$

Hence,

$$W = 356 \times 0.049 \times 40 \times \left(\frac{28.9}{520}\right)^{0.5} = 164 \text{ lb/hr}$$

1.20a

Q: Derive an expression for the leakage of gas across a damper, stating the assumptions made.

A: Most of the dampers used for isolation of gas or air in ducts are not 100% leakproof. They have a certain percentage of leakage area, which causes a flow of gas across the area. Considering the conditions to be similar to those of flow across an orifice, we have

$$V_g = C_d \sqrt{2gH_w \frac{\rho_w}{2\,\rho_g}} \tag{29}$$

where

V_g = gas velocity through the leakage area, fps
H_w = differential pressure across the damper, in. WC

ρ_g, ρ_w = density of gas and water, lb/cu ft
    g = acceleration due to gravity, ft/sec^2
   C_d = coefficient of discharge, 0.61

The gas flow W in lb/hr can be obtained from

$$W = 3600 \, \rho_g A(100 - E) \, \frac{V_g}{100} \qquad (30)$$

where E is the sealing efficiency on an area basis (%). Most dampers have an E value of 95 to 99%. This figure is provided by the damper manufacturer. A is the duct cross section, ft^2. Substituting $C_d = 0.61$ and $\rho_g = 40/(460 + t)$ into (29) and (30) and simplifying, we have

$$W = 2484A(100 - E) \sqrt{\frac{H_w}{460 + t}} \qquad (31)$$

where t is the gas or air temperature, °F.

1.20b

Q: A boiler flue gas duct with a diameter of 5 ft has a damper whose sealing efficiency is 99.5%. It operates under a differential pressure of 7 in. WC when closed. Gas temperature is 540°F. Estimate the leakage across the damper. If energy costs $3/MM Btu, what is the hourly heat loss and the cost of leakage?

A: Substitute $A = 3.14 \times 5^2/4$, $H_w = 7$, $t = 540$, and $E = 99.5$ into Eq. (31). Then

$$W = 2484 \times 3.14 \times \frac{5^2}{4} \times (100 - 99.5) \times \frac{7}{\sqrt{1000}}$$

$$= 2040 \text{ lb/hr}$$

The hourly heat loss can be obtained from

$$Q = WC_p(t - t_a) = 2040 \times 0.26 \times (540 - 80)$$

$$= 240,000 \text{ Btu/hr} = 0.24 \text{ MM Btu/hr}$$

where C_p is the gas specific heat, Btu/lb °F. Values of 0.25 to 0.28 can be used for quick estimates, depending on gas temperature. t_a is the ambient temperature in °F. 80°F was assumed in this case. The cost of this leakage = 0.24 × 3 = $0.72/hr.

1.20c

Q: How is the sealing efficiency of a damper defined?

A: The sealing efficiency of a damper is defined on the basis of the area of cross section of the damper and also as a percentage of flow. The latter method of definition is a function of the actual gas flow condition.

In Q 1.20b, the damper had an efficiency of 99.5% on an area basis. Assume that the actual gas flow was 230,000 lb/hr. Then, on a flow basis, the efficiency would be

$$100 - \frac{2040}{230,000} = 99.12\%$$

If the flow were 115,000 lb/hr and the differential pressure were maintained, the efficiency on an area basis would still be 99.5%, while that on a flow basis would be

$$100 - \frac{2040}{115,000} = 98.24\%$$

Plant engineers should be aware of these two methods of stating the efficiency of dampers.

1.21

Q: 50,000 lb/hr of flue gas from a boiler exists at 800°F. If a waste heat recovery system is added to reduce it to 350°F, how much energy is saved? If energy costs $3/MM Btu and the plant operates for 6000 hr/year, what is the annual savings? If the cost of the heat recovery system is $115,000, what is the simple payback?

A: The energy savings $Q = WC_p(t_1 - t_2)$ · t_1 and t_2 are gas temperatures before and after installation of the heat recovery system, °F. C_p is the gas specific heat, Btu/lb °F. Use a value of 0.265 when the gas temperature is in the range 400 to 600°F.

$$Q = 50,000 \times 0.265 \times (800 - 350) = 5.85 \times 10^6$$
$$= 5.85 \text{ MM Btu/hr}$$

Annual savings $= 5.85 \times 6000 \times 3 = \$105,000$

Hence

$$\text{Simple payback} = \frac{115,000}{105,000} = 1.1 \text{ years, or 13 months}$$

1.22

Q: What is life-cycle costing? Two bids are received for a fan as shown below. Which bid is better?

	Bid 1	Bid 2
Flow (acfm)	10,000	10,000
Head (in. WC)	8	8
Efficiency (%)	60	75
Total cost, fan and motor ($)	17,000	21,000

A: Life-cycle costing is a methodology that computes the total cost of owning and operating the equipment over its life. Several financing methods and tax factors would make this a complicated evaluation. However, let us use a simple approach to illustrate the concept. To begin with, the following data should be obtained.

Cost of electricity, $C_e = \$0.25/\text{kWh}$
Annual period of operation, $N = 8000$ hr
Life of equipment, $T = 15$ years
Interest rate, $i = 0.13$ (13%)
Escalation rate, $e = 0.08$ (8%)

If the annual cost of operation is C_a, the life-cycle cost (LCC) is

$$\text{LCC} = C_c + C_a F \tag{32}$$

where C_c is the cost of equipment. F is a factor that capitalizes the operating cost over the life of the equipment. It can be shown [4,5] that

$$F = \frac{1 + e}{1 + i} \times \frac{1 - \left(\dfrac{1 + e}{1 + i}\right)^T}{1 - \left(\dfrac{1 + e}{1 + i}\right)}$$

The annual cost of operation is given by

$$C_a = PC_eN \tag{34}$$

where P is the electric power consumed, kW.

$$P = 1.17 \times 10^{-4} \times \frac{qH_w}{\eta_f} \tag{35}$$

where

H_w = head, in. WC
η_f = efficiency, fraction
q = flow, acfm

Let us use the subscripts 1 and 2 for bids 1 and 2.

$$P_1 = 1.17 \times 10^{-4} \times 10{,}000 \times \frac{8}{0.60} = 15.6 \text{ kW}$$

$$P_2 = 1.17 \times 10^{-4} \times 10{,}000 \times \frac{8}{0.75} = 12.48 \text{ kW}$$

From Eq. (33), substituting $e = 0.08$, $i = 0.13$, and $T = 15$, we get $F = 10.64$. Calculate C_a from Eq. (34):

C_{a1} = 15.6 × 8000 × 0.025 = $3120
C_{a2} = 12.48 × 8000 × 0.025 = $2500

Using Eq. (32), calculate the life-cycle cost.

LCC_1 = 17,000 + 3120 × 10.64 = $50,196
LCC_2 = 21,000 + 2500 × 10.64 = $47,600

We note that bid 2 has a lower LCC and thus may be chosen. However, we have to analyze other factors such as period of operation, future cost of energy, and so on, before deciding. If N were lower, it is likely that bid 1 would be better.

Hence, choosing equipment should not be based only on the initial investment but on an evaluation of the life-cycle cost, especially as the cost of energy is continually increasing.

1.23

Q: A process kiln exits 50,000 lb/hr of flue gas at 800°F. Two bids were received for heat recovery systems, as follows:

	Bid 1	Bid 2
Gas temperature leaving system, °F	450	300
Investment, $	215,000	450,000

If the plant operates for 6000 hr/yr, and interest, escalation rates, and life of plant are as in Q1.22, evaluate the two bids if energy costs $4/MM Btu.

A: Let us calculate the capitalized savings and compare them with the investments. For bid 1:

Energy recovered = $50,000 \times 0.25 \times (800 - 450)$
= 4.375 MM Btu/hr

This energy is worth

$4.375 \times 4 = \$17.5/hr$

Annual savings = $6000 \times 17.5 = \$105,000$

The capitalization factor from Q1.22 is 10.64. Hence capitalized savings (savings through the life of the plant) = $105,000 \times 10.64 = \$1.12 \times 10^6$. A similar calculation for bid 2 shows that the capitalized savings will be $\$1.6 \times 10^6$. The difference in capitalized savings of $\$0.48 \times 10^6$, or $480,000, exceeds the difference in the investment of $235,000. Hence bid 2 is more attractive.

If, however, energy costs $3/MM Btu and the plant works for 2500 hr/year, capitalized savings on bid 1 will by $465,000 and that of bid 2 = $665,000. The difference of $200,000 is less than the difference in investment of $235,000. Hence under these conditions, bid 1 is better.

Hence cost of energy and period of operation are important factors in arriving at the best choice.

1.24

Q: Determine the thickness of the tubes required for a boiler super-heater. The material is SA 213 T 11; the metal temperature is 900°F (see Q4.16a for a discussion of metal temperature calculation), and the tube outer diameter is 1.75 in. The design pressure is 1000 psig.

A: Per ASME *Boiler and Pressure Vessel Code*, Sec. 1, 1980, p. 27, the following equation can be used to obtain the thickness or the allowable pressure for tubes. (A tube is specified by the outer diameter and minimum wall thickness, while a pipe is specified by the nominal diameter and average wall thickness.) Typical pipe and tube materials used in boiler applications are shown in Tables 1.3 and 1.7.

$$t_w = \frac{Pd}{2S_a + P} + 0.005d + e \tag{36}$$

$$P = S_a \times \frac{2t_w - 0.01d - 2e}{d - (t_w - 0.005d - e)} \tag{37}$$

where

t_w = minimum wall thickness, in.
P = design pressure, psig
d = tube outer diameter, in.
e = factor that accounts for compensation in screwed tubes, generally zero
S_a = allowable stress, psi

Table 1.7   Allowable Stress Values, Ferrous Tubing, 1000 psi

Material specifications	Temperatures not exceeding (°F):									
	20 to 650	700	750	800	850	900	950	1000	1200	1400
SA 178 gr A	10.0	9.7	9.0	7.8	6.7	5.5	3.8	2.1	—	—
	12.8	12.2	11.0	9.2	7.4	5.5	3.8	2.1	—	—
SA 192 gr C	11.8	11.5	10.6	9.2	7.9	6.5	4.5	2.5	—	—
SA 210 gr A-1 SA 53 B	15	14.4	13.0	10.8	8.7	6.5	4.5	2.5	—	—
gr C	17.5	16.6	14.8	12.0	7.8	5.0	3.0	1.5	—	—
SA 213 T 11, P 11	15.0	15.0	15.0	15.0	14.4	13.1	11.0	7.8	1.2	—
T 22, P 22	15.0	15.0	15.0	15.0	14.4	13.1	11.0	7.8	1.6	—
T 9	—	13.4	13.1	12.5	12.5	12.0	10.8	8.5	—	—
SA 213 TP 304 H	—	15.9	15.5	15.2	14.9	14.7	14.4	13.8	6.1	2.3
TP 316 H	—	16.3	16.1	15.9	15.7	15.5	15.4	15.3	7.4	2.3
TP 321 H	—	15.8	15.7	15.5	15.4	15.3	15.2	14.0	5.9	1.9
TP 347 H	—	14.7	14.7	14.7	14.7	14.7	14.6	14.4	7.9	2.5

Source: ASME, Boiler and Pressure Vessel Code, Sec. 1, Power boilers, 1980.

From Table 1.7, S_a is 13,100. Substituting into Eq. (36) yields

$$t_w = \frac{1000 \times 1.75}{2 \times 13,100 + 1000} + 0.005 \times 1.75 = 0.073 \text{ in.}$$

The tube with the next higher thickness would be chosen. A corrosion allowance, if required, may be added to t_w.

1.25

Q: Determine the maximum pressure that an SA 53 B carbon steel pipe of size 3 in. schedule 80 can be subjected to at a metal temperature of 550°F. Use a corrosion allowance of 0.02 in.

A: By the ASME Code, Sec. 1, 1980, p. 27, the formula for determining allowable pressures or thickness of pipes, drums, and headers is

$$t_w = \frac{Pd}{2S_a E + 0.8P} + c \tag{38}$$

where

E = ligament efficiency, 1 for seamless pipes
c = corrosion allowance

From Table 1.3, a 3-in. schedule 80 pipe has an outer diameter of 3.5 in. and a nominal wall thickness of 0.3 in. Considering the manufacturing tolerance of 12.5%, the minimum thickness available is $0.875 \times 0.3 = 0.2625$ in.

Substituting $S_a = 15,000$ psi (Table 1.7) and $c = 0.02$ into Eq. (38), we have

$$0.2625 = \frac{3.5P}{2 \times 15,000 + 0.8P} + 0.02$$

Solving for P, we have $P = 2200$ psig.

For alloy steels, the factor 0.8 in the denominator would be different. The ASME Code may be referred to for details [6]. Table 1.8 gives the maximum allowable pressures for carbon steel pipes up to a temperature of 650°F [7].

Table 1.8   Maximum Allowable Pressure[a]

Nominal pipe size (in.)	Schedule 40	Schedule 80	Schedule 160
1/4	4830	6833	—
1/2	3750	5235	6928
1	2857	3947	5769
1 1/2	2112	3000	4329
2	1782	2575	4225
2 1/2	1948	2702	3749
3	1693	2394	3601
4	1435	2074	3370
5	1258	1857	3191
6	1145	1796	3076
8	1006	1587	2970

[a]Based on allowable stress of 15,000 psi; corrosion allowance is zero.
Source: *Chemical Engineering*, July 25, 1983.

1.26

Q: How is the maximum allowable external pressure for boiler tubes determined?

A: According to ASME Code Sec. 1, 1989, para PFT 51, the external pressures of tubes or pipes can be determined as follows.

1. For cylinders having $d_o/t > 10$ [9],

$$P_a = \frac{4B}{3 \, (d_o/t)} \tag{39}$$

where

P_a = maximum allowable external pressure, psi
A, B = factors obtained from ASME Code, Sec. 1, depending on values of d_o/t and L/d_o, where L, d_o, and t refer to tube length, external diameter, and thickness.

2. When $d_o/t <10$, A and B are determined from tables or charts as in Q1.25. For $d_o/t <4$, $A = 1.1/(d_o/t)^2$. Two values of allowable pressures are then computed, namely, P_{a1} and P_{a2}.

$$P_{a1} = \left(\frac{2.167}{d_o/t} - 0.0833 \right) \times B \quad \text{and}$$

$$P_{a2} = 2S_b \times \frac{1 - (t/d_o)}{d_o/t}$$

where S_b is the lesser of 2 times the maximum allowable stress values at design metal temperature from the code stress tables or 1.8 times the yield strength of the material at design metal temperature. Then the smaller of the P_{a1} or P_{a2} is used for P_a.

EXAMPLE

Determine the maximum allowable external pressure at 600°F for 120-in. SA 192 tubes of outer diameter 2 in. and length 15 ft used in fire tube boilers.

Solution.

$$\frac{d_o}{L} = \frac{15 \times 12}{2.0} = 90 \quad \text{and}$$

$$\frac{d_o}{t} = \frac{2}{0.120} = 16.7$$

From Figure 1.3, factor $A = 0.004$. From Figure 1.4, $B = 9500$. Since $d_o/t > 10$,

$$P_a = \frac{4B}{3(d_o/t)} = 4 \times \frac{9500}{3/16.7} = 758 \text{ psi}$$

1.27

Q: What is a decibel? How is it expressed?

A: The decibel (dB) is the unit of measure used in noise evaluation. It is a ratio (not an absolute value) of a sound level to a reference level and is stated as a sound pressure level (SPL) or a sound

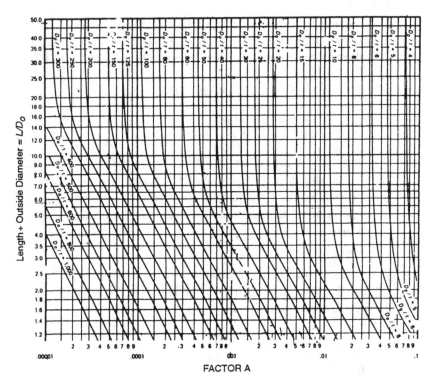

Figure 1.3 Factor *A* for use in external pressure calculation [9].

power level (PWL). The reference level for SPL is 0.0002 μbar. A human ear can detect from about 20 dB to sound pressures 100,000 times higher, 120 dB.

Audible frequencies are divided into octave bands for analysis. The center frequencies in hertz (Hz) of the octave bands are 31.5, 63, 125, 250, 500, 1000, 2000, 4000, and 8000 Hz. The human ear is sensitive to frequencies between 500 and 3000 Hz and less sensitive to very high and low frequencies. At 1000 Hz, for example, 90 dB is louder than it is at 500 Hz.

The sound meter used in noise evaluation has three scales, A, B, and C, which selectively discriminate against low and high frequencies. The A scale (dBA) is the most heavily weighted

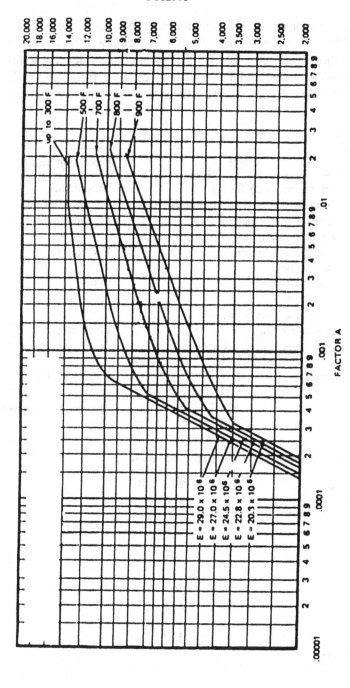

Figure 1.4 Factor B for use in external pressure calculation (SA 178A, SA 192 tubes) [9].

43

scale and approximates the human ear's response to noise (500 to 6000 Hz). It is used in industry and in regulations regarding the evaluation of noise. Table 1.9 gives typical dBA levels of various noise sources, and Table 1.10 gives the permissible OSHA (Occupational Safety and Health Act) noise exposure values.

Table 1.9   Typical A-Weighted Sound Levels

dBA	Source	Perception/hearing
140	Jet engine at 25 ft	Unbearable
130	High-pressure safety vent at 25 ft	Threshold of pain
120	Large FD fan plenum area	Uncomfortably loud
110	8000-hp engine exhaust at 25 ft	
100	Compressor building	Very loud
90	Boiler room	
80	Pneumatic drill	Loud
70	Commercial area	
60	Normal conversation	
50	Average home	Comfortable
40	Nighttime residential area	
30	Broadcast studio	
20	Whisper	Barely audible
10		
0		Threshold of hearing

Table 1.10   Permissible Noise Exposures (OSHA)

Duration per day (hr)	Sound level (dBA) (slow response)
8	90
6	92
4	95
3	97
2	100
1 1/2	102
1	105
1/2	110
1/4 or less	115

1.28

Q: How are decibels added? A noise source has the following dB values at center frequencies:

Hz	31.5	63	125	250	500	1000	2000	4000	8000
dB	97	97	95	91	84	82	80	85	85

What is the overall noise level?

A: Decibels are added logarithmically and not algebraically. 97 dB plus 97 dB is not 194 dB but 100 dB.

$$P = 10 \log(10^{P_1/10} + 10^{P_2/10} + 10^{P_3/10} + \cdots)$$
$$= 10 \log(10^{9.7} + 10^{9.7} + 10^{9.5} + 10^{9.1} + 10^{8.4}$$
$$+ 10^{8.2} + 10^8 + 10^{8.5} + 10^{8.5})$$
$$= 102 \text{ dB}$$

1.29

Q: What are SPL and PWL?

A: SPL is sound pressure level, which is dependent on the distance and environment and is easily measured with a sound-level meter. SPL values should be referred to distance. PWL is sound power level and is a measure of the total acoustic power radiated by a given source. It is defined as

$$\text{PWL} = 10 \log \frac{W}{10^{12}} \text{ dB} \tag{40}$$

PWL is a constant for a given source and is independent of the environment. It cannot be measured directly but must be calculated. PWL can be roughly described as being equal to the watt rating of a bulb. Manufacturers of fans and gas turbines publish the values of PWL of their machines. While selecting silencers for these equipment, PWL may be converted to SPL depending on distance, and the attenuation desired at various frequencies may be obtained. A silencer that gives the desired attenuation can then be chosen.

1.30

Q: A sound level of 120 dB is measured at a distance of 3 ft from a source. Find the value at 100 ft.

A: The following formula relates the PWL and SPL with distance:

$$SPL = PWL - 20 \log L + 2.5 \text{ dB} \qquad (41)$$

where

L = distance, ft

PWL is a constant for a given source. Hence

$SPL + 20 \log L =$ a constant

$120 + 20 \log 3 = SPL_2 + 20 \log 100$

Hence

$SPL_2 = 89.5 \text{ dB}$

Thus we see that SPL has decreased by 30 dB with a change from 3 ft to 100 ft. While selecting silencers, one should be aware of the desired SPL at the desired distance. Neglecting the effect of distance can lead to specifying a large and costly silencer.

1.31

Q: How is the noise level from exhaust of engines computed?

A: A gas turbine exhaust has the noise spectrum given in Table 1.11 at various octave bands. The exhaust gases flow through a heat recovery boiler into a stack that is 100 ft high. Determine the noise level 150 ft from the top of the stack (of diameter 60 in.) and in front of the boiler.

Assume that the boiler attenuation is 20 dB at all octave bands. In order to arrive at the noise levels at the boiler front, three corrections are required: (1) boiler attenuation, (2) effect of directivity, and (3) divergence at 150 ft. The effect of directivity

Table 1.11   Table of Noise Levels

	63	125	250	500	1K	2K	4K	8K
1. Frequency (Hz)	63	125	250	500	1K	2K	4K	8K
2. PWL ($\times 10^{-12}$ W dB) (gas turbine)	130	134	136	136	132	130	131	133
3. Boiler attenuation (dB)	-20	-20	-20	-20	-20	-20	-20	-20
4. Directivity (dB)	0	-1	-2	-5	-8	-10	-13	-16
5. Divergence (dB)	-41	-41	-41	-41	-41	-41	-41	-41
6. Resultant	69	72	73	70	63	59	57	56
7. A scale (dB)	-25	-16	-9	-3	0	1	1	-1
8. Net	44	56	64	67	63	60	58	55
	⎣___56___⎦		⎣___69___⎦		⎣___65___⎦		⎣___60___⎦	
		⎣_____69_____⎦				⎣_____66.5_____⎦		
				⎣_____71_____⎦				

is shown in Table 1.12. The divergence effect is given by 20 log $L - 2.5$, where L is the distance from the noise source.

Row 8 values are converted to dBA by adding the dB at various frequencies. The final value is 71 dBA.

1.32

Q: How is the holdup or volume of water in boiler drums estimated? A boiler generating 10,000 lb/hr of steam at 400 psig has a 42-in. drum 10 ft long with 2:1 ellipsoidal ends. Find the duration available between normal water level (NWL) and low-level cutoff (LLCO) if NWL is at 2 in. below drum centerline and LLCO is 2 in. below NWL.

A: The volume of the water in the drum must include the volume due to the straight section plus the dished ends.
Volume in straight section, V_s, is given by

$$V_s = L \times R^2 \times \left(\frac{a}{57.3} - \sin a \times \cos a \right)$$

where a is the angle shown in Figure 1.5. The volume of liquid in each end is given by

Table 1.12   Effect of Directivity Based on Angle to Direction of Flow and Size of Silencer Outlet

Angle to direction of flow	Silencer outlet diameter (in.)	Octave band center frequency (Hz)							
		63	125	250	500	1K	2K	4K	8K
0°	72–96	+4	+5	+5	+6	+6	+7	+7	+7
	54–66	+3	+4	+4	+5	+5	+5	+5	+5
	36–48	+2	+3	+3	+4	+4	+4	+4	+4
	26–32	+1	+1	+2	+2	+2	+2	+2	+2
	16–24	0	0	+1	+1	+1	+1	+1	+1
	8–14	0	0	0	0	0	0	0	0
	6	0	0	0	0	0	0	0	0
45°	72–96	+2	+3	+3	+4	+4	+5	+5	+5
	54–66	+1	+2	+2	+3	+3	+3	+3	+3
	36–48	0	+1	+1	+2	+2	+2	+2	+2
	26–32	0	0	0	+1	+1	+1	+1	+1
	16–24	0	0	0	0	0	0	0	0
	8–14	0	0	0	0	0	0	0	0
	6	0	0	0	0	0	0	0	0
90 and 135°	72–96	-1	-2	-5	-7	-10	-12	-15	-17
	54–66	0	-1	-2	-5	-8	-10	-13	-16
	36–48	0	0	-1	-3	-6	-7	-11	-15
	26–32	0	0	0	-1	-3	-5	-9	-14
	16–24	0	0	0	0	-1	-3	-7	-13
	8–14	0	0	0	0	-1	-2	-5	-11
	5–6	0	0	0	0	0	-1	-3	-6
	4	0	0	0	0	0	0	-1	-3

Source: Burgess Manning.

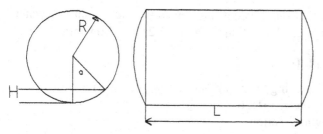

Figure 1.5   Partial volume of water in boiler drum.

$$V_e = 0.261 \times H^2 \times (3R - H)$$

where

H = straight length of drum
R = drum radius

In this case, $H = 120$ in. and $R = 21$ in.

Let us compute V_{s1} and V_{e1}, the volume of the straight section and each end corresponding to the 19-in. level from the bottom of the drum.

$$\cos a = \frac{2}{21} = 0.09523. \qquad \text{Hence } a = 84.5° \text{ and } \sin a$$

$$= 0.9954.$$

$$V_{s1} = 120 \times 21 \times 21 \times \left(\frac{84.53}{57.3} - 0.09523 \times 0.9954 \right)$$

$$= 73,051 \text{ cu in.}$$

$$V_{e1} = 0.261 \times 19 \times 19 \times (3 \times 21 - 19) = 4146 \text{ cu in.}$$

Hence total volume of liquid up to 19-in. level = 73,051 + 2 × 4146 = 81,343 cu in. = 47.08 cu ft.

Similarly, we can show that total volume of water up to the 17-in. level = 34.1 cu ft. Hence the difference is 13 cu ft.

Specific volume of water at 400 psig = 0.0193 cu ft/lb

$$\text{Normal evaporation rate} = 10,000 \times \frac{0.0193}{60}$$

$$= 3.2 \text{ cu ft/min}$$

Hence the duration between the levels assuming that the water supply has been discontinued $= 13/3.21 = 4.05$ min.

NOMENCLATURE

A	Area of opening, in.^{2}, or duct cross section, ft^2
A, B	factors used in Q1.26
BD	Blowdown, fraction
BHP	Boiler horsepower
c	Corrosion allowance, in.
C_c	Initial investment, $
C_d	Coefficient of discharge
C_e	Cost of electricity, $/kWh
C_p	Specific heat, Btu/lb °F
d	Tube or outer diameter, in.
d_i	Tube or pipe inner diameter, in.
e	Escalation factor
E	Sealing efficiency, %; ligament efficiency, fraction
F	Factor defined in Eq. (33)
G	Gas mass velocity, lb/ft^2 hr
h	Enthalpy, Btu/lb
H	Height of liquid column, in.
h_g, h_f	Enthalpy of saturated vapor and liquid, Btu/lb
H_g	Head of gas column, ft
H_l	Head of liquid, ft
H_w	Differential pressure across damper, in. WC
i	Interest rate
L	Distance, ft
LCC	Life-cycle cost, $
M	Moisture in air, lb/lb
MW	Molecular weight
N	Annual period of operation, hr
P_w	Partial pressure of water vapor, psia
P	Gas pressure, psia; design pressure, psig
ΔP	Differential pressure, psi
PWL	Sound power level
q	Volumetric flow, gpm or cfm
Q	Energy, Btu/hr

R Radius of drum, in.

RH   Relative humidity

s Specific gravity

S_a Allowable stress, psi

SBV   Steam by volume

SPL   Sound pressure level

SVP   Saturated vapor pressure, psia

t Fluid temperature, °F

t_w Minimum wall thickness of pipe or tube, in.

T Life of plant, years

v Specific volume, cu ft; subscripts g and f stand for saturated vapor and liquid

V_e, V_s Volume of drum ends, straight section, cu in.

V_g Velocity of gas

W Mass flow, lb/hr

x Steam quality

y Volume fraction

ρ Density, lb/cu ft; subscript g stands for gas

REFERENCES

1. V. Ganapathy, Determining operating parameters for hot exhaust gas cooling systems, *Plant Engineering*, Mar. 3, 1983, p. 182.
2. V. Ganapathy, Nomograph estimates steam leakage and cost, *Heating Piping and Air-Conditioning*, Nov. 1982, p. 101.
3. V. Ganapathy, Quick estimates of damper leakage and cost energy loss, *Oil and Gas Journal*, Sept. 21, 1981, p. 124.
4. V. Ganapathy, *Applied Heat Transfer*, PennWell Books, Tulsa, Okla., 1982, p. 186.
5. Brown and Yanuck, *Life Cycle Costing*, Fairmont Press, Atlanta, Ga., 1980, p. 188.
6. ASME, *Boiler and Pressure Vessel Code*, Sec. 1, New York, 1980, p. 119.
7. V. Ganapathy, Estimate maximum allowable pressures for steel piping, *Chemical Engineering*, July 25, 1983, p. 99.
8. ASME, *Boiler and Pressure Vessel Code*, Sec. 8, Div. 1, Para. UG 131, 1980.
9. ASME, *Boiler and Pressure Vessel Code*, Sec. 1, Para. PFT 51, 1989.

Fuels, Combustion, and Efficiency of Boilers and Heaters

2.07:   Determining energy, steam quantity, and electric heater capacity required for heating air

2.08:   Determining energy, steam quantity, and electric heater capacity required for heating fuel oils

2.09:   Combustion calculations from ultimate analysis of fuels; determining wet and dry air and flue gas quantities; volumetric analysis of flue gas on wet and dry basis; partial pressures of water vapor and carbon dioxide in flue gas; molecular weight and density of flue gas

2.10:   Combustion calculations on MM Btu basis; determining air and flue gas quantities in the absence of fuel data

2.11:   Estimating excess air from flue gas CO_2 readings

2.12:   Estimating excess air from CO_2 and O_2 readings; estimating excess air from O_2 readings alone

2.13:   Effect of reducing oxygen in flue gas; calculating flue gas produced; calculating energy saved and reduction in fuel cost

2.14:   Effect of fuel heating values on air and flue gas produced in boilers

2.15:   Determining combustion temperature of different fuels in the absence of fuel analysis

2.16a: Calculating ash concentration in flue gases

2.16b: Relating ash concentration between mass and volumetric units

2.17:   Determining melting point of ash knowing ash analysis

2.18:   Determining SO_2 and SO_3 in flue gases in lb/MM Btu and in ppm (volume)

2.19:   Determining efficiency of boilers and heaters; efficiency on HHV basis; dry gas loss; loss due to moisture and combustion of hydrogen; loss due to moisture in air; radiation loss; efficiency on LHV basis; wet flue gas loss; relating efficiencies on HHV and LHV basis

2.20:   Determining efficiency of boilers and heaters on HHV and LHV basis from flue gas analysis

2.21:   Loss due to CO formation

2.22:   Simple formula for efficiency determination

2.23:   Determining radiation losses in boilers and heaters if casing temperature and wind velocity are known

2.24:   Variation of heat losses and efficiency with boiler load

2.25a: Sulfur dew point of flue gases

2.25b: Computing acid dew points for various acid vapors

2.25c: Effect of gas temperature on corrosion potential

2.26a: Converting NO_x and CO from lb/hr to ppm for turbine exhaust gases

2.26b: Converting NO_x and CO from lb/hr to ppm for fired boilers

2.27:   Oxygen consumption versus fuel input for gas turbine exhaust gases

2.28:   Relating heat rates of engines to fuel consumption

2.01

Q: How are the HHV (higher heating value) and LHV (lower heating value) of fuels estimated when the ultimate analysis is known?

A: We can use the expressions [1]

$$HHV = 14{,}500 \times C + 62{,}000 \times \left(H_2 - \frac{O_2}{8}\right)$$
$$+ \ 4000 \times S \tag{1}$$
$$LHV = HHV - 9720 \times H_2 - 1110W \tag{2}$$

where W is the fraction by weight of moisture in fuel, and C, H_2, O_2, and S are fractions by weight of carbon, hydrogen, oxygen, and sulfur in the fuel.

If a coal has $C = 0.80$, $H_2 = 0.003$, $O_2 = 0.005$, $W = 0.073$, $S = 0.006$, and the rest ash, find its HHV and LHV. Substituting in Eqs. (1) and (2), we have

$$HHV = 14{,}500 \times 0.80 + 62{,}000 \times \left(0.003 - \frac{0.005}{8}\right)$$
$$+ \ 4000 \times 0.006 = 11{,}771 \ Btu/lb$$
$$LHV = 11{,}771 - 9720 \times 0.003 - 1110 \times 0.073$$
$$= 11{,}668 \ Btu/lb$$

Fuel inputs to furnaces and boilers and efficiencies are often specified without reference to the heating values, whether HHV or LHV, which is misleading.

If a burner has a capacity of Q MM Btu/hr (million Btu/hr) on an HHV basis, its capacity on an LHV basis would be

$$Q_{LHV} = Q_{HHV} \times \frac{LHV}{HHV} \tag{3a}$$

Similarly, if η_{HHV} and η_{LHV} are the efficiencies of a boiler on an HHV and an LHV basis, respectively, they are related as follows:

$$\eta_{HHV} \times HHV = \eta_{LHV} \times LHV \tag{3b}$$

2.02a

Q: How can we estimate the HHV and LHV of a fuel oil in the absence of its ultimate analysis?

A: Generally, the degree API of a fuel oil will be known, and the following expressions can be used:

$$HHV = 17,887 + 57.5 \times °API - 102.2 \times \%S \qquad (4a)$$
$$LHV = HHV - 91.23 \times \%H_2 \qquad (4b)$$

where $\%H_2$ is the % hydrogen, by weight.

$$\%H_2 = F - \frac{2122.5}{°API + 131.5} \qquad (5)$$

where

$F = 24.50$ for $0 \le °API \le 9$
$F = 25.00$ for $9 < °API \le 20$
$F = 25.20$ for $20 < °API \le 30$
$F = 25.45$ for $30 < °API \le 40$

HHV and LHV are in Btu/lb.

2.02b

Q: Determine the HHV and LHV of 30° API fuel oil in Btu/gal and in Btu/lb. Assume that %S is 0.5.

A: From Eq. (4a),

$$HHV = 17,887 + 57.5 \times 30 - 102.2 \times 0.5$$
$$= 19,651 \text{ Btu/lb}$$

To calculate the density or specific gravity of fuel oils we can use the expression

$$s = \frac{141.5}{131.5 + °API} = \frac{141.5}{131.5 + 30} = 0.876 \qquad (6)$$

Hence

Density $= 0.876 \times 8.335 = 7.3$ lb/gal

8.335 is the density of liquids in lb/gal when $s = 1$.

HHV in BTU gas $= 19,561 \times 7.3 = 142,795$

From Eq. (5),

$$\%H_2 = 25.2 - \frac{2122.5}{131.5 + 30} = 12.05$$

$$LHV = 19,561 - 91.23 \times 12.05 = 18,460 \text{ Btu/lb}$$
$$= 18,460 \times 7.3 = 134,758 \text{ Btu/gal}$$

2.03a

Q: A good way to compare fuel costs is to check their values per MM Btu fired. If coal having HHV $= 9500$ Btu/lb costs \$25/long ton, what is the cost in \$/MM Btu?

A: 1 long ton $= 2240$ lb. 1 MM Btu has $10^6/9500 = 105$ lb of coal. Hence 105 lb would cost

$$105 \times \frac{25}{2240} = \$1.17/\text{MM Btu}$$

2.03b

Q: If No. 6 fuel oil costs 30 cents/gal, is it cheaper than the coal in Q2.03a?

A: Table 2.1 gives the HHV of fuel oils. It is 152,400 Btu/gal. Hence 1 MM Btu would cost

$$\frac{10^6}{152,400} \times 0.30 = \$1.96/\text{MM Btu}$$

2.03c

Q: Which is less expensive, electricity at 1.5 cents/kWh or gas at \$3/MM Btu?

A: 3413 Btu $= 1$ kWh. At 1.5 cents/kWh, 1 MM Btu of electricity costs $(10^6/3413) \times 1.5/100 = \$4.4$. Hence in this case, electric-

Table 2.1 Typical Heat Contents of Various Oils

Typical of:	°API	Sp. gr. 60°F (15.6°C)	lb/gal	kg/m³	Gross Btu/gal	Gross kcal/liter	Wt % H	Net Btu/gal	Net kcal/liter	Sp. heat at 40°F	Sp. heat at 300°F	Temp. corr. (°API/°F)	ft³ 60°F air/gal	Ult. % CO₂
	0	1.076	8.969	1,075	160,426	10,681	8.359	153,664	10,231	0.391	0.504	0.045	1,581	—
	2	1.060	8.834	1,059	159,038	10,589	8.601	152,183	10,133	0.394	0.508	—	—	—
	4	1.044	8.704	1,043	157,692	10,499	8.836	150,752	10,037	0.397	0.512	—	—	18.0
	6	1.029	8.577	1,0028	156,384	10,412	9.064	149,368	9,945	0.400	0.516	0.048	1,529	17.6
	8	1.014	8.454	1,013	155,115	10,328	9.285	148,028	9,856	0.403	0.519	0.050	1,513	17.1
No. 6 oil	10	1.000	8.335	1,000	153,881	10,246	10.00	146,351	9,744	0.406	0.523	0.051	1,509	16.7
	12	0.986	8.219	985.0	152,681	10,166	10.21	145,100	9,661	0.409	0.527	0.052	1,494	16.4
	14	0.973	8.106	971.5	151,515	10,088	10.41	143,888	9,580	0.412	0.530	0.054	1,478	16.1
	16	0.959	7.996	958.3	150,380	10,013	10.61	142,712	9,502	0.415	0.534	0.056	1,463	15.8
	18	0.946	7.889	945.5	149,275	9,939	10.80	141,572	9,426	0.417	0.538	0.058	1,448	15.5
No. 5 oil	20	0.934	7.785	933.0	148,200	9,867	10.99	140,466	9,353	0.420	0.541	0.060	1,433	15.2
	22	0.922	7.683	920.9	147,153	9,798	11.37	139,251	9,272	0.423	0.545	0.061	1,423	14.9
No. 4 oil	24	0.910	7.585	909.9	146,132	9,730	11.55	138,210	9,202	0.426	0.548	0.063	1,409	14.7
	26	0.898	7.488	897.5	145,138	9,664	11.72	137,198	9,135	0.428	0.552	0.065	1,395	14.5
	28	0.887	7.394	886.2	144,168	9,599	11.89	136,214	9,069	0.431	0.555	0.067	1,381	14.3
No. 2 oil	30	0.876	7.303	875.2	143,223	9,536	12.06	135,258	9,006	0.434	0.559	0.089	1,368	14.0
	32	0.865	7.213	864.5	142,300	9,475	12.47	134,163	8,933	0.436	0.562	0.072	1,360	13.8
	34	0.855	7.126	854.1	141,400	9,415	12.63	133,259	8,873	0.439	0.566	0.074	1,347	13.6
	36	0.845	7.041	843.9	140,521	9,356	12.78	132,380	8,814	0.442	0.569	0.076	1,334	13.4
	38	0.835	6.958	833.9	139,664	9,299	12.93	131,524	8,757	0.444	0.572	0.079	1,321	13.3
No. 1 oil	40	0.825	6.877	824.2	138,826	9,243	13.07	130,689	8,702	0.447	0.576	0.082	1,309	13.1
	42	0.816	6.798	814.7	138,007	9,189	—	—	—	0.450	0.579	0.085	—	13.0
	44	0.806	6.720	805.4	137,207	9,136	—	—	—	0.452	0.582	0.088	—	12.8

Source: Adapted from Ref. 2.

ity is costlier than gas. This example serves to illustrate the conversion of units and does not imply that this situation will prevail in all regions.

2.04

Q: Estimate the annual fuel cost for a 300-MW coal-fired power plant if the overall efficiency is 40% and the fuel cost is $1.1/ MM Btu. The plant operates for 6000 hr/year.

A: Power plants have efficiencies in the range 35 to 42%. Another way of expressing this is to use the term *heat rate*, defined as

$$\text{Heat rate} = \frac{3413}{\text{efficiency, in fraction}} \text{ Btu/kWh}$$

In this case it is $3413/0.4 = 8530$ Btu/kWh.

Annual fuel cost = $1000 \times$ megawatt $\times$ heat rate $\times$ (hr/ year) $\times$ cost of fuel in $/MM Btu

$$= 1000 \times 300 \times 8530 \times 6000$$
$$\times \frac{1.1}{10^6} = \$16.9 \times 10^6$$

The fuel cost for any other type of power plant could be found in a similar fashion. Heat rates are provided by power plant suppliers.

2.05

Q: A 20-MM Btu/hr burner was firing natural gas of HHV = 1050 Btu/scf with a specific gravity of 0.6. If it is now required to burn propane having HHV = 2300 Btu/scf with a specific gravity of 1.5, and if the gas pressure to the burner was set at 4 psig earlier for the same duty, estimate the new gas pressure. Assume that the gas temperature in both cases is 60°F.

A: The heat input to the burner is specified on an HHV basis. The fuel flow rate would be Q/HHV, where Q is the duty in Btu/hr. The gas pressure differential between the gas pressure regulator and the furnace is used to overcome the flow resistance according to the equation

$$\Delta P = \frac{KW_f^2}{\rho} \tag{7}$$

where

ΔP = pressure differential, psi
K = a constant
ρ = gas density = $0.075s$ (s is the gas specific gravity; $s = 1$ for air)
W_f = fuel flow rate in lb/hr = flow in scfh $\times$ $0.075s$

Let the subscripts 1 and 2 denote natural gas and propane, respectively.

$$W_{f1} = \frac{20 \times 10^6}{1050} \times 0.075 \times 0.6$$

$$W_{f2} = \frac{20 \times 10^6}{2300} \times 0.075 \times 1.5$$

$\Delta P_1 = 4$, $\rho_1 = 0.075 \times 0.6$, and $\rho_2 = 0.075 \times 1.5$. Hence, from Eq. (7),

$$\frac{\Delta P_1}{\Delta P_2} = \frac{W_{f1}^2 \rho_2}{W_{f2}^2 \rho_1} = \frac{4}{\Delta P_2} = \frac{0.6}{(1050)^2} \times \frac{(2300)^2}{1.5}$$

or

$$\Delta P_2 = 2.08 \text{ psig}$$

Hence, if the gas pressure is set at about 2 psig, we can obtain the same duty. The calculation assumes that the back pressure has not changed.

2.06

Q: Gas flow measurement using displacement meters indicates actual cubic feet of gas consumed. However, gas is billed, generally, at reference conditions of 60°F and 14.65 psia (4 oz). Hence gas flow has to be corrected for actual pressure and temperature. Plant engineers should be aware of this conversion.

In a gas-fired boiler plant, 1000 cu ft of gas per hour was measured, gas conditions being 60 psig and 80°F. If the gas has a higher calorific value of 1050 Btu/scf, what is the cost of fuel consumed if energy costs $4/MM Btu?

A: The fuel consumption at standard conditions is found as follows.

$$V_s = V_a P_a \times \frac{T_s}{P_s T_a} \tag{8}$$

where

V_s, V_a = fuel consumption, standard and actual, cu ft/hr
T_s = reference temperature of 520°R
T_a = actual temperature, °R
P_s, P_a = standard and actual pressures, psia

$$V_s = 100 \times (30 + 14.22) \times \frac{520}{14.65 \times 540}$$

$$= 2900 \text{ scfh}$$

Hence

Energy used = 2900 × 1050 = 3.05 MM Btu/hr
Cost of fuel = 3.05 × 4 = $12.2/hr

If pressure and temperature corrections are not used, the displacement meter reading can lead to wrong fuel consumption data.

2.07

Q: Estimate the energy in Btu/hr and in kilowatts for heating 75,000 lb/hr of air from 90°F to 225°F. What is the steam quantity required if 200 psia saturated steam is used to accomplish the duty noted above? What size of electric heater would be used?

A: The energy required to heat the air can be expressed as

$$Q = W_a C_p \Delta T \tag{9}$$

where

Q = duty, Btu/hr
W_a = air flow, lb/hr

C_p = specific heat of air, Btu/lb °F

ΔT = temperature rise, °F

C_p may be taken as 0.25 for the specified temperature range.

$$Q = 75,000 \times 0.25 \times (225 - 90) = 2.53 \times 10^6 \text{ Btu/hr}$$

Using the conversion factor 3413 Btu = 1 kWh, we have

$$Q = 2.53 \times \frac{10^6}{3413} = 741 \text{ kW}$$

A 750-kW heater or the next higher size could be chosen.

If steam is used, the quantity can be estimated by dividing Q in Btu/hr by the latent heat obtained from the steam tables (see the Appendix). At 200 psia, the latent heat is 843 Btu/lb. Hence

$$\text{Steam required} = 2.55 \times \frac{10^6}{843} = 3046 \text{ lb/hr}$$

2.08

Q: Estimate the steam required at 25 psig to heat 20 gpm of 15° API fuel oil from 40°F to 180°F. If an electric heater is used, what should be its capacity?

A: Table 2.2 gives the heat content of fuel oils in Btu/gal [2]. At 180°F, enthalpy is 529 Btu/gal, and at 40°F it is 26 Btu/gal. Hence the energy absorbed by the fuel oil is

$$Q = 20 \times 60 \times (529 - 26) = 0.6 \times 10^6 \text{ Btu/hr}$$
$$= 0.6 \times \frac{10^6}{3413} = 175 \text{ kW}$$

Latent heat of steam (from the steam tables) is 934 Btu/lb at 25 psig or 40 psia. Hence

$$\text{Steam required} = 0.6 \times \frac{10^6}{934} = 646 \text{ lb/hr}$$

If an electric heater is used, its capacity will be a minimum of 175 kW. Allowing for radiation losses, we may choose a 200-kW heater.

Table 2.2   Heat Content (Btu/gal) of Various Oils[a]

Temp. (°F)	Gravity, °API at 60°F (15.6°C)							
	10	15	20	25	30	35	40	45
	Specific gravity, 60°F/60°F							
	1.0000	0.9659	0.9340	0.9042	0.8762	0.8498	0.8251	0.8017
32	0	0	0	0	0	0	0	0
	0	**0**	**0**	**0**	**0**	**0**	**0**	**0**
60	95	93	92	90	89	87	86	85
								965
100	237	233	229	226	222	219	215	
							1065	**1062**
120	310	305	300	295	290	286	281	
							1116	**1112**
140	384	378	371	366	360	355	349	
							1169	**1164**
160	460	453	445	438	431	425	418	
					1236		**1223**	**1217**
180	538	529	520	511	503	496	488	
					1293		**1278**	**1272**
200	617	607	596	587	577	569	560	
		1371			**1352**		**1335**	**1327**
220	697	686	674	663	652	643	633	
		1434			**1412**		**1393**	**1384**
240	779	766	753	741	729	718	707	
		1498			**1474**		**1452**	**1442**
260	862	848	833	820	807	795	783	
		1563			**1537**		**1513**	**1502**
300	1034	1017	999	984	968	954	939	
		1699			**1668**		**1639**	**1626**
400	1489	1463	1439	1416	1393	1372	1352	1333
		2088	**2064**	**2041**	**2018**	**1997**	**1977**	**1958**
500	1981	1947	1914	1884	1854	1826	1799	1774
		2497	**2464**	**2434**	**2404**	**2376**	**2349**	**2324**
600	2511	2467	2426	2387	2350	2314	2281	2248
		2942	**2901**	**2862**	**2825**	**2789**	**2756**	**2723**
700	3078	3025	2974	2927	2881	2837	2796	2756
	3478	**3425**	**3374**	**3327**	**3281**	**3237**	**3196**	**3156**
800	3683	3619	3559	3502	3447	3395	3345	3297
	4008	**3944**	**3884**	**3827**	**3772**	**3720**	**3670**	**3622**

[a]Values in regular type are for liquid; **bold values** are for vapor.

In the absence of information on fuel oil enthalpy, use a specific gravity of 0.9 and a specific heat of 0.5 Btu/lb °F. Hence the duty will be

$$Q = 20 \times 60 \times 62.40 \times \frac{0.9}{7.48} \times .05 \times (180 - 40)$$

$$= 0.63 \times 10^6 \text{ Btu/hr}$$

(7.48 is the conversion factor from cubic feet to gallons.)

2.09a

Q: Natural gas having CH_4 = 83.4%, C_2H_6 = 15.8%, and N_2 = 0.8% by volume is fired in a boiler. Assuming 15% excess air, 70°F ambient, and 80% relative humidity, perform detailed combustion calculations and determine flue gas analysis.

A: From Chapter 1 we know that air at 70°F and 80% RH has a moisture content of 0.012 lb/lb dry air. Table 2.3 can be used to figure air requirements of various fuels. For example, we see that CH_4 requires 9.53 mol of air per mole of CH_4, and C_2H_6 requires 16.68 mol.

Let us base our calculations on 100 mol of fuel. The theoretical dry air required will be

83.4 × 9.53 + 16.68 × 15.8 = 1058.3 mol

Considering 15% excess,

Actual dry air = 1.15 × 1058.3 = 1217 mol
Excess air = 0.15 × 1058.3 = 158.7 mol
Excess O_2 = 158.7 × 0.12 = 33.3 mol
Excess N_2 = 1217 × 0.79 = 961 mol

(Air contains 21% by volume O_2, and the rest is N_2.)

$$\text{Moisture in air} = 1217 \times 29 \times \frac{0.012}{18} = 23.5 \text{ mol}$$

(We multiplied moles of air by 29 to get its weight, and then the water quantity was divided by 18 to get moles of water.)

Table 2.3 can also be used to get the moles of CO_2, H_2O, and O_2 [3].

Table 2.3 Combustion Constants

No. Substance	Formula	Mol. Wt[a]	Lb per Cu ft[b]	Cu ft per lb[b]	Sp gr air = 1.000[b]	Heat of combustion[c] Btu per cu ft Gross	Net[d]	Btu per lb Gross	Net[d]
1 Carbon	C	12.01	—	—	—	—	—	14,093[g]	14,093[g]
2 Hydrogen	H$_2$	2.016	0.005327	187.723	0.06959	325.0	275.0	61,100	51,623
3 Oxygen	O$_2$	32.000	0.08461	11.819	1.1053	—	—	—	—
4 Nitrogen (atm)	N$_2$	28.016	0.07439[e]	13.443[e]	0.9718[e]	—	—	—	—
5 Carbon monoxide	CO	28.01	0.07404	13.506	0.9672	321.8	321.8	4,347	4,347
6 Carbon dioxide	CO$_2$	44.01	0.1170	8.548	1.5282	—	—	—	—
Paraffin series C$_n$H$_{2n+2}$									
7 Methane	CH$_4$	16.041	0.04243	23.565	0.5543	1013.2	913.1	23,879	21,520
8 Ethane	C$_2$H$_6$	30.067	0.08029[e]	12.455[e]	1.04882[e]	1792	1641	22,320	20,432
9 Propane	C$_3$H$_8$	44.092	0.1196[e]	8.365[e]	1.5617[e]	2590	2385	21,661	19,944
10 n-Butane	C$_4$H$_{10}$	58.118	0.1582[e]	6.321[e]	2.06654[e]	3370	3113	21,308	19,680
11 Isobutane	C$_4$H$_{10}$	58.118	0.1582[e]	6.321[e]	2.06654[e]	3363	3105	21,257	19,629
12 n-Pentane	C$_5$H$_{12}$	72.144	0.1904[e]	5.252[e]	2.4872[e]	4016	3709	21,091	19,517
13 Isopentane	C$_5$H$_{12}$	72.144	0.1904[e]	5.252[e]	2.4872[e]	4008	3716	21,052	19,478
14 Neopentane	C$_5$H$_{12}$	72.144	0.1904[e]	5.252[e]	2.4872[e]	3993	3693	20,970	19,396
15 n-Hexane	C$_6$H$_{14}$	86.169	0.2274[e]	4.398[e]	2.9704[e]	4762	4412	20,940	19,403
Olefin series C$_n$H$_{2n}$									
16 Ethylene	C$_2$H$_4$	28.051	0.07456	13.412	0.9740	1613.8	1513.2	21,644	20,295
17 Propylene	C$_3$H$_6$	42.077	0.1110[e]	9.007[e]	1.4504[e]	2336	2186	21,041	19,691
18 n-Butene (butylene)	C$_4$H$_8$	56.102	0.1480[e]	6.756[e]	1.9336[e]	3084	2885	20,840	19,496
19 Isobutene	C$_4$H$_8$	56.102	0.1480[e]	6.756[e]	1.9336[e]	3068	2869	20,730	19,382
20 n-Pentene	C$_5$H$_{10}$	70.128	0.1852[e]	5.400[e]	2.4190[e]	3836	3586	20,712	19,363
Aromatic series C$_n$H$_{2n-6}$									
21 Benzene	C$_6$H$_6$	78.107	0.2060[e]	4.852[e]	2.6920[e]	3751	3601	18,210	17,480
22 Toluene	C$_7$H$_8$	92.132	0.2431[e]	4.113[e]	3.1760[e]	4484	4284	18,440	17,620
23 Xylene	C$_8$H$_{10}$	106.158	0.2803[e]	3.567[e]	3.6618[e]	5230	4980	18,650	17,760
Miscellaneous gases									
24 Acetylene	C$_2$H$_2$	26.036	0.06971	14.344	0.9107	1499	1448	21,500	20,776
25 Naphthalene	C$_{10}$H$_8$	128.162	0.3384[e]	2.955[e]	4.4208[e]	5854[f]	5654[f]	17,298[f]	16,708[f]
26 Methyl alcohol	CH$_3$OH	32.041	0.0846[e]	11.820[e]	1.1052[e]	867.9	768.0	10,259	9,078
27 Ethyl alcohol	C$_2$H$_5$OH	46.067	0.1216[e]	8.221[e]	1.5890[e]	1600.3	1450.5	13,161	11,929
28 Ammonia	NH$_3$	17.031	0.0456[e]	21.914[e]	0.5961[e]	441.1	365.1	9,668	8,001
29 Sulfur	S	32.06	—	—	—	—	—	3,983	3,983
30 Hydrogen sulfide	H$_2$S	34.076	0.09109[e]	10.979[e]	1.1898[e]	647	596	7,100	6,545
31 Sulfur dioxide	SO$_2$	64.06	0.1733	5.770	2.264	—	—	—	—
32 Water vapor	H$_2$O	18.016	0.04758[e]	21.017[e]	0.6215[e]	—	—	—	—
33 Air	—	28.9	0.07655	13.063	1.0000	—	—	—	—

All gas volumes corrected to 60°F and 30 in Hg dry. For gases saturated with water at 60°F, 1.73% of the Btu value must be deducted.

[a]Calculated from atomic weights given in *Journal of the American Chemical Society*, February 1937.

[b]Densities calculated from values given in grams per liter at 0°C and 760 mmHs in the International Critical Tables allowing for the known deviations from the gas laws. Where the coefficient of expansion was not available, the assumed value was taken as 0.0037 per °C. Compare this with 0.003662 which is the coefficient for a perfect gas. Where no densities were available, the volume of the mole was taken as 22.4115 liters.

[c]Converted to mean Btu per lb (1/180 of the heat per lb of water from 32 to 212°F) from data by Frederick D. Rossini, National Bureau of Standards, letter of April 10, 1937, except as noted.

Cu ft per cu ft of combustible						Lb per lb of combustible						Experimental error in heat of combustion ($\pm\%$)
Required for combustion			Flue products			Required for combustion			Flue products			
O_2	N_2	Air	CO_2	H_2O	N_2	O_2	N_2	Air	CO_2	H_2O	N_2	
—	—	—	—	—	—	2.664	8.863	11.527	3.664	—	8.863	0.012
0.5	1.882	2.382	—	1.0	1.882	7.937	26.407	34.344	—	8.937	26.407	0.015
—	—	—	—	—	—	—	—	—	—	—	—	—
0.5	1.882	2.382	1.0	—	1.882	0.571	1.900	2.471	1.571	—	1.900	0.045
—	—	—	—	—	—	—	—	—	—	—	—	—
2.0	7.528	9.528	1.0	2.0	7.528	3.990	13.275	17.265	2.744	2.246	13.275	0.033
3.5	13.175	16.675	2.0	3.0	13.175	3.725	12.394	16.119	2.927	1.798	12.394	0.030
5.0	18.821	23.821	3.0	4.0	18.821	3.629	12.074	15.703	2.994	1.634	12.074	0.023
6.5	24.467	30.967	4.0	5.0	24.467	3.579	11.908	15.487	3.029	1.550	11.908	0.022
6.5	24.467	30.967	4.0	5.0	24.467	3.579	11.908	15.487	3.029	1.550	11.908	0.019
8.0	30.114	38.114	5.0	6.0	30.114	3.548	11.805	15.353	3.050	1.498	11.805	0.025
8.0	30.114	38.114	5.0	6.0	30.114	3.548	11.805	15.353	3.050	1.498	11.805	0.071
8.0	30.114	38.114	5.0	6.0	30.114	3.548	11.805	15.353	3.050	1.498	11.805	0.11
9.5	35.760	45.260	6.0	7.0	35.760	3.528	11.738	15.266	3.064	1.464	11.738	0.05
3.0	11.293	14.293	2.0	2.0	11.293	3.422	11.385	14.807	3.138	1.285	11.385	0.021
4.5	16.939	21.439	3.0	3.0	16.939	3.422	11.385	14.807	3.138	1.285	11.385	0.031
6.0	22.585	28.585	4.0	4.0	22.585	3.422	11.385	14.807	3.138	1.285	11.385	0.031
6.0	22.585	28.585	4.0	4.0	22.585	3.422	11.385	14.807	3.138	1.285	11.385	0.031
7.5	28.232	35.732	5.0	5.0	28.232	3.422	11.385	14.807	3.138	1.285	11.385	0.037
7.5	28.232	35.732	6.0	3.0	28.232	3.073	10.224	13.297	3.381	0.692	10.224	0.12
9.0	33.878	32.878	7.0	4.0	33.878	3.126	10.401	13.527	3.344	0.782	10.401	0.21
10.5	39.524	50.024	8.0	5.0	39.524	3.165	10.530	13.695	3.317	0.849	10.530	0.36
2.5	9.411	11.911	2.0	1.0	9.411	3.073	10.224	13.297	3.381	0.692	10.224	0.16
12.0	45.170	57.170	10.0	4.0	45.170	2.996	9.968	12.964	3.434	0.562	9.968	—r
1.5	5.646	7.146	1.0	2.0	5.646	1.498	4.984	6.482	1.374	1.125	4.984	0.027
3.0	11.293	14.293	2.0	3.0	11.293	2.084	6.934	9.018	1.922	1.170	6.934	0.030
0.75	2.823	3.573	—	1.5	3.323	1.409	4.688	6.097	— (SO_2)	1.587	5.511	0.088
—	—	—	— (SO_2)	—	—	0.998	3.287	4.285	1.998	—	3.287	0.071
1.5	5.646	7.146	1.0 (SO_2)	1.0	5.646	1.409	4.688	6.097	1.880	0.529	4.688	0.30
—	—	—	—	—	—	—	—	—	—	—	—	—

dDeduction from gross to net heating value determined by deducting 18,919 Btu per pound mol of water in the products of combustion. Osborne, Stimson and Ginnings, *Mechanical Engineering*, p. 163, March 1935, and Osborne, Stimson, and Flock, National Bureau of Standards Research Paper 209.

eDenotes that either the density or the coefficient of expansion has been assumed. Some of the materials cannot exist as gases at 60°F and 30 in. Hg pressure, in which cases the values are theoretical ones given for ease of calculation of gas problems. Under the actual concentrations in which these materials are present their partial pressure is low enough to keep them as gases.

fFrom third edition of *Combustion*.

gNtional Bureau of Standards, RP 1141.

Source: Reprinted from *Fuel Flue Gases*, 1941 edition, courtesy of American Gas Association.

$$CO^2 = 1 \times 83.4 + 2 \times 15.8 = 115 \text{ mol}$$
$$H_2O = 2 \times 83.4 + 3 \times 15.8 + 23.5 = 237.7 \text{ mol}$$
$$O_2 = 33.3 \text{ mol}$$
$$N_2 = 961 + 0.8 = 961.8 \text{ mol}$$

The total moles of flue gas produced is $115 + 237.7 + 33.3 + 961.8 = 347.8$. Hence

$$\%CO_2 = \frac{115}{1347.8} \times 100 = 8.5$$

Similarly,

$$\%H_2O = 17.7, \qquad \%O_2 = 2.5, \qquad \%N_2 = 71.3$$

The analysis above is on a wet basis. On a dry flue gas basis,

$$\%CO_2 = 8.5 \times \frac{100}{100 - 17.7} = 10.3\%$$

Similarly,

$$\%O_2 = 3.0\%, \qquad \%N_2 = 86.7\%$$

To obtain w_{da}, w_{wa}, w_{dg}, and w_{wg}, we need the density of the fuel or the molecular weight, which is

$$\frac{1}{100} \times (83.4 \times 16 + 15.8 \times 30 + 0.8 \times 28) = 18.30$$

$$w_{da} = 1217 \times \frac{29}{100 \times 18.3} = 19.29 \text{ lb dry air/lb fuel}$$

$$w_{wa} = 19.29 + \frac{23.5 \times 18}{18.3 \times 100} = 19.52 \text{ lb wet air/lb fuel}$$

$$w_{dg} = \frac{115 \times 44 + 33.3 \times 32 + 961 \times 28}{1830} = 18 \text{ lb dry gas/lb fuel}$$

$$w_{wg} = \frac{115 \times 44 + 33.3 \times 32 + 237.7 \times 18 + 961.8 \times 28}{1830}$$

$$= 20.40 \text{ lb wet gas/lb fuel}$$

This procedure can be used when the fuel analysis is given. More often, plant engineers will be required to perform estimates of air for combustion without a fuel analysis. In such situations, the MM Btu basis of combustion and calculations will come in handy. This is discussed in Q2.10.

2.09b

Q: For the case stated in Q2.09a, estimate the partial pressure of water vapor, p_w, and of carbon dioxide, p_c, in the flue gas. Also estimate the density of flue gas at 300°F.

A: The partial pressures of water vapor and carbon dioxide are important in the determination of nonluminous heat transfer coefficients.

$$p_w = \frac{\text{volume of water vapor}}{\text{total flue gas volume}} = 0.177 \text{ atm} = 2.6 \text{ psia}$$

$$p_c = \frac{\text{volume of carbon dioxide}}{\text{total flue gas volume}} = 0.085 \text{ atm} = 1.27 \text{ psia}$$

To estimate the gas density, its molecular weight must be obtained (see Q1.05).

$$MW = \Sigma (MW_i \times y_i)$$

$$= \frac{28 \times 71.3 + 18 \times 17.7 + 32 \times 2.5 + 44 \times 8.5}{100}$$

$$= 27.7$$

Hence, from Eq. (6),

$$\rho_g = 27.7 \times 492 \times \frac{14.7}{359 \times 760 \times 14.7} = 0.05 \text{ lb/cu ft}$$

The gas pressure was assumed to be 14.7 psia. In the absence of flue gas analysis, we can obtain the density as discussed in Q1.03.

$$\rho_g = \frac{40}{760} = 0.052 \text{ lb/cu ft}$$

2.10a

Q: Discuss the basis for the million Btu method of combustion calculations.

A: Each fuel such as natural gas, coal, or oil requires a certain amount of stoichiometric air per MM Btu fired (on an HHV basis). This quantity does not vary much with the fuel analysis and hence has become a valuable method of evaluating combustion air and flue gas quantities produced when fuel gas analysis is not available.

For solid fuels such as coal and oil, the dry stoichiometric air w_{da} in lb/lb fuel can be obtained from

$$w_{da} = 11.53 \times C + 34.34 \times \left(H_2 - \frac{O_2}{8}\right) + 4.29 \times S$$

where C, H_2, O_2, and S are carbon, hydrogen, oxygen, and sulfur in the fuel in fraction by weight.

For gaseous fuels, w_{da} is given by

$$w_{da} = 2.47 \times CO + 34.34 \times H_2 + 17.27 \times CH_4$$
$$+ 13.3 \times C_2H_2 + 14.81 \times C_2H_4$$
$$+ 16.12 \times C_2H_6 - 4.32 \times O_2$$

EXAMPLE 1

Let us compute the amount of air required per MM Btu fired for fuel oil. C = 0.875, H = 0.125, and °API = 28.

Solution. From (4a),

HHV = $17,887 + 57.5 \times 28 - 102.2 \times 0$
      = 19,497 Btu/lb

Amount of air in lb/lb fuel from the above equation is

$$w_{da} = 11.53 \times 0.875 + 34.34 \times 0.125 = 14.38 \text{ lb/lb}$$
$$\text{fuel}$$

1 MM Btu of fuel fired requires $(1 \times 10^6)/19,497 = 51.28$ lb of fuel. Hence, from the above, 51.28 lb of fuel requires

$$51.28 \times 14.38 = 737 \text{ lb of dry air}$$

Table 2.4 shows a range of 735 to 750. To this must be added excess air; the effect of moisture in the air should also be considered.

EXAMPLE 2

Let us take the case of natural gas with the following analysis: methane = 83.4%, ethane = 15.8%, and nitrogen = 0.8%. *Solution.* Converting this to percent weight basis, we have

Table 2.4 Combustion Constant A For Fuels

No.	Fuel	A
1	Blast furnace gas	575
2	Bagasse	650
3	Carbon monoxide gas	670
4	Refinery and oil gas	720
5	Natural gas	730
6	Furnace oil and lignite	745–750
7	Bituminous coals	760
8	Anthracite	780
9	Coke	800

Fuel	% Vol	MW	Col 2 × col 3	% Wt
CH_4	83.4	16	1334.4	72.89
C_2H_6	15.8	30	474	25.89
N_2	0.8	28	22.4	1.22

Let us compute the air required in lb/lb fuel.
From Table 2.3,

$$\text{Air required} = 17.265 \times 0.7289 + 16.119 \times 0.2589$$
$$= 16.75 \text{ lb/lb fuel}$$
$$\text{HHV of fuel} = 0.7289 \times 23{,}876 + 0.2589 \times 22{,}320$$
$$= 23{,}181 \text{ Btu/lb}$$

where 23,876 and 22,320 are HHV of methane and ethane from Table 2.3.

The amount of fuel equivalent to 1 MM Btu would be $(1 \times 10^6)/23{,}181 = 43.1$ lb, which requires $43.1 \times 16.75 = 722$ lb of air, or 1 MM Btu fired would need 722 lb of dry air; this is close to the value indicated in Table 2.4.

Let us take the case of 100% methane and see how much air it needs for combustion. From Table 2.3, air required per lb of methane is 17.265 lb, and its heating value is 23,879 Btu/lb. In this case 1 MM Btu is equivalent to $(1 \times 10^6)/23{,}879 = 41.88$ lb of fuel, which requires $41.88 \times 17.265 = 723$ lb of dry air.

Taking the case of propane, 1 lb requires 15.703 lb of air. 1 MM Btu equals $1 \times 10^6/21{,}661 = 46.17$ lb of fuel. This would require $46.17 \times 15.703 = 725$ lb of air.

Thus for all fossil fuels we can come up with a good estimate of theoretical dry air per MM Btu fired on an HHV basis, and gas analysis does not affect this value significantly. The amount of air per MM Btu is termed A and is shown in Table 2.4 for various fuels.

2.10b

Q: A fired heater is firing natural gas at an input of 75 MM Btu/hr on an HHV basis. Determine the dry combustion air required at 10% excess air and the amount of flue gas produced if the HHV of fuel is 20,000 Btu/lb.

A: From Table 2.4, A is 730 lb/MM Btu. Hence the total air required is

$$W_a = 75 \times 1.1 \times 730 = 60{,}200 \text{ lb/hr}$$

The flue gas produced is

$$W_g = W_a + W_f = 60{,}200 + \frac{10^6}{20{,}000} = 60{,}250 \text{ lb/hr}$$

These values can be converted to volume at any temperature using the procedure described in Chapter 1.

The MM Btu method is quite accurate for engineering purposes such as fan selection and sizing of ducts and air and gas systems. Its advantage is that fuel analysis need not be known, which is generally the case in power and process plants. The efficiency of heaters and boilers can also be estimated using the MM Btu method of combustion calculations.

2.10c

Q: A coal-fired boiler is firing coal of HHV = 9500 Btu/lb at 25% excess air. If ambient conditions are 80°F, relative humidity 80%, and flue gas temperature 300°F, estimate the combustion air in lb/lb fuel, the volume of combustion air in cu ft/lb fuel, the flue gas produced in lb/lb fuel, and the flue gas volume in cu ft/lb fuel.

A: Since the fuel analysis is not known, let us use the MM Btu method. From Table 2.4, $A = 760$ for coal. 1 MM Btu requires $760 \times 2.15 = 950$ lb of dry air. At 80% humidity and 80°F, air contains 0.018 lb of moisture per pound of air (Chapter 1). Hence the wet air required per MM Btu fired is 950×1.018 lb. Also, 1 MM Btu fired equals $10^6/9500 = 105$ lb of coal. Hence

$$w_{da} = \text{dry air, lb/lb fuel} = \frac{950}{105} = 9.05$$

$$w_{wa} = \text{wet air, lb/lb fuel} = 950 \times \frac{1.018}{950} = 9.21$$

$$\rho_a = \text{density of air at 80°F} = 29 \times \frac{492}{359 \times 540}$$

$$= 0.0736 \text{ lb/cu ft (see Ch. 1, Q. 1.03)}$$

Hence

$$\text{Volume of air} = \frac{9.21}{0.0736} = 125 \text{ cu ft/lb fuel}$$

$$\rho_g = \text{density of flue gas} = \frac{40}{760} = 0.0526 \text{ lb/cu ft}$$

$$w_{dg} = \text{dry flue gas in lb/lb fuel} = \frac{950 + 105}{105} = 10.05$$

$$\text{Volume of flue gas, cu ft/lb fuel} = \frac{10.05}{0.0526} = 191$$

2.11

Q: Is there a way to figure the excess air from flue gas CO_2 readings?

A: Yes. A good estimate of excess air E in percent can be obtained from the equation

$$E = 100 \times [(K_1/\%CO_2) - 1] \tag{10a}$$

Table 2.5 K_1 Factors for Fuels

Fuel type	K_1
Bituminous coals	18.6
Coke	20.5
Oil	15.5
Refinery gas and gas oil	13.4
Natural gas	12.5
Blast furnace gas	25.5

Source: Ref. 1.

%CO_2 is the percent of the carbon dioxide in dry flue gas by volume, and K_1 is a constant depending on the type of fuel, as seen in Table 2.5. For example, if %CO_2 = 15 in flue gas in a coal-fired boiler, then for bituminous coal (K_1 = 18.6),

$$E = 100 \times \left(\frac{18.6}{15} - 1 \right) = 24\%$$

2.12

Q: Discuss the significance of %CO_2 and %O_2 in flue gases.

A: Excess air levels in flue gas can be estimated if the %CO_2 and %O_2 in dry flue gas by volume is known. The higher the excess air, the higher the flue gas quantity and the greater the losses. Plant engineers should control excess air levels to help control plant operating costs. The cost of operation with high excess air is discussed in Q2.13.

A formula that is widely used to figure the excess air is [1]

$$E = 100 \times \frac{O_2 - CO/2}{0.264 \times N_2 - (O_2 - CO/2)} \tag{10b}$$

where O_2, CO, and N_2 are the oxygen, carbon monoxide, and nitrogen in dry flue gas, vol %, and E is the excess air, %.

Another formula that is quite accurate is [1]

Table 2.6   Constant K_2 Used in Eq. (10c)

Fuel	K_2
Carbon	100
Hydrogen	80
Carbon monoxide	121
Sulfur	100
Methane	90
Oil	94.5
Coal	97
Blast furnace gas	223
Coke oven gas	89.3

Source: Ref. 1.

$$E = K_2 \times \frac{O_2}{21 - O_2} \tag{10c}$$

where K_2 is a constant that depends on the type of fuel (see Table 2.6).

2.13

Q: In a natural gas boiler of capacity 50 MM Btu/hr (HHV basis), the oxygen level in the flue gas is reduced from 3.0% to 2.0%. What is the annual savings in operating costs if fuel costs \$4/ MM Btu? The HHV of the fuel is 19,000 Btu/lb. The exit gas temperature is 500°F, and the ambient temperature is 80°F.

A: The original excess air is $90 \times 3/(21 - 3) = 15\%$ (see Q2.12). The excess air is now

$$E = 90 \times \frac{2.0}{21 - 2} = 9.47\%$$

With 15% excess, the approximate air required (see Q2.10a) is $50 \times 746 \times 1.15 = 42,895$ lb/hr.

$$\text{Flue gas} = 42,895 + 50 \times \frac{10^6}{19,000} = 45,256 \text{ lb/hr}$$

With 9.47% excess air,

Air required $= 50 \times 746 \times 1.0947 = 40{,}832$ lb/hr

Flue gas produced $= 40{,}832 + 50 \times \dfrac{10^6}{19{,}000}$

$= 43{,}463$ lb/hr

Reduction in heat loss $= (45{,}526 - 43{,}463) \times 0.25$
$\times (500 - 80) = 0.22$ MM Btu/hr

This is equivalent to an annual savings of $0.22 \times 4 \times 300 \times 24$ = \$6336. (We assumed 300 days of operation a year.) This could be a significant savings considering the life of the plant. Hence plant engineers should operate the plant realizing the implications of high excess air and high exit gas temperature. Oxygen levels can be continuously monitored and recorded and hooked up to combustion air systems in order to operate the plant more efficiently. (It may be noted that exit gas temperature will also be reduced if excess air is reduced. The calculation above indicates the minimum savings that can be realized.)

2.14

Q: Fuels are often interchanged in boiler plants because of relative availability and economics. It is desirable, then, to analyze the effect on the performance on the system. Discuss the implications of burning coal of 9800 Btu/lb in a boiler originally intended for 11,400-Btu/lb coal.

A: Let us assume that the duty does not change and that the efficiency of the unit is not altered. However, the fuel quantity will change. Combustion air required, being a function of MM Btu fired, will not change, but the flue gas produced will increase. Let us prepare a table.

We can use the same fans, as the variation in flue gas produced is not significant to warrant higher gas pressure drops. We must look into other aspects, such as the necessity of higher

	Coal 1	Coal 2
Fuel HHV, Btu/lb	11,400	9800
Fuel fired per MM Btu (10^6/HHV)	87	102
Air required per MM Btu (25% excess air)	$760 \times 1.25 = 950$	$760 \times 1.25 = 950$
Flue gas, lb	1037	1052
Ratio of flue gas	1	1.015

combustion air temperature (due to higher moisture in the fuel), ash concentration, and fouling characteristics of the new fuel. If a different type of fuel is going to be used, say oil, this will be a major change, and the fuel-handling system burners and furnace design will have to be reviewed. The gas temperature profiles will change owing to radiation characteristics, and absorption of surfaces such as superheaters and economizers will be affected. A discussion with the boiler design engineers will help.

2.15

Q: What is meant by combustion temperature of fuels? How is it estimated?

A: The adiabatic combustion temperature is the maximum temperature that can be attained by the products of combustion of fuel and air. However, because of dissociation and radiation losses, this maximum is never attained. Estimation of temperature after dissociation involves solving several equations. For purposes of estimation, we may decrease the adiabatic combustion temperature by 3 to 5% to obtain the actual combustion temperature. From energy balance it can be shown that

$$t_c = \frac{\text{LHV} + A\alpha \times \text{HHV} \times C_{pa} \times (t_a - 80)/10^6}{(1 - \% \text{ ash}/100 + A\alpha \times \text{HHV}/10^6) \times C_{pg}} \quad (11)$$

where

$\qquad$ LHV, HHV $=$ lower and higher calorific value of fuel, Btu/lb

$\qquad\qquad$ $A =$ theoretical air required per million Btu fired, lb

$\qquad\qquad$ $\alpha =$ excess air factor $= 1 + E/100$

$\qquad\qquad$ $t_a, t_c =$ temperature of air and combustion temperature, °F

$\qquad$ $C_{pa}, C_{pg} =$ specific heats of air and products of combustion, Btu/lb °F

For example, for fuel oil with combustion air at 300°F, LHV $=$ 17,000 Btu/lb, HHV $=$ 18,000 Btu/lb, $\alpha = 1.15$, and $A = 745$ (see Table 2.4). We have

$$t_c = \frac{17,000 + 745 \times 1.15 \times 18,000 \times 0.25 \times (300 - 80)/10^6}{(1 + 745 \times 1.15 \times 18,000/10^6) \times 0.32}$$

$$= 3400°F$$

C_{pa} and C_{pg} were taken as 0.25 and 0.32, respectively.

2.16a

Q: How is the ash concentration in flue gases estimated?

A: Particulate emission data are needed to size dust collectors for coal-fired boilers. In coal-fired boilers, about 75% of the ash is carried away by the flue gases and 25% drops into the ash pit. The following expression may be derived using the MM Btu method of combustion calculation [5]:

$$C_a = \frac{240,000 \times (\% \, ash/100)}{T \times [7.6 \times 10^{-6} \times HHV \times (100 + E) + 1 - (\% \, ash/100)]} \qquad (12a)$$

where

$\qquad$ $C_a =$ ash concentration, grains/cu ft

$\qquad$ $E =$ excess air, %

$\qquad$ $T =$ gas temperature, °R

$\qquad$ HHV $=$ heating value, Btu/lb

EXAMPLE

If coals of HHV = 11,000 Btu/lb having 11% ash are fired in a boiler with 25% excess air and the flue gas temperature is 850°R, determine the ash concentration.

Solution. Substituting into Eq. (12a), we have

$$C_a = \frac{240,000 \times 0.11}{850 \times (7.6 \times 10^{-6} \times 11,000 \times 125 + 1 - 0.11)}$$

$$= 2.75 \text{ grains/cu ft}$$

2.16b

Q: How do you convert the ash concentration in the flue gas in % weight to grains/acf or grains/scf?

A: Flue gases from incineration plants or solid fuel boilers contain dust or ash, and often these components are expressed in mass units such as lb/hr or % by weight while engineers involved in selection of pollution control equipment prefer to work in terms of grains/acf or grains/scf (actual and standard cubic feet). The relation is as follows:

$$C_a = 0.01 \times A \times 7000 \times \rho = 70A \qquad (12b)$$

where

ρ = gas density, lb/cu ft = $39.5/(460 + t)$

t = gas temperature, °F

C_a = ash content, grains/acf or grains/scf depending on whether density is computed at actual temperature or at 60°F.

A = ash content, % by weight

The expression for density is based on atmospheric flue gases having a molecular weight of 28.8 (see Q1.03).

Flue gases contain 1.5% by weight of ash. The concentration in grains/acf at 400°F is

$$C_a = 70 \times 1.5 \times \frac{39.5}{860} = 4.8 \text{ grains/acf}$$

and at 60°F,

$$C_a = 70 \times 1.5 \times \frac{39.5}{520} = 7.98 \text{ grains/scf}$$

2.17

Q: Discuss the importance of the melting point of ash in coal-fired boilers. How is it estimated?

A: In the design of steam generators and ash removal systems, the ash fusion temperature is considered an important variable. Low ash fusion temperature may cause slagging and result in deposition of molten ash on surfaces like superheaters and furnaces. The furnace will then absorb less energy, leading to higher furnace exit gas temperatures and overheating of superheaters.

A quick estimate of ash melting temperature in °C can be made using the expression [6]

$$
\begin{aligned}
t_m = {} & 19 \times Al_2O_3 + 15 \times (SiO_2 \times TiO_2) \\
& + 10 \times (CaO + MgO) \\
& + 6 \times (Fe_2O_3 + Na_2O + K_2O) \qquad (13)
\end{aligned}
$$

where t_m is the fusion temperature in °C, and the rest of the terms are percent ash content of oxides of aluminum, silicon, titanium, calcium, magnesium, iron, sodium, and potassium.

EXAMPLE

Analysis of a given ash indicates the following composition:

$$Al_2O_3 = 20\%, \quad SiO_2 + TiO_2 = 30\%$$
$$Fe_2O_3 + Na_2O + K_2O = 20\%, \quad CaO + MgO = 15\%$$

Find the fusion temperature.

Solution Substituting into Eq. (13), we find that $t_m = 1100°C$.

2.18a

Q: What is the emission of SO_2 in lb/MM Btu if coals of HHV = 11,000 Btu/lb and having 1.5% sulfur are fired in a boiler?

A: The following expression gives e, the emission of SO_2 in lb/MM Btu:

$$e = 2 \times 10^4 \ \frac{S}{HHV} \tag{14}$$

where S is the % sulfur in the fuel.

$$e = 2 \times 10^4 \times \frac{1.5}{11,000} = 2.73 \text{ lb/MM Btu}$$

If an SO_2 scrubbing system of 75% efficiency is installed, the exiting SO_2 concentration will be $0.25 \times 2.73 = 0.68$ lb/MM Btu.

2.18b

Q: What is the SO_2 level in ppm (parts per million) by volume if the coals in Q2.18a are fired with 25% excess air?

A: We have to estimate the flue gas produced. Using the MM Btu method,

$$w_g = \frac{10^6}{11,000} + 1.25 \times 760 = 1041 \text{ lb/MM Btu}$$

Let the molecular weight be 30, which is a good estimate in the absence of flue gas analysis. Then,

$$\text{Moles of flue gas} = \frac{1041}{30} = 34.7 \text{ per MM Btu fired}$$

$$\text{Moles of } SO_2 = \frac{2.73}{64} = 0.042 \quad \text{(from Q2.18a and Table 1.1)}$$

(64 is the molecular weight of SO_2. Dividing weight by molecular weight gives the moles.)

Hence ppm of SO_2 in flue gas will be $0.042 \times 34.7/10^6 = 1230$ ppm.

2.18c

Q: If 5% of the SO_2 gets converted to SO_3, estimate the ppm of SO_3 in the flue gas.

A:
$$\text{Moles of } SO_3 = 0.05 \times \frac{2.73}{80} = 0.0017 \text{ per MM Btu}$$

Hence

$$\text{ppm by volume of } SO_3 = \frac{0.0017}{34.7} \times 10^6 = 49 \text{ ppm}$$

(80 is the molecular weight of SO_3).

2.19a

Q: How is the efficiency of a boiler or a fired heater determined?

A: The estimation of the efficiency of a boiler or heater involves computation of several losses such as those due to flue gases leaving the unit, unburned fuel, radiation losses, heat loss due to molten ash, and so on. Readers may refer to the ASME *Power Test Code* [7] for details. Two methods are widely used, one based on the measurement of input and output and the other based on heat losses. The latter is preferred, as it is easy to use.

There are two ways of stating the efficiency, one based on the HHV and the other on LHV. As discussed in Q2.01,

$$\eta_{HHV} \times HHV = \eta_{LHV} \times LHV$$

The various losses are [1], on an HHV basis,

1. Dry gas loss, L_1:

$$L_1 = 24w_{dg} \frac{t_g - t_a}{HHV} \tag{15a}$$

2. Loss due to combustion of hydrogen and moisture in fuel, L_2:

$$L_2 = (9 \times H_2 + W) \times (1080 - 0.46t_g - t_a) \times \frac{100}{HHV} \tag{15b}$$

3. Loss due to moisture in air, L_3:

$$L_3 = 46 \times Mw_{da} \times \frac{t_g - t_a}{HHV} \tag{15c}$$

4. Radiation loss, L_4. The American Boiler Manufacturers Association (ABMA) chart [7] may be referred to to obtain this value. A quick estimate of L_4 is

$$L_4 = 10^{0.62 - 0.42 \ \log \ Q} \tag{15d}$$

For Eqs. (15a) to (15d),

$$
\begin{aligned}
w_{dg} &= \text{dry flue gas produced, lb/lb fuel} \\
w_{da} &= \text{dry air required, lb/lb fuel} \\
H_2, W &= \text{hydrogen and moisture in fuel, fraction} \\
M &= \text{moisture in air, lb/lb dry air (see Q1.09b)} \\
t_g, t_a &= \text{temperatures of flue gas, air, °F} \\
Q &= \text{duty in MM Btu/hr}
\end{aligned}
$$

5. To losses L_1 to L_4 must be added a margin or unaccounted loss, L_5. Hence efficiency becomes

$$\eta_{HHV} = 100 - (L_1 + L_2 + L_3 + L_4 + L_5) \tag{15e}$$

Note that combustion calculations are a prerequisite to efficiency determination. If the fuel analysis is not available, plant engineers can use the MM Btu method to estimate w_{dg} rather easily and then estimate the efficiency (see Q2.20).

The efficiency can also be estimated on an LHV basis. The various losses considered are the following.

1. Wet flue gas loss:

$$w_{wg} \times C_p \times \frac{t_g - t_a}{LHV} \tag{15f}$$

(C_p, gas specific heat, will be in the range of 0.25 to 0.265 for wet flue gases.)

2. Radiation loss (see Q2.23)
3. Unaccounted loss, margin

Then

$$\eta_{LHV} = 100 - \text{(sum of the above three losses)}$$

One can also convert η_{HHV} to η_{LHV} using Eq. (3b) (see Q2.01).

2.19b

Q: Coals of HHV = 13,500 and LHV = 12,600 Btu/lb are fired in a boiler with excess air of 25%. If the exit gas temperature is 300°F and ambient temperatures is 80°F, determine the efficiency on an HHV basis and on an LHV basis.

A: From the MM Btu method of combustion calculations, assuming that moisture in air is 0.013 lb/lb dry air,

$$w_{wg} = \frac{1.013 \times 760 \times 1.25 + 10^6/13{,}500}{10^6/13{,}500}$$

$$= \frac{1036}{74} = 14.0$$

(760 is the constant obtained from Table 2.4.) Hence

Wet flue gas loss $= 100 \times 14.0 \times 0.26$

$$\times \frac{300 - 80}{12{,}600} = 6.35\%$$

Let radiation and unaccounted losses by 1.3%. Then

$$\eta_{LHV} = 100 - (6.35 + 1.3) = 92.34\%$$

$$\eta_{HHV} = 92.34 \times \frac{12{,}600}{13{,}500} = 86.18\%$$

(Radiation losses vary from 0.5 to 1.0% in large boilers and may go up to 2.0% in smaller units. The major loss is the flue gas loss.)

2.19c

Q: Determine the efficiency of a boiler firing the fuel given in Q2.09a at 15% excess air. Assume radiation loss = 1%, exit gas temperature $= 400°F$, and ambient $= 70°F$. Excess air and relative humidity are the same as in Q2.09a (15% and 80%).

A: Results of combustion calculations are already available.

Dry flue gas $= 18$ lb/lb fuel

Moisture in air $= 19.52 - 19.29 = 0.23$ lb/lb/ fuel

Water vapor formed due to combustion of fuel $= 20.4 - 18 - 0.23 = 2.17$ lb/lb fuel

$$HHV = \frac{83.4 \times 1013.2 + 15.8 \times 1792}{100} = 1128 \, Btu/cu \, ft$$

Fuel density at $60°F = 18.3/379 = 0.483$ lb/cu ft, so

$$HHV = \frac{1128}{0.0483} = 23,364 \, Btu/lb$$

The losses are

1. Dry gas loss,

$$L_1 = 100 \times 18 \times 0.24 \times \frac{400 - 70}{23,364} = 6.1\%$$

2. Loss due to combustion of hydrogen and moisture in fuel,

$$L_2 = 100 \times 2.17 \times \frac{1080 + 0.46 \times 400 - 70}{23,364} = 11.1\%$$

3. Loss due to moisture in air,

$$L_3 = 100 \times 0.23 \times 0.46 \times \frac{400 - 70}{23,364} = 0.15\%$$

4. Radiation loss $= 1.0\%$
5. Unaccounted losses and margin $= 0\%$

Total losses $= 6.1 + 11.1 + 0.15 + 1.0 = 18.35\%$

Hence

Efficiency on HHV basis $= 100 - 18.35 = 81.65\%$

One can convert this to LHV basis after computing the LHV.

2.19d

Q: How do excess air and boiler exit gas temperature affect the various losses and boiler efficiency?

A: Table 2.7 shows the results of combustion calculations for various fuels at different excess air levels and boiler exit gas temperature. It also shows the amount of CO_2 generated per MM Btu fired.

It can be seen that natural gas generates the lowest amount of CO_2.

$$CO_2/\text{MM Btu natural gas} = \frac{10^6}{23,789} \times 19.17 \times$$
$$\frac{9.06 \times 44}{27.57/100} = 116.5 \text{ lb}$$

(The above is obtained by converting the volumetric analysis to weight basis using the molecular weights of CO_2 and the flue gas.) For oil, CO_2 generated $= 162.4$ lb, and for coal, 202.9 lb.

2.20

Q: A fired heater of duty 100 MM Btu/hr (HHV basis) firing No. 6 oil shows the following dry flue gas analysis:

$CO_2 = 13.5\%$, $O_2 = 2.5\%$, $N_2 = 84\%$

The exit gas temperature and ambient temperature are 300 and 80°F, respectively. If moisture in air is 0.013 lb/lb dry air, estimate the efficiency of the unit on an LHV and an HHV basis. LHV $= 18,400$ Btu/lb and HHV $= 19,500$ Btu/lb.

A: Since the fuel analysis is not known, let us estimate the flue gas produced by the MM Btu method. First, compute the excess air, which is

$$E = 94.5 \times \frac{2.5}{21 - 2.5} = 12.8\%$$

The factor 94.5 is from Table 2.6 (see Q2.12). The wet flue gas produced is

Table 2.7 Combustion Calculations for Various Fuels

	Gas				Oil				Coal	
T_{go}, °F	350	450	350	450	350	450	350	450	450	550
EA, %	5	5	15	15	5	5	15	15	25	25
CO_2	9.06		8.34		12.88		11.82		13.38	
H_2O	19.11		17.70		12.37		11.47		7.10	
N_2	70.93		71.48		73.83		74.19		75.43	
O_2	0.90		2.48		0.92		2.53		3.94	
SO_2									0.15	
W_g/W_f	19.17		20.9		16.31		17.77		13.42	
L_1, %	4.74	6.44	5.23	7.09	5.13	6.96	5.62	7.63	8.91	11.25
L_2, %	0.09	0.12	0.10	0.13	0.09	0.12	0.10	0.14	0.15	0.19
L_3, %	10.89	11.32	10.89	11.32	6.63	6.89	6.63	6.89	4.3	4.46

	Gas				Oil				Coal	
T_{go}, °F	350	450	350	450	350	450	350	450	450	550
EA, %	5	5	15	15	5	5	15	15	25	25
L_4, %						1.0				
E_h, %	83.2	81.1	82.9	80.5	87.1	85.0	86.7	84.3	85.6	83.0
E_l, %	92.3	89.9	91.7	89.2	92.8	90.0	92.3	89.9	89.0	86.4
MW	27.57		27.66		28.86		28.97		29.64	

Coal: C=72.8, H_2=4.8, N_2=1.5, O_2=6.2, S=2.2, H_2O=3.5, ash=9.0 (% weight); HHV=1313 Btu/lb; LHV=12,634 Btu/lb
Oil: C=87.5, H_2=12.5 (% weight); °API=32; HHV=19,727 Btu/lb; LHV=18,512 Btu/lb
Gas: Ch_4=97, C_2H_6=2, C_3H_8=1 (% vol); HHV=23,789 Btu/lb; LHV=21,462 Btu/lb

$$\frac{\left(\dfrac{745 \times 1.128 \times 1.013}{10^6} + \dfrac{10^6}{19{,}500} \right)}{(10^6/19{,}500)} = 17.6 \text{ lb/lb fuel}$$

Hence

$$\text{Wet gas loss} = 100 \times 17.6 \times 0.26 \times \frac{300 - 80}{18{,}400} = 5.47\%$$

The radiation loss on an HHV basis can be approximated by Eq. (15d):

$$\text{Radiation loss} = 10^{0.62 - 0.42 \log Q} = 0.60\%;$$

$$Q = 100 \text{ MM Btu/hr}$$

Let us use 1.0% on an LHV basis, although this may be a bit high. Hence the efficiency on an LHV basis is $100 - 6.47 = 93.53\%$. The efficiency on an HHV basis would be [Eq. (3b)]

$$\eta_{HHV} \times HHV = \eta_{LHV} \times LHV$$

or

$$\eta_{HHV} = 93.53 \times \frac{18{,}400}{19{,}500} = 88.25$$

Thus, even in the absence of fuel ultimate analysis, the plant personnel can check the efficiency of boilers and heaters based on operating data.

2.21

Q: How is the loss due to incomplete combustion such as formation of CO determined?

A: Efforts must be made by the boiler and burner designers to ensure that complete combustion takes place in the furnace. However, because of various factors such as size of fuel particles, turbulence, and availability of air to fuel and the mixing process, some carbon monoxide will be formed, which means losses. If CO is formed from carbon instead of CO_2, 10,600 Btu/lb is lost. This is the difference between the heat of reaction of the two processes

$$C + O_2 \to CO_2 \quad \text{and} \quad C + O_2 \to CO$$

The loss in Btu/lb is given by [1]

$$L = \frac{CO}{CO + CO_2} \times 10{,}160 \times C$$

where C is the carbon in fuel, fraction by weight, and CO and CO_2 are vol % of the gases.

EXAMPLE

Determine the losses due to formation of CO if coal with HHV of 12,000 Btu/lb is fired in a boiler, given that CO and CO_2 in the flue gas are 1.5% and 17% and the fuel has a carbon content of 56%.

Solution.   Substituting into the equation given above,

$$L = \frac{1.5}{18.5} \times 10{,}160 \times \frac{0.56}{12{,}000} = 0.038,$$

or 3.8% on HHV basis (dividing loss in Btu/lb by HHV)

2.22

Q: Is there a simple formula to estimate the efficiency of boilers and heaters if the excess air and exit gas temperature are known and the fuel analysis is not available?

A: Boiler efficiency depends mainly on excess air and the difference between the flue gas exit temperature and the ambient temperature. The following expressions have been derived from combustion calculations of typical natural gas and oil fuels. These may be used for quick estimations.

For natural gas:

$$\eta_{HHV}, \% = 89.4 - (0.001123 + 0.0195 \times EA) \times \Delta T \qquad (16a)$$

$$\eta_{LHV}, \% = 99.0 - (0.001244 + 0.0216 \times EA) \times \Delta T \qquad (16b)$$

For fuel oils:

$$\eta_{HHV}, \% = 92.9 - (0.001298 + 0.01905 \times EA) \times \Delta T$$

$$\eta_{LHV}, \% = 99.0 - (0.001383 + 0.0203 \times EA) \times \Delta T$$

where

EA = excess air factor (EA = 1.15 means 15% excess air)

ΔT = difference between exit gas and ambient temperatures

EXAMPLE

Natural gas at 15% excess air is fired in a boiler, with exit gas
temperature 280°F and ambient temperature 80°F. Determine the
boiler efficiency. EA = 1.15 and ΔT = (280 − 80) = 200°F.
Solution.

$$\eta_{HHV} = 89.4 - (0.001123 + 0.0195 \times 1.15)$$
$$\times (280 - 80) = 84.64\%$$
$$\eta_{LHV} = 99.0 - (0.001244 + 0.0216 \times 1.15)$$
$$\times (280 - 80) = 93.78\%$$

The above equations are based on 1% radiation plus unaccounted
losses.

2.23

Q: The average surface temperature of the aluminum casing of a
gas-fired boiler was measured to be 180°F when the ambient
temperature was 85°F and the wind velocity was 5 mph. The
boiler was firing 50,000 scfh of natural gas with LHV = 1075
Btu/scf. Determine the radiation loss on an LHV basis if the total
surface area of the boiler was 2500 ft^2. Assume that the emissivity of the casing = 0.1.

A: This example shows how radiation loss can be obtained from the
measurement of casing temperatures. The wind velocity is 5 mph
= 440 fpm. From Q4.51 we see that the heat loss q in Btu/ft^2 hr
will be

$$q = 0.173 \times 10^{-8} \times 0.1 \times [(460 + 180)^4 - (460 + 85)^4]$$
$$+ 0.296 \times (180 - 85)^{1.25} \times \sqrt{\frac{440 + 69}{69}}$$
$$= 252 \text{ Btu/ft}^2 \text{ hr} \tag{17}$$

The total heat loss will be 2500 × 252 = 0.63 × 10^6 Btu/hr.
The radiation loss on an LHV basis will be 0.63 × 10^6 × 100/

(50,000 × 1075) = 1.17%. If the HHV of the fuel were 1182 Btu/scf, the radiation loss on an HHV basis would be 0.63 × 1182/1075 = 1.06%.

2.24

Q: How does the radiation loss vary with boiler duty or load? How does this affect the boiler efficiency?

A: The heat losses from the surface of a boiler will be nearly the same at all loads if the ambient temperature and wind velocity are the same. Variations in heat losses can occur owing to differences in the gas temperature profile in the boiler, which varies with load. However, for practical purposes this variation can be considered minor. Hence the heat loss as a percent will increase as the boiler duty decreases.

The boiler exit gas temperature decreases with a decrease in load or duty and contributes to some improvement in efficiency, which is offset by the increase in radiation losses. Hence there will be a slight increase in efficiency as the load increases, and after a certain load, efficiency decreases.

The above discussion pertains to fired water tube or fire tube boilers and not waste heat boilers, which have to be analyzed for each load because the gas flow and inlet gas temperature can vary significantly with load depending on the type of process or application.

2.25a

Q: Discuss the importance of dew point corrosion in boilers and heaters fired with fuels containing sulfur.

A: During the process of combustion, sulfur in fuels such as coal, oil, and gas is converted to sulfur dioxide. Some portion of it (1 to 5%) is converted to sulfur trioxide, which can combine with water vapor in the flue gas to form gaseous sulfuric acid. If the surface in contact with the gas is cooler than the acid dew point, sulfuric acid can condense on it, causing corrosion. ADP (acid dew point) is dependent on several factors, such as excess air, percent sulfur in fuel, percent conversion of SO_2 to SO_3, and partial pressure of water vapor in the flue gas. Manufacturers of economizers and air heaters suggest minimum cold and temperatures that are required to avoid corrosion. Figures 2.1 and 2.2 are typical. Sometimes, the minimum fluid temperature, which affects the tube metal temperature, is suggested. The following equation gives a conservative estimate of the acid dew point [8]:

$$T_{dp} = 1.7842 + 0.0269 \log p_w - 0.129 \log p_{SO_3}$$
$$+ 0.329 \log p_w \times \log p_{SO_3} \qquad (18)$$

where

T_{dp} = acid dew point, K
p_w = partial pressure of water vapor, atm
p_{SO_3} = partial pressure of sulfur trioxide, atm

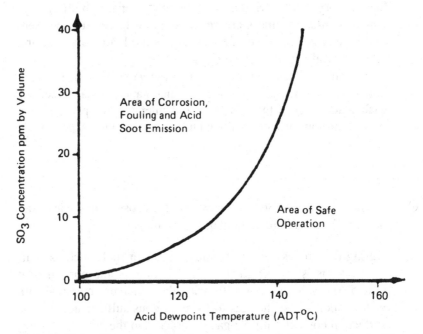

Figure 2.1 The relationship between SO_3 and ADT. (Courtesy of Land Combustion Inc.)

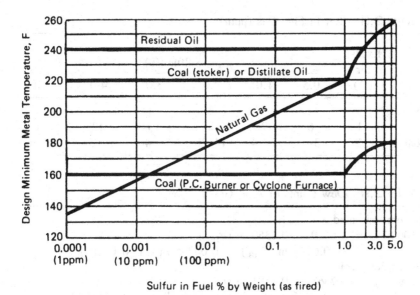

Figure 2.2   Limiting tube-metal temperatures to avoid external corrosion in economizers and air heaters when burning fuels containing sulfur. (From *Steam: Its Generation and Use*, 39th ed., Babcock and Wilcox, courtesy of the publisher.)

Table 2.8 gives typical p_{SO_3} values for various fuels and excess air. Q2.18c shows how p_{SO_3} can be computed.

A practical way to determine T_{dp} is to use a dew point meter. An estimation of the cold-end metal temperature can give an indication of possible corrosion.

2.25b

Q: How is the dew point of an acid gas computed?

A: Table 2.9 shows the dew point correlations for various acid gases [9, 11].

Flue gas from an incinerator has the following analysis: H_2O = 12, SO_2 = 0.02, HCl = 0.0015% by volume and the rest oxygen and nitrogen. Gas pressure = 10 in. wg. Compute the dew points of sulfuric and hydrochloric acids given that 2% of

Table 2.8 SO_3 in Flue Gas (ppm)

Fuel	Excess air (%)	Sulfur (%)					
		0.5	1.0	2.0	3.0	4.0	5.0
Oil	5	2	3	3	4	5	6
	11	6	7	8	10	12	14
Coal	25	3–7	7–14	14–28	20–40	27–54	33–66

Table 2.9 Dew Points of Acid Gases[a]

Hydrobromic acid
$$1000/T_{dp} = 3.5639 - 0.1350 \ln (P_{H_2O})$$
$$- 0.0398 \ln (P_{HBr}) + 0.00235 \ln (P_{H_2O}) \ln (P_{HBr})$$

Hydrochloric acid
$$1000/T_{dp} = 3.7368 - 0.1591 \ln (P_{H_2O})$$
$$- 0.0326 \ln (P_{HCl}) + 0.00269 \ln (P_{H_2O}) \ln (P_{HCl})$$

Nitric acid
$$1000/T_{dp} = 3.6614 - 0.1446 \ln (P_{H_2O})$$
$$- 0.0827 \ln (P_{HNO_3}) + 0.00756 \ln (P_{H_2O}) \ln (P_{HNO_3})$$

Sulfurous acid
$$1000/T_{dp} = 3.9526 - 0.1863 \ln (P_{H_3O}) +$$
$$0.000867 \ln (P_{SO_2}) - 0.000913 \ln (P_{H_2O}) \ln (P_{SO_2})$$

Sulfuric acid
$$1000/T_{dp} = 2.276 - 0.0294 \ln (P_{H_2O}) -$$
$$0.0858 \ln (P_{H_3SO_4}) + 0.0062 \ln (P_{H_2O}) \ln (P_{H_2SO_4})$$

[a]T_{dp} is dew point temperature (K), and P is partial pressure (mm Hg). Compared with published data, the predicted dew points are within about 6 K of actual values except for H_2SO_4, which is within about 9 K.
Source: HCl, HBr, HNO_3 and SO_2 correlations were derived from vapor–liquid equilibrium data. The H_2SO_4 correlation is from Ref. 5.

SO_2 converts to SO_3. In order to use the correlations, the gas pressures must be converted to mm Hg. Atmospheric pressure = 10 in. wg = 10/407 = 0.02457 atmg or 1.02457 atm abs.

$$P_{H_2O} = 0.12 \times 1.02457 \times 760 = 93.44 \text{ mm Hg},$$
$$\ln(P_{H_2O}) = 4.537$$

$$P_{HCl} = 0.00015 \times 1.0245 \times 760 = 0.1168 \text{ mm Hg,}$$
$$\ln(P_{HCl}) = -2.1473$$

Partial pressures of sulfuric acid and SO_3 are equal. Hence

$$P_{SO_3} = 0.02 \times 0.0002 \times 760 \times 1.0245 = 0.0031 \text{ mm Hg,}$$
$$\ln(P_{SO_3}) = -5.7716$$

Substituting into the equations, we obtain the following.
For hydrochloric acid:

$$\frac{1000}{T_{dp}} = 3.7368 - 0.1591 \times 4.537 + 0.0326 \times 2.1473$$
$$- 0.00269 \times 4.537 \times 2.1473 = 3.0588$$
$$\text{or } T_{dp} = 327 \text{ K} = 54°C = 129°F$$

For sulfuric acid:

$$\frac{1000}{T_{dp}} = 2.276 - 0.0294 \times 4.537 + 0.0858 \times 5.7716$$
$$- 0.0062 \times 4.537 \times 5.7716 = 2.4755$$
$$\text{or } T_{dp} = 404 \text{ K} = 131°C = 268°F$$

The dew points of other gases can be obtained in a similar manner.

2.25c

Q: Does the potential for acid dew point corrosion decrease if the gas temperature at the economizer is increased?

A: Acid dew points were computed in Q2.25a. If the tube wall temperatures can be maintained above the dew point, then condensation of vapors is unlikely. However, the tube wall temperature in a gas to liquid heat exchanger such as the economizer is governed by the gas film heat transfer coefficient rather than the tube-side water coefficient, which is very high.

It can be shown by using the electrical analogy and neglecting the effects of fouling that [9]

$$t_m = t_o - (t_o - t_i) \times \frac{h_i}{h_i + h_o}$$

where

t_m = tube wall temperature
t_o = gas and tube-side fluid temperature
h_i = tube-side heat transfer coefficient
h_o = gas-side heat transfer coefficient

In an economizer, h_i is typically about 1000 Btu/ft^2 hr °F and h_o is about 15 Btu/ft^2 hr °F.

Let us assume that water temperature t_i = 250°F and compute the wall temperature t_m for two gas temperatures, 350°F and 750°F.

$$t_{m1} = 350 - (350 - 250) \times \frac{1000}{1015} = 252°F,$$

$$t_{m2} = 750 - (750 - 250) \times \frac{1000}{1015} = 258°F$$

Hence for a variation of 400°F in gas temperature, the tube wall temperature changes by only 6°F because the gas film heat transfer coefficient is so low compared to the water-side coefficient. Even with finned tubes the difference would be marginal.

We see that by specifying a higher stack gas temperature when selecting or designing an economizer we cannot avoid corrosion concerns if the water temperature is low or close to the acid dew point. A better way is to increase the water temperature entering the economizer by raising the deaerator pressure or by using a heat exchanger to preheat the water.

2.26a

Q: How do you convert pollutants such as NO$_x$ and CO from gas turbine exhaust gases from mass units such as lb/hr to ppm?

A: With strict emission regulations, plant engineers and consultants often find it necessary to relate mass and volumetric units of pollutants such as NO$_x$ and CO. In gas turbine cogeneration and combined cycle plants, in addition to the pollutants from the gas turbine itself, one has to consider the contributions from duct burners or auxiliary burners that are added to increase the steam generation from the HRSGs (heat recovery steam generators).

One can easily obtain the total lb/hr of NO_x or CO in the exhaust gas. However, regulations refer to NO_x and CO in ppmvd (parts per million volume dry) referred to 15% oxygen in the gas. The conversion can be done as follows.

If w lb/hr is the flow rate of NO_x (usually reported as NO_2) in a turbine exhaust flow of W lb/hr, the following expression gives NO_x in volumetric units on dry basis [9].

$$V = 100 \times \frac{(w/46)/(W/MW)}{100 - \%H_2O} \tag{19}$$

where

$\%H_2O$ = volume of water vapor
MW = molecular weight of the exhaust gases

The value of V obtained with Eq. (19) must be converted to 15% oxygen on dry basis to give ppmvd of NO_x:

$$V_n = \frac{V \times (21 - 15) \times 10^6}{21 - 100 \times \%O_2/(100 - \%H_2O)} = V \times F \tag{20}$$

where $\%O_2$ is the oxygen present in the wet exhaust gases and factor F converts V to 15% oxygen basis, which is the usual basis of reporting emissions. Similarly, CO emission in ppmvd can be obtained as

$$V_c = 1.642 \times V_n \text{ (for the same } w \text{ lb/hr rate)}$$

because the ratio of the molecular weights of NO_2 and CO is 1.642.

EXAMPLE
Determine the NO_x and CO concentrations in ppmvd, 15% oxygen dry basis if 25 lb/hr of NO_x and 15 lb/hr of CO are present in 550,000 lb/hr of turbine exhaust gas that has the following analysis by % volume (usually argon is added to nitrogen content):

$CO_2 = 3.5,$ $H_2O = 10,$ $N_2 = 75,$ $O_2 = 11.5$
Solution.   First,

$$MW = (3.5 \times 44 + 10 \times 18 + 75 \times 28 + 11.5 \times 32)/100 = 28$$

Let us compute NO_x on dry basis in the exhaust.

$$V = \frac{100 \times (25/46)}{(550,000/28)/(100 - 10)} = 0.00003074.$$

$$F = \frac{10^6 \times (21 - 15)}{21 - [100/(100 - 10)] \times 11.5} = 0.73 \times 10^6$$

Hence

$$V_n = 0.00003074 \times 0.73 \times 10^6 = 16.4 \text{ ppmvd}$$

Similarly, $V_c = (15/25) \times 1.642 \times 16.4 = 16.2$ ppmvd.

2.26b

Q: How can the emissions due to NO_x and CO be converted from ppm to lb/MM Btu or vice versa [10]?

A: Packaged steam generators firing gas or oil must limit emissions of pollutants in order to meet state and federal regulations. Criteria on emissions of common pollutants such as carbon monoxide (CO) and oxides of nitrogen (NO_x) are often specified in parts per million volume dry (ppmvd) at 3% oxygen. On the other hand, burner and boiler suppliers often cite or guarantee values in pounds per million Btu fired.

Table 2.10 demonstrates a simple method for calculating the conversion. It should be noted that excess air has little effect on the conversion factor.

Table 2.10 shows the results of combustion calculations for natural gas and No. 2 oil at various excess air levels. The table shows the flue gas analysis, molecular weight, and amount of flue gas produced per million Btu fired on higher heating value (HHV) basis. Using these, we will arrive at the relationship between ppmvd values of NO_x or CO to the corresponding values in lb/MM Btu fired.

CALCULATIONS FOR NATURAL GAS
From simple mass to mole conversions we have

$$V_n = 10^6 \times Y \times \frac{N}{46} \times \frac{MW}{Wgm} \times \frac{21 - 3}{21 - O_2 \times Y} \qquad (21)$$

Table 2.10   Results of Combustion Calculations

Component (vol %)	\multicolumn{4}{c}{Percent excess air}							
	0	10	20	30	0	10	20	30
	Natural gas[a]				No. 2 oil[b]			
CO_2	9.47	8.68	8.02	7.45	13.49	12.33	11.35	10.51
H_2O	19.91	18.38	17.08	15.96	12.88	11.90	11.07	10.36
N_2	70.62	71.22	71.73	72.16	73.63	74.02	74.34	74.62
O_2	0	1.72	3.18	4.43	0	1.76	3.24	4.50
MW	27.52	27.62	27.68	27.77	28.87	28.85	28.84	28.82
Wgm	768	841	914	966	790	864	938	1011

[a]Natural gas analysis assumed: $C_1 = 97$, $C_2 = 2$, $C_3 = 1$ vol %.
  (Higher and lower heating values = 23,759 and 21,462 Btu/lb, respectively.)
[b]No. 2 oil analysis assumed: $C = 87.5\%$, $H_2 = 12.5\%$, deg API $= 32$.
  (Higher and lower heating values = 19,727 and 18,512 Btu/lb, respectively.)

where

MW = molecular weight of wet flue gases
N = pounds of NO_x per million Btu fired
O_2 = % volume of oxygen in wet flue gases
V_n = parts per million volume dry NO_x
Wgm = flue gas produced per MM Btu fired, lb
Y = $100/(100 - \%H_2O)$, where H_2O is the volume of water vapor in wet flue gases.

From Table 2.10, for zero excess air:

$$Wgm = (10^6/23,789) \times 18.3 = 769$$
$$Y = 100/(100 - 19.91) = 1.248$$
$$MW = 27.53, \qquad O_2 = 0$$

Substituting these into Eq. (21) we have

$$V_n = 106 \times 1.248 \times N \times 27.52$$
$$\times \frac{18}{(46 \times 769 \times 21)} = 832 N$$

Similarly, to obtain ppmvd CO (parts per million volume dry CO), one would use 28 instead of 46 in the denominator. Thus the molecular weight of NO_x would be 46 and the calculated molecular weight of CO would equal 28.

$$V_e = 1367 \ CO$$

where CO is the pounds of CO per MM Btu fired on higher heating value basis.

Now repeat the calculations for 30% excess air:

$$Wgm = 986.6, \quad Y = \frac{100}{100 - 15.96} = 1.189,$$

$$MW = 27.77, \quad O_2 = 4.43$$

$$V_n = 10^6 \times 1.189 \times \frac{N}{46} \times \frac{27.77}{986.6}$$

$$\times \frac{18}{21 - (4.43 \times 1.189)} = 832 \times N$$

Thus, independent of excess air, we obtain 832 as the conversion factor for NO_x and 1367 for CO.

Similarly, the No. 2 oil and using values from Table 2.10,

$$V_n = 783 \ N \quad \text{and} \quad V_c = 1286 \ CO$$

EXAMPLE

If a natural gas burner generates 0.1 lb of NO_x per MM Btu fired, then the equivalent would equal $832 \times 0.1 = 83$ ppmvd.

2.27

Q: In gas turbine cogeneration and combined cycle projects the heat recovery steam generator may be fired with auxiliary fuel in order to generate additional steam. One of the frequently asked questions concerns the consumption of oxygen in the exhaust gas versus fuel quantity fired. Would there be sufficient oxygen in the exhaust to raise the exhaust gas to the desired temperature?

A: Gas turbine exhaust gases typically contain 14 to 16% oxygen by volume compared to 21% in air. Hence generally there is no need for additional oxygen to fire auxiliary fuel such as gas or oil or even coal while raising its temperature. (If the gas turbine is injected with large amounts of steam, the oxygen content will be lower, and we should refer the analysis to a burner supplied.) Also if the amount of fuel fired is very large, then we can run out of oxygen in the gas stream. Supplementary firing or auxiliary firing can double or even quadruple the steam generation in the boiler compared to its unfired model of operation [1]. The energy Q in Btu/hr required to raise W_g lb/hr of exhaust gases from a temperature of t_1 to t_2 is given by

$$Q = W_g \times (h_2 - h_1)$$

where

h_1, h_2 = enthalpy of the gas at t_1 and t_2, respectively.

The fuel quantity in lb/hr is W_f in Q/LHV, where LHV is the lower heating value of the fuel in Btu/lb.

If 0% volume of oxygen is available in the exhaust gases, the equivalent amount of air W_a in the exhaust is [9]

$$W_a = \frac{100 \times W_g \times 0 \times 32}{23 \times 100 \times 29.5}$$

In this equation we are merely converting the moles of oxygen from volume to weight basis. A molecular weight of 29.5 is used for the exhaust gases, and 32 for oxygen. The factor 100/23 converts the oxygen to air.

$$W_a = 0.0417 \times W_g \times O \tag{22}$$

Now let us relate the air required for combustion with fuel fired. From Q1.03 to Q1.05 we know that each MM Btu of fuel fired on an HHV basis requires a constant amount A of air. A is 745 for oil and 730 for natural gas; thus, $10^6/\text{HHV}$ lb of fuel requires A lb of air. Hence Q/LHV lb of fuel requires

$$\frac{Q}{\text{LHV}} \times A \times \frac{\text{HHV}}{10^6} \text{ lb of air}$$

and this equals W_a from (22).

$$\frac{Q}{\text{LHV}} \times A \times \frac{\text{HHV}}{10^6} = W_a = 0.0417 \, W_g \times O \tag{23}$$

or

$$Q = 0.0417 \times W_g \times O \times 10^6 \times \frac{LHV}{A \times HHV} \quad (24)$$

Now for natural gas and fuel oils, it can be shown that LHV/(A × HHV) = 0.00124. Hence substituting into Eq. (24), we get

$$Q = 58.4 \times W_g \times O \quad (25)$$

This is a very important equation, as it relates the energy input by the fuel (on an LHV basis) with oxygen consumed.

EXAMPLE

It is desired to raise the temperature of 150,000 lb/hr of turbine exhaust gases from 950°F to 1575°F in order to double the output of the waste heat boiler. If the exhaust gases contain 15% volume of oxygen, and the fuel input is 29 MM Btu/hr (LHV basis), determine the oxygen consumed.

Solution. From (4),

$$O = \frac{29 \times 10^6}{150,000 \times 58.4} = 3.32\%$$

Hence if the incoming gases had 15% volume of oxygen, even after the firing of 29 MM Btu/hr, we would have 15 − 3.32 = 11.68% oxygen in the exhaust gases.

A more accurate method would be to use a computer program [9], but the above equation clearly tells us if there is likely to be a shortage of oxygen.

2.28

Q: How can the fuel consumption for power plant equipment such as gas turbines and diesel engines be determined if the heat rates are known?

A: The heat rate (HR) of gas turbines or engines in Btu/kWh refers indirectly to the efficiency.

$$\text{Efficiency} = \frac{3413}{HR}$$

where 3413 is the conversion factor from Btu/hr to kW.

One has to be careful about the basis for the heat rate, whether it is on an HHV or LHV basis. The efficiency will be on the same basis.

EXAMPLE

If the heat rate for a gas turbine is 9000 Btu/kWh on an LHV basis, and the higher and lower heating values of the fuel are 20,000 and 22,000 Btu/lb, then

$$\text{Efficiency on LHV basis} = \frac{3413}{9000} = 0.379, \text{ or } 37.9\%$$

To convert this efficiency to an HHV basis, simply multiply it by the ratio of the heating values:

$$\text{Efficiency on HHV basis} = 37.9 \times \frac{20,000}{22,000} = 34.45\%$$

NOMENCLATURE

A	Theoretical amount of air for combustion per MM Btu fired, lb
C, CO, CO_2	Carbon, carbon monoxide, and carbon dioxide
C_a	Ash concentration in flue gas, grains/cu ft
C_p	Specific heat, Btu/lb °F
e	Emission rate of sulfur dioxide, lb/MM Btu
E	Excess air, %
EA	Excess air factor
HHV	Higher heating value, Btu/lb or Btu/scf
HR	Heat rate, Btu/kWh
h_i, h_o	Inside and outside heat transfer coefficients, Btu/ft^2 hr °F
K	Constant used in Eq. (7)
K_1, K_2	Constants used in Eqs. (10a) and (10c)
L_1 to L_5	Losses in steam generator, %
LHV	Lower heating value, Btu/lb or Btu/scf
MW	Molecular weight
P_c, P_w, P_{H_2O}	Partial pressures of carbon dioxide and water vapor, atm
P_{SO_3}	Partial pressure of sulfur trioxide, atm
P_a, P_s	Actual and standard pressures, psia
ΔP	Differential pressure, psi

q	Heat loss, Btu/ft^2 hr
Q	Energy, Btu/hr or kW
s	Specific gravity
S	Sulfur in fuel
t_a, t_g	Temperatures of air, gas, °F
t_m	Melting point of ash, °C; tube wall temperature, °C
T_{dp}	Acid dew point temperature, K
T_s, T_a	Standard and actual temperatures, °R
V_s, V_a	Standard and actual volumes, cu ft
V_c, V_n	CO and NO$_x$, ppmvd
w	Weight of air, lb/lb fuel; subscript da stands for dry air; wa, wet air; wg, wet gas; dg, dry gas
W	Moisture, lb/hr
W_a, W_g, W_f	Flow rates of air, gas, and fuel, lb/hr
η	Efficiency; subscripts HHV and LHV denote the basis
ρ	Density, lb/cu ft; subscript g stands for gas, f for fuel

REFERENCES

1. V. Ganapathy, *Applied Heat Transfer*, PennWell Books, Tulsa, Okla., 1982, pp. 14–24.
2. *North American Combustion Handbook*, 2nd ed., North American Mfg. Co., Cleveland, Ohio, 1978, pp. 9–40.
3. Babcock and Wilcox, *Steam: Its Generation and Use*, 38th ed., New York, 1978, p. 6–2.
4. V. Ganapathy, Use chart to estimate furnace parameters, *Hydrocarbon Processing*, Feb. 1982, p. 106.
5. V. Ganapathy, Figure particulate emission rate quickly, *Chemical Engineering*, July 26, 1982, p. 82.
6. V. Ganapathy, Nomogram estimates melting point of ash, *Power Engineering*, March 1978, p. 61.
7. ASME, *Power Test Code*, Performance test code for steam generating units, PTC 4.1, ASME, New York, 1974.
8. V. Ganapathy, Estimate combustion gas dewpoint, *Oil and Gas Journal*, April 1978, p. 105.
9. V. Ganapathy, *Waste Heat Boiler Deskbook*, Fairmont Press, Atlanta, Ga., 1991.
10. V. Ganapathy, Converting ppm to lb/MM Btu; an easy method, *Power Engineering*, April 1992, p. 32.
11. K. Y. Hsiung, Predicting dew points of acid gases. *Chemical Engineering*, Feb. 9, 1981, p. 127.

3

Fluid Flow, Valve Sizing, and Pressure Drop Calculations

3.08:   Selecting safety valves for boiler superheater; actual and re-
        quired relieving capacities

3.09:   Relieving capacities of a given safety valve on different gases

3.10:   Relieving capacity of safety relief valve for liquid service

3.11:   Determining relieving capacity of a given safety valve on air
        and steam service

3.12:   Sizing control valves; valve coefficient C_v

3.13:   Calculating C_v for steam service; saturated and superheated
        steam; critical and noncritical flow

3.14:   Calculating C_v for liquid service

3.15:   On cavitation: recovery factors

3.16:   Selecting valves for laminar flow

3.17:   Calculating pressure loss in water line; determining friction
        factor for turbulent flow; equivalent length of piping; viscosity
        of water

3.18:   Pressure loss in boiler superheater; estimating friction factor in
        smooth tubes; pressure drop in smooth tubing; Reynolds num-
        ber for gases

3.19:   Determining pressure drop under laminar conditions; pressure
        drop in fuel oil lines; effect of temperature on specific volume,
        viscosity of oils

3.20:   Pressure drop for viscous liquids; friction factor under turbu-
        lent conditions

3.21: Calculating flow in gpm and in lb/hr for fuel oils; expansion factors for fuel oils at different temperatures

3.22: Pressure loss in natural gas lines using Spitzglass formula

3.23: Calculating pressure drop of flue gas and air in ducts; friction factors; equivalent diameter for rectangular ducts; Reynolds number estimation

3.24: Determining Reynolds number for superheated steam in tubes; viscosity of steam; Reynolds number for air flowing over tube bundles

3.25: Determining flow in parallel passes of a superheater

3.26: Equivalent length of piping system; equivalent length of valves and fittings

3.27: Pressure drop of air and flue gases over plain tube bundles; friction factor for in-line and staggered arrangements

3.28: Pressure drop of air and flue gases over finned tube bundles

3.29: Factors influencing boiler circulation

3.30: Purpose of determining circulation ratio

3.31: Determining circulation ratio in water tube boilers

3.32: Determining circulation ratio in fire tube boilers

3.33: Determining steam flow in blowoff lines

3.34: Sizing boiler blowdown lines

3.35: Stack height and friction losses

3.01a

Q: How are flow meters sized?

A: The basic equation for pressure differential in head meters (venturi, nozzles, orifices) is [1]

$$W = 359YC_d d_o^2 \frac{\sqrt{h\rho}}{\sqrt{1 - \beta^4}} \qquad (1a)$$

where

W = flow of the fluid, lb/hr

d_o = orifice diameter, in.

Y = expansion factor, which allows for changes in density of compressible fluids (for liquids $Y = 1$, and for most gases it varies from 0.92 to 1.0)

β = ratio of orifice to pipe inner diameter = d_o/d_i

ρ = density of fluid, lb/cu ft

C_d = a coefficient of charge

C_d may be taken as 0.61 for orifices and 0.95 to 0.98 for venturis and nozzles. It is a complicated function of Reynolds number and orifice size. The permanent pressure drop, Δp, across a flow meter is important, as it means loss in power or additional consumption of energy. It is the highest for orifices [2]:

$$\Delta P = h \times (1 - \beta^2) \qquad (1b)$$

For nozzles,

$$\Delta P = h \times \frac{1 - \beta^2}{1 + \beta^2} \qquad (1c)$$

and for venturis it depends on the angle of divergence but varies from 10 to 15% of h. Q3.04 discusses the significance of permanent pressure drop and the cost associated with it.

3.01b

Q: The differential pressure across an orifice of a steam flow meter shows 180 in. WC when the upstream conditions are 1600 psia and 900°F. The steam flow was calibrated at 80,000 lb/hr under these conditions. Because of different plant load requirements, the steam parameters are now 900 psia and 800°F. If the differential pressure is 200 in. WC, what is the steam flow?

A: From Eq. (1a),

$$W \propto \sqrt{\rho h} \propto \sqrt{\frac{h}{v}}$$

where

W = steam flow, lb/hr
v = specific volume, cu ft/lb
h = differential pressure, in. WC
ρ = density, lb/cu ft

From the steam tables (see the Appendix),

v_1 = 0.4553 cu ft/lb at 1600 psia, 900°F

v_2 = 0.7716 cu ft/lb at 900 psia, 800°F

h_1 = 180, h_2 = 200,   and W_1 = 80,000. We need to find W_2.

$$\frac{80,000}{W_2} = \sqrt{\frac{180 \times 0.7716}{200 \times 0.4553}} = 1.235$$

Hence W_2 = 64,770 lb/hr.

3.02

Q: Determine the orifice size to limit the differential pressure to 100 in. WC when 700 lb/sec of water at 60°F flows in a pipe of inner diameter 18 in. The density of water is 62.4 lb/cu ft.

A: Equation (1a) is not handy to use when it is required to solve for the orifice diameter d_o. Hence, by substituting for $\beta = d_o/d_i$ and simplifying, we have

$$W = 359 C_d Y d_i^2 \frac{\sqrt{\rho h}\ \beta^2}{\sqrt{1 - \beta^4}} \tag{2}$$

This equation is easy to use either when orifice size is needed or when flow through a given orifice is required. The term β

$\sqrt{1 - \beta^4}$ is a function of β and can be looked up from Table 3.1. Substituting for $W = 700 \times 3600$, $C_d = 0.61$, $Y = 1$, $\rho = 62.4$, and $h = 100$, we have

$$700 \times 3600 = 359 \times 0.61 \times 1 \times 18^2 \sqrt{62.4 \times 100}$$
$$\times F(\beta)$$

or

$$F(\beta) = 0.45$$

From Table 3.1, by interpolation, we note that $\beta = 0.64$. Thus the orifice diameter $d_o = 0.64 \times 18 = 11.5$ in.

3.03

Q: What size of orifice is needed to pass a saturated steam flow of 26,480 lb/hr when the upstream pressure is 1000 psia and line size is 2.9 in. and the differential is not to exceed 300 in. WC?

A: Using Eq. (2) and substituting $Y = 0.95$, $\rho = 1/v = 1/0.4456 = 2.24$ lb/cu ft, and $d_i = 2.9$, we have

$$W = 26,480 = 359 \times 0.61 \times 0.95 \times 2.9^2 \times F(\beta)$$
$$\times \sqrt{2.24 \times 300}$$

Hence

$$F(\beta) = 0.58$$

From Table 3.1, $\beta = 0.71$. Hence

$$d_o = 0.71 \times 2.9 = 2.03 \text{ in.}$$

Table 3.1 $F(\beta)$ Values for Solving Eq. (2)

β	$F(\beta) = \beta^2/\sqrt{1 - \beta^4}$
0.3	0.09
0.4	0.162
0.5	0.258
0.6	0.39
0.7	0.562
0.8	0.83

3.04

Q: What is the significance of a permanent pressure drop across the flow measurement device? 1.3 million scfh of natural gas with a specific gravity of 0.62 at 125 psia is metered using an orifice plate with a differential head of 100 in. WC. The line size is 12 in. What are the operating costs involved? Assume that electricity costs 20 mills/kWh.

A: The first step is to size the orifice. Use a molecular weight of $0.62 \times 29 = 18$ to compute the density. (The molecular weight of any gas = specific gravity $\times$ 29.) From Q1.03,

$$\rho = 18 \times 492 \times \frac{125}{359 \times 520 \times 15} = 0.39 \text{ lb/cu ft}$$

(A temperature of 60°F was assumed.) The density at standard conditions of 60°F, 15 psia, is

$$\rho = 18 \times \frac{492}{359 \times 520} = 0.047 \text{ lb/cu ft}$$

Hence mass flow is

$$W = 1.3 \times 10^6 \times 0.047 = 359 \times 0.61 \times 12^2$$
$$\times \sqrt{0.39 \times 100} \times F(\beta)$$
$$F(\beta) = 0.31$$

From Table 3.1, $\beta = 0.55$, so $\beta^2 = 0.3$. The permanent pressure drop, from Q3.01, is

$$\Delta P = (1 - \beta^2)h = (1 - 0.3) \times 100 = 70 \text{ in. WC}$$

The horsepower consumed in developing this head is

$$HP = \text{scfh} \times (460 + t) \times \frac{\Delta P}{P \times 10^7} \qquad (3)$$

It was assumed in the derivation of Eq. (3) that compressor efficiency was 75%. Substitution yields

$$HP = 1.3 \times 10^6 \times 520 \times \frac{70}{10^7 \times 125} = 38$$

The annual cost of operation is

$$38 \times 0.746 \times 8000 \times 0.02 = \$4535$$

(8000 hours of operation was assumed per year; 0.746 is the
factor converting horsepower to kilowatts.)

3.05

Q: Often, pitot tubes are used to measure air velocities in ducts in
order to compute the air flow. A pitot tube in a duct handling air
at 200°F shows a differential of 0.4 in. WC. If the duct cross
section is 4 ft^2, estimate the air velocity and the flow rate.

A: It can be shown [3] by substituting $\rho = 40/(460 + t)$ that for a
pitot,

$$V = 2.85 \times \sqrt{h \times (460 + t)} \tag{4}$$

where

V = velocity, fps
h = differential pressure, in. WC
t = air or flue gas temperature, °F
$V = 2.85 \times \sqrt{0.4 \times 660} = 46$ fps

The air flow rate in acfm will be $46 \times 4 \times 60 = 11,040$ acfm.
The flow W in lb/hr = $11,040 \times 60 \times 40/660 = 40,145$ lb/hr.
[W = acfm $\times$ 60 $\times$ density, and density = $40/(460 + t)$.]

3.06

Q: How are safety valves for boilers sized?

A: The ASME Code for boilers and pressure vessels (Secs. 1 and 8)
describes the procedure for sizing safety or relief valves. For
boilers with 500 ft^2 or more of heating surface, two or more
safety valves must be provided. Boilers with superheaters must
have at least one valve on the superheater. The valves on the
drum must relieve at least 75% of the total boiler capacity.
Superheater valves must relieve at least 20%. Boilers that have
reheaters must have at least one safety valve on the reheater
outlet capable of handling a minimum of 15% of the flow. The
remainder of the flow must be handled by valves at the reheater
inlet.

If there are only two valves for a boiler, the capacity of the smaller one must be at least 50% of that of the larger one. The difference between drum pressure and the lowest set valve may be at least 5% above drum pressure but never more than the design pressure and not less than 10 psi. The range between the lowest set boiler valve and the highest set value is not to be greater than 10% of the set pressure of the highest set valve. After blowing, each valve is to close at 97% of its set pressure. The highest set boiler valve cannot be set higher than 3% over the design pressure.

The guidelines above are some of those used in selecting safety valves. For details the reader should refer to the ASME Code [4].

3.07

Q: How are the capacities of safety valves for steam service determined?

A: The relieving capacities of safety valves are given by the following expressions. ASME Code, Sec. 1 uses a 90% rating, while Sec. 8 uses a 100% rating [5].

$$W = 45AP_aK_{sh} \tag{5a}$$
$$W = 50AP_aK_{sh} \tag{5b}$$

where

W = lb/hr of steam relieved
A = nozzle or throat area of valve, in.^{2}
P_a = accumulated inlet pressure = $P_s \times (1 + \text{acc}) + 15$, psia (The factor acc is the fraction of pressure accumulation.)
P_s = set pressure, psig
K_{sh} = correction factor for superheat (see Figure 3.1)

The nozzle areas of standard orifices are specified by letters D to T and are given in Table 3.2. For saturated steam, the degree of superheat is zero, so $K_{sh} = 1$. The boiler safety valves are sized for 3% accumulation.

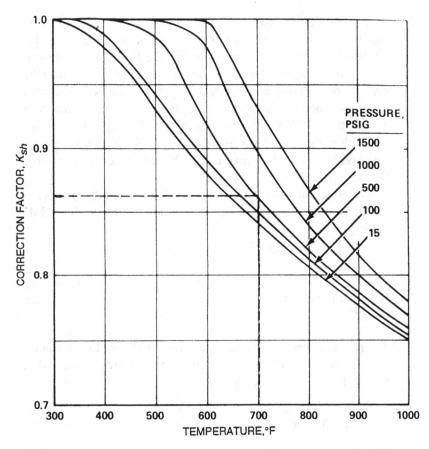

Figure 3.1 Correction factors for superheat.

3.08

Q: Determine the sizes of valves to be used on a boiler that has a superheater; the parameters are the following.

Total steam generation = 650,000 lb/hr
Design pressure = 1500 psig
Drum operating pressure = 1400 psig
Steam outlet temperature = 950°F
Pressure accumulation = 3%
Superheater outlet operating pressure = 1340 psig

Table 3.2   Orifice Designation

Type	Area (in.^{2})
D	0.110
E	0.196
F	0.307
G	0.503
H	0.785
J	1.287
K	1.838
K2	2.545
L	2.853
M	3.600
M2	3.976
N	4.340
P	6.380
Q	11.05
R	16.00

A: The set pressure must be such that the superheater valve opens before the drum valves. Hence the set pressure can be $(1500 - 60 - 40) = 1400$ psig (60 is the pressure drop and 40 is a margin). The inlet pressure $P_a = 1.03 \times 1400 + 15 = 1457$ psia. From Figure 3.1, $K_{sh} = 0.79$.

$$A = \frac{W}{45K_{sh}P_a} = \frac{130,000}{45 \times 0.79 \times 1457} = 2.51 \text{ in.}^2$$

We used a value of 130,000 lb/hr, which is 20% of the total boiler capacity. A K2 orifice is suitable. This relieves $(2.545/2.51) \times 130,000 = 131,550$ lb/hr. The drum valves must relieve $650,000 - 131,550 = 518,450$ lb/hr. About 260,000 lb/hr may be handled by each drum valve if two are used. Let the first valve be set at 1475 psig, or

$$P_a = 1.03 \times 1475 + 15 = 1535 \text{ psia}$$

and the next at $P_a = 1575$ psia.

$$\text{Area of first valve: } A = \frac{260,000}{45 \times 1535} = 3.76 \text{ in.}^2$$

$$\text{Area of second valve: } A = \frac{260,000}{45 \times 1575} = 3.67 \text{ in.}^2$$

Use two M2 orifices, which each have an area of 3.976 in.^{2}
Relieving capacities are

$$\frac{3.976}{3.76} + \frac{3.976}{3.67} \times 260,000 = 556,000 \text{ lb/hr}$$

which exceeds our requirement of 520,000 lb/hr.

3.09a

Q: How is the relieving capacity of safety valves for gaseous service found?

A: The expression used for estimating the relieving capacity for gases and vapors [6] is

$$W = CKAP_a \sqrt{\frac{MW}{T}} \tag{6}$$

where

$C =$ a function of the ratio (k) of specific heats of gases (Table 3.3)

$K =$ valve discharge coefficient, varies from 0.96 to 0.98

$P_a =$ accumulated inlet pressure $= P_s(1 + \text{acc}) + 15$, psia

$P_s =$ set pressure, psig

$MW =$ molecular weight of gas

$T =$ absolute temperature, °R

3.09b

Q: A safety valve is set for 100 psig for air service at 100°F and uses a G orifice. What is the relieving capacity if it is used on ammonia service at 50°F, pressure being the same?

A: Assume that k is nearly the same for both air and ammonia. Hence for the valve, $CKAP_a$ is a constant. For air, use $C = 356$, $K = 0.98$, $A = 0.503$, $MW = 29$, and $T = 560$.

Table 3.3   Constant C for Gas or Vapor Related to Ratio of Specific Heats ($k = C_p/C_v$)

k	Constant C	k	Constant C	k	Constant C
1.00	315	1.26	343	1.52	366
1.02	318	1.28	345	1.54	368
1.04	320	1.30	347	1.56	369
1.06	322	1.32	349	1.58	371
1.08	324	1.34	351	1.60	372
1.10	327	1.36	352	1.62	374
1.12	329	1.38	354	1.64	376
1.14	331	1.40	356	1.66	377
1.16	333	1.42	358	1.68	379
1.18	335	1.44	359	1.70	380
1.20	337	1.46	361	2.00	400
1.22	339	1.48	363	2.20	412
1.24	341	1.50	364		

Source: Ref. 5.

$$W_a = 356 \times 0.98 \times 0.503 \times (1.1 \times 100 + 15) \sqrt{\frac{29}{560}}$$

$$= 4990 \text{ lb/hr}$$

(An acc value of 0.10 was used above.) From Eq. (6), substituting MW $= 17$ and $T = 510$ for ammonia, we have

$$\frac{W_a}{W_{amm}} = \sqrt{\frac{29 \times 510}{17 \times 560}} = 1.246$$

Hence

$$W_{amm} = \frac{4990}{1.246} = 4006 \text{ lb/hr}$$

3.10a

Q: How are the relieving capacities for liquids determined?

A: An expression for relieving capacity at 25% accumulation [5] is

$$q = 27.2 A K_s \sqrt{P_1 - P_b} \tag{7}$$

where

P_1 = set pressure, psig
P_b = back pressure, psig
K_s = $\sqrt{1/s}$, s being the specific gravity
A = orifice area, in.^{2}
q = capacity, gpm

3.10b

Q: Determine the relieving capacity of a relief value on an economizer if the set pressure is 300 psig, back pressure is 15 psig, and s = 1. The valve has a G orifice (A = 0.503 in.^{2}).

A: Using Eq. (7), we have

$$q = 27.2 \times 0.503 \times 1 \times \sqrt{300 - 15} = 231 \text{ gpm}$$
$$= 231 \times 500 = 115,000 \text{ lb/hr}$$

At 10% accumulation, q would be 0.6 × 231 = 140 gpm and the flow W = 70,000 lb/hr. (500 is the conversion factor from gpm to lb/hr when s = 1.)

3.11

Q: A safety valve bears a rating of 20,017 lb/hr at a set pressure of 450 psig for saturated steam. If the same valve is to be used for air at the same set pressure and at 100°F, what is its relieving capacity?

A: For a given valve, $CKAP_a$ is a constant if the set pressure is the same. (See Q3.09a for definition of these terms.)
For steam,

$$20,017 = 50 \times KAP_a$$

Hence

$$KAP_a = \frac{20,017}{50} = 400.3$$

For air,

$$W = CKAP_a \sqrt{\frac{MW}{T}}$$

$C = 356$, $MW = 29$, and $T = 560°R$ for the case of air. Hence,

$$W_a = 356 \times 400.3 \times \sqrt{\frac{29}{560}} = 32,430 \text{ lb/hr}$$

Converting to acfm, we have

$$q = 32,430 \times 560 \times \frac{15}{0.081 \times 492 \times 465 \times 60}$$

$$= 244 \text{ acfm}$$

(The density of air was estimated at 465 psia and 100°F.)

3.12

Q: How is the size of control valves for steam service determined?

A: Control valves are specified by C_v or valve coefficients. The manufacturers of control valves provide these values (see Table 3.4). The C_v provided must exceed the C_v required. Also, C_v at several points of possible operation of the valve must be found, and the best C_v characteristics that meet the load requirements must be used, as controllability depends on this. For example, a quick-opening characteristic (see Figure 3.2) is desired for on–off service. A linear characteristic is desired for general flow control and liquid-level control systems, while equal percentage trim is desired for pressure control or in systems where pressure varies. The control valve supplier must be contacted for the selection and for proper actuator sizing.

For the noncritical flow of steam ($P_1 < 2P_2$) [7],

$$C_v = \frac{W \times [1 + 0.00065 \times (t - t_s)]}{2.11 \times \sqrt{\Delta P} \times P_t} \tag{8}$$

For critical flow ($P_1 \geq 2P_2$),

$$C_v = \frac{W \times [1 + 0.00065 \times (t - t_s)]}{1.85 \times P_1} \tag{9}$$

Table 3.4 Flow Coefficient C_v

Body size (in.)	Port diameter (in.)	Total travel (in.)	Valve opening (% total travel)										K_m and C_f
			10	20	30	40	50	60	70	80	90	100	
3/4	1/4	3/4	0.075	0.115	0.165	0.230	0.321	0.448	0.625	0.870	1.15	1.47	0.70
	3/8	3/4	0.120	0.190	0.305	0.450	0.628	0.900	1.24	1.68	2.18	2.69	0.80
	1/2	3/4	0.235	0.400	0.600	0.860	1.16	1.65	2.15	2.85	3.40	3.66	0.70
1	1/4	3/4	0.075	0.115	0.165	0.230	0.321	0.448	0.625	0.870	1.20	1.56	0.80
	3/8	3/4	0.120	0.190	0.305	0.450	0.630	0.910	1.35	1.97	2.78	3.68	0.70
	1/2	3/4	0.235	0.410	0.610	0.900	1.26	1.80	2.50	3.45	4.50	5.36	0.70
	3/4	3/4	0.380	0.700	1.10	1.57	2.36	3.40	5.00	6.30	6.67	6.95	0.75
1½	1/4	3/4	0.075	0.115	0.165	0.230	0.321	0.448	0.625	0.870	1.20	1.56	0.80
	3/8	3/4	0.120	0.190	0.305	0.450	0.630	0.910	1.35	1.97	2.78	3.68	0.70
	1/2	3/4	0.265	0.420	0.620	0.915	1.31	1.90	2.64	3.65	4.56	6.04	0.80
	3/4	3/4	0.380	0.700	1.10	1.65	2.45	3.70	5.30	7.10	8.88	10.2	0.75
	1	3/4	0.930	1.39	2.12	3.10	4.44	6.12	8.13	10.1	11.5	12.2	0.75
2	1/4	3/4	0.075	0.115	0.165	0.230	0.321	0.448	0.625	0.870	1.20	1.56	0.80
	3/8	3/4	0.120	0.190	0.305	0.450	0.630	0.910	1.35	1.97	2.78	3.68	0.70
	1/2	3/4	0.265	0.420	0.620	0.915	1.31	1.90	2.64	3.65	4.89	6.44	0.70
	3/4	3/4	0.380	0.700	1.10	1.65	2.45	3.70	5.53	8.00	10.3	12.3	0.70
	1	3/4	0.930	1.39	2.12	3.10	4.50	6.45	9.31	12.9	15.7	17.8	0.75
	1½	3/4	0.957	1.45	2.31	3.70	6.05	9.86	15.2	20.2	22.0	22.0	0.79
3	1/4	3/4	0.075	0.115	0.165	0.230	0.321	0.448	0.625	0.870	1.20	1.56	0.80
	3/8	3/4	0.120	0.190	0.305	0.450	0.630	0.910	1.35	1.97	2.78	3.68	0.70
	1/2	3/4	0.265	0.420	0.620	0.915	1.31	1.90	2.64	3.65	4.89	6.44	0.70
	3/4	3/4	0.380	0.700	1.10	1.65	2.45	3.70	5.70	8.66	12.3	14.8	0.65
	1	3/4	0.930	1.39	2.12	3.10	4.50	6.70	9.90	13.2	17.9	23.6	0.65
	1½	1⅛	1.15	2.29	3.41	4.77	6.44	8.69	12.5	19.2	26.7	32.2	0.74
	2	1⅛	1.92	3.13	4.83	7.93	12.6	24.6	35.9	40.5	43.4	44.3	0.72

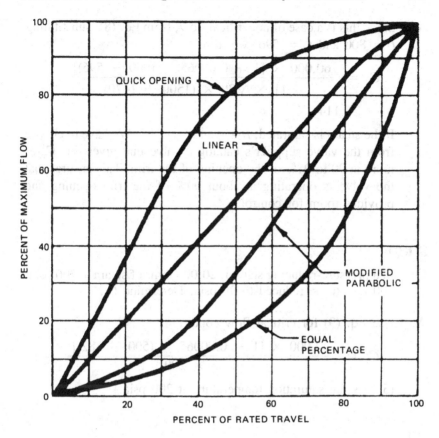

Figure 3.2 Typical control valve characteristics.

where

t, t_s = steam temperature and saturation temperature (for
saturated steam, $t = t_s$)

W = steam flow, lb/hr

P_t = total pressure ($P_1 + P_2$), psia

3.13a

Q: Estimate the C_v required when 60,000 lb/hr of superheated steam
at 900°F, 1500 psia flows in a pipe. The allowable pressure drop
is 30 psi.

A: Since this is a case of noncritical flow, from Eq. (8), substituting $t = 800$ and $t_s = 596$, we have

$$C_v = \frac{60{,}000 \times [1 + 0.00065 \times (900 - 596)]}{2.11 \times \sqrt{30} \times (1500 + 1470)}$$

$$= 114$$

If the steam is saturated, $t = t_s$ and $C_v = 95$. We have to choose from the valve supplier's catalog a valve that gives this C_v or more at 90 to 95% of the opening of the trim. This ensures that the valve is operating at about 90% of the trim opening and provides room for control.

3.13b

Q: In a pressure-reducing station, 20,000 lb/hr of steam at 200 psia, 500°F is to be reduced to 90 psia. Determine C_v.

A: Use Eq. (9) for critical flow conditions:

$$C_v = \frac{20{,}000 \times [1 + 0.00065 \times (500 - 382)]}{1.85 \times 200} = 58$$

(382 is the saturation temperature at 200 psia.)

3.14

Q: Determine the valve coefficient for liquids. A liquid with density 45 lb/cu ft flows at the rate of 100,000 lb/hr. If the allowable pressure drop is 50 psi, determine C_v.

A: The valve coefficient for liquid, C_v, is given by [8]

$$C_v = q\sqrt{\frac{s}{\Delta P}} \tag{10}$$

where

q = flow, gpm
ΔP = pressure drop, psi
s = specific gravity

From Q1.01,

$$W = 8q\rho$$

$$q = \frac{100,000}{8 \times 45} = 278 \text{ gpm}$$

$$s = \frac{45}{62.4} = 0.72$$

$$\Delta P = 50$$

Hence

$$C_v = 278 \times \sqrt{\frac{0.72}{50}} = 34$$

3.15

Q: How is cavitation caused? How is the valve sizing done to consider this aspect?

A: Flashing and cavitation can limit the flow in a control valve for liquid. The pressure distribution through a valve explains the phenomenon. The pressure at the vena contracta is the lowest, and as the fluid flows it gains pressure but never reaches the upstream pressure. If the pressure at the port or vena contracta should drop below the vapor pressure corresponding to upstream conditions, bubbles will form. If the pressure at the exit remains below the vapor pressure, bubbles remain in the stream and flashing occurs.

A valve has a certain recovery factor associated with it. If the recovery of pressure is high enough to raise the outlet pressure above the vapor pressure of the liquid, the bubbles will collapse or implode, producing cavitation. High-recovery valves tend to be more subject to cavitation [9]. The formation of bubbles tends to limit the flow through the valve. Hence the pressure drop used in sizing the valve should allow for this reduced capacity. ΔP_{all} is used in sizing,

$$\Delta P_{\text{all}} = K_m (P_1 - r_c P_v) \tag{11}$$

where

K_m = valve recovery coefficient (depends on valve make)
P_1 = upstream pressure, psia
r_c = critical pressure ratio (see Figure 3.3)
P_v = vapor pressure at inlet liquid temperature, psia

Full cavitation will occur if the actual ΔP is greater than ΔP_{all} and if the outlet pressure is higher than the fluid vapor pressure. If the actual ΔP is less than ΔP_{all}, the actual ΔP should be used for valve sizing. To avoid cavitation, select a valve with a low recovery factor (a high K_m factor).

3.16

Q: How are valves selected for laminar flow and viscous liquids?

A: Calculate the turbulent flow C_v from Eq. (10) and the laminar C_v from [10]

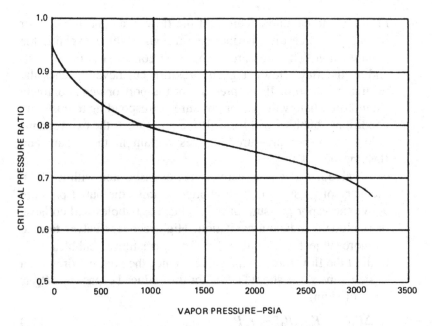

Figure 3.3   Critical pressure ratios for water.

$$\text{lam } C_v = 0.072 \times \left(\frac{\mu q}{\Delta P} \right)^{2/3} \tag{12}$$

Use the larger C_v in the valve selection. (μ is the liquid viscosity in centipoise.)

3.17a

Q: Determine the pressure loss in a 3-in. schedule 80 line carrying water at 100°F and 2000 psia if the total equivalent length is 1000 ft. Flow is 38,000 lb/hr.

A: The expression for turbulent flow pressure drop of fluids (Reynolds number > 2100) is [11]

$$\Delta P = 3.36 \times 10^{-6} \times fW^2 L_e \frac{v}{d_i^5} \tag{13}$$

where

ΔP = pressure loss, psi
f = Darcy friction factor
W = flow, lb/hr
L_e = equivalent length, ft (Q3.26 shows how the equivalent length can be computed)
v = specific volume of fluid, cu ft/lb
d_i = tube inner diameter, in.

For water at 100°F and 2000 psia, from Table 1.1, $v = 0.016$. In industrial heat transfer equipment such as boilers, superheaters, economizers, and air heaters, the fluid flow is generally turbulent, and hence we need not check for Reynolds number. (Q3.24 shows how Re can be found.) However, let us quickly check Re here:

$$Re = 15.2 \frac{W}{d_i \mu} \tag{14}$$

Referring to Table 3.5, water viscosity, μ, at 100°F is 1.645 lb/ft hr.

$$Re = 15.2 \times \frac{38,000}{2.9 \times 1.645} = 121,070$$

Table 3.5 Viscosity of Steam and Water (lb m/hr ft)

Temp. (°F)	Pressure (psia)											
	1	2	5	10	20	50	100	200	500	1000	2000	5000
1500	0.0996	0.0996	0.0996	0.0996	0.0996	0.0996	0.0996	0.0996	0.1008	0.1008	0.1019	0.1066
1400	0.0938	0.0938	0.0938	0.0938	0.0938	0.0938	0.0952	0.0952	0.0952	0.0961	0.0973	0.1019
1300	0.0892	0.0982	0.0892	0.0892	0.0892	0.0892	0.0892	0.0892	0.0892	0.0903	0.0915	0.0973
1200	0.0834	0.0834	0.0834	0.0834	0.0834	0.0834	0.0834	0.0834	0.0846	0.0846	0.0867	0.0926
1100	0.0776	0.0776	0.0776	0.0776	0.0776	0.0776	0.0776	0.0776	0.0788	0.0799	0.0811	0.0892
1000	0.0730	0.0730	0.0730	0.0730	0.0730	0.0730	0.0730	0.0730	0.0730	0.0741	0.0764	0.0857
900	0.0672	0.0672	0.0672	0.0672	0.0672	0.0672	0.0672	0.0672	0.0683	0.0683	0.0707	0.0846
800	0.0614	0.0614	0.0614	0.0614	0.0614	0.0614	0.0614	0.0614	0.0625	0.0637	0.0660	0.0973
700	0.0556	0.0556	0.0556	0.0556	0.0556	0.0556	0.0568	0.0568	0.0568	0.0579	0.0625	0.171
600	0.0510	0.0510	0.0510	0.0510	0.0510	0.0510	0.0510	0.0510	0.0510	0.0510	0.210	0.221
500	0.0452	0.0452	0.0452	0.0452	0.0452	0.0452	0.0452	0.0440	0.0440	0.250	0.255	0.268
400	0.0394	0.0394	0.0394	0.0394	0.0394	0.0394	0.0394	0.0382	0.317	0.320	0.323	0.335
300	0.0336	0.0336	0.0336	0.0336	0.0336	0.0336	0.441	0.442	0.444	0.445	0.448	0.460
250	0.0313	0.0313	0.0313	0.0313	0.0313	0.551	0.551	0.551	0.552	0.554	0.558	0.569
200	0.0290	0.0290	0.0290	0.0290	0.725	0.725	0.725	0.726	0.729	0.729	0.732	0.741
150	0.0255	0.0255	1.032	1.032	1.032	1.032	1.032	1.032	1.033	1.034	1.037	1.044
100	1.645	1.645	1.645	1.645	1.645	1.645	1.645	1.645	1.645	1.646	1.646	1.648
50	3.144	3.144	3.144	3.144	3.144	3.144	3.144	3.142	3.141	3.139	3.134	3.119
32	4.240	4.240	4.240	4.240	4.240	4.240	4.240	4.239	4.236	4.231	4.222	4.192

Table 3.6 Tube Diameter Versus
Friction Factor (Darcy) for Turbulent Flow

d_i (in.)	f
0.5	0.028
0.75	0.0245
1.0	0.0230
1.5	0.0210
2.0	0.0195
2.5	0.0180
3.0	0.0175
4.0	0.0165
5.0	0.0160
8.0	0.0140
10.0	0.013

The inner diameter of 2.9 for the pipe was obtained from Table 1.6. For turbulent flow and carbon or alloy steels of commercial grade, f may be obtained from Table 3.6. Here f for a tube inner diameter of 2.9 in. is 0.0175. Substituting in Eq. (13) yields

$$\Delta P = 3.36 \times 10^{-6} \times 0.0175 \times (38,000)^2 \times 100$$
$$\times \frac{0.016}{2.9^5} = 6.6 \text{ psi}$$

3.17b

Q: Estimate the pressure drop in a superheater of a boiler that has an equivalent length of 200 ft. The tube inner diameter is 2.0 in., the flow per pass is 8000 lb/hr, the steam pressure is 800 psia, and the temperature is 700°F.

A: Using Eq. (13) and substituting $v = 0.78$ cu ft/lb and $f = 0.0195$ for turbulent flow from Table 3.6 (generally flow in superheaters, economizers, and piping would be turbulent), we obtain

$$\Delta P = 3.36 \times 10^{-6} \times 0.0195 \times 200 \times 8000^2$$
$$\times \frac{0.78}{2^5} = 21 \text{ psi}$$

3.18a

Q: How does friction factor depend on pipe roughness?

A: For smooth tubes such as copper and other heat exchanger tubes, f is given by [12]

$$f = 0.133 \times Re^{-0.174} \tag{15}$$

Substituting this in Eq. (13) gives us

$$\frac{\Delta P}{L_e} = 0.0267 \times \rho^{0.8267} \times \mu^{0.174} \times \frac{V^{1.826}}{d_i^{1.174}} \tag{16}$$

(μ is the viscosity, lb/ft hr; V is the velocity, fps.)

3.18b

Q: Determine the pressure drop per 100 ft in a drawn copper tube of inner diameter 1.0 in. when 250 lb/hr of air at a pressure of 30 psig and at 100°F flows through it.

A: Calculate the density (see Chapter 1):

$$\rho = 29 \times 492 \times \frac{45}{359 \times 560 \times 15} = 0.213 \text{ lb/cu ft}$$

The effect of pressure can be neglected in the estimation of viscosity of gases up to 40 psig. For a detailed computation of viscosity as a function of pressure, readers may refer to the author's book, Ref. 11. From Table 3.7, $\mu = 0.047$ lb/ft hr. The velocity is

$$V = 250 \times \frac{576}{3600 \times 3.14 \times 0.213} = 60 \text{ fps}$$

$$\frac{\Delta P}{100} = 0.0267 \times 0.213^{0.8267} \times 0.047^{0.174}$$

$$\times \frac{60^{1.826}}{1} = 7.7 \text{ psi}$$

Table 3.7 Viscosity of Air

Temperature (°F)	Viscosity (lb/ft hr)
100	0.0459
200	0.0520
400	0.062
600	0.0772
800	0.0806
1000	0.0884
1200	0.0957
1400	0.1027
1600	0.1100
1800	0.1512

3.19a

Q: Derive the expression for ΔP for laminar flow of fluids.

A: For laminar flow of fluids in pipes such as that occurring with oils, the friction factor is

$$f = \frac{64}{Re} \tag{17a}$$

Substituting into Eq. (13) and using Eq. (14) gives us

$$\Delta P = 3.36 \times 10^{-6} \times 64 \times d_i \mu W^2 \frac{L_e \times v}{15.2 \times Wd_i^5}$$

$$= 14.4 \times 10^{-6} \times W \times L_e \times \frac{v\mu}{d_i^4} \tag{17b}$$

Converting lb/hr to gph (gallons per hour), we can rewrite this as

$$\Delta P = 4.5 \times 10^{-6} \times L_e \times cS \times s \times \frac{gph}{d_i^4} \tag{18}$$

where

  cS = viscosity, centistokes
  s = specific gravity

Equation (18) is convenient for calculations for oil flow situations.

3.19b

Q: Estimate the pressure drop per 100 ft in an oil line when the oil has a specific gravity of 16°API and is at 180°F. The line size is 1.0 in., and the flow is 7000 lb/hr.

A: We must estimate Re. To do this we need the viscosity [13] in centistokes:

$$cS = 0.226 \, SSU - \frac{195}{SSU} \quad \text{for SSU 32 to 100} \quad (19)$$

$$cS = 0.220 \, SSU - \frac{135}{SSU} \quad \text{for SSU} > 100 \quad (20)$$

SSU represents the Saybolt seconds, a measure of viscosity. Also, $cS \times s = cP$, where cP is the viscosity in centipoise, and 0.413 cP = 1 lb/ft hr.

The specific gravity is to be found. At 180°F, from Eq. (23) (see Q3.21) it can be shown that the specific volume at 180°F is 0.0176 cu ft/lb. Then

$$s = \frac{1}{0.0176 \times 62.4} = 0.91$$

Hence cP = 0.91 × 24.83, where

$$cS = 0.22 \times 118 - \frac{135}{118} = 24.83$$

and

$$\mu = 2.42 \times 0.91 \times 24.83 = 54.6 \text{ lb/ft hr}$$

$$Re = 15.2 \times \frac{7000}{0.91 \times 24.83 \times 2.42} = 1948$$

(2.42 was used to convert cP to lb/ft hr.) From Eq. (17a),

$$f = \frac{64}{1948} = 0.0328$$

Substituting into Eq. (17b) yields

$$\Delta P = 14 \times 10^{-6} \times 54.6 \times 7000 \times 100$$
$$\times \frac{1}{0.91 \times 62.4} = 9.42 \text{ psi}$$

3.20a

Q: For viscous fluids in turbulent flow, how is the pressure drop determined?

A: For viscous fluids, the following expression can be used for the friction factor:

$$f = \frac{0.316}{Re^{0.22}} \tag{21}$$

Substituting into Eq. (13) gives us

$$\Delta P = 3.36 \times 10^{-6} \times 0.361 \times (d_i\mu)^{0.22}L_eW^2$$

$$\times \frac{v}{(15.2W)^{0.22}d_i^5}$$

$$= 0.58 \times 10^{-6} \times \mu^{0.22}W^{1.78}L_e \times \frac{v}{d_i^{4.78}} \tag{22}$$

3.20b

Q: A fuel oil system delivers 4500 lb/hr of light oil at 70°F in a pipe. What is the flow that can be delivered at 30°F, assuming that $\mu_{70}/\mu_{30} = 0.5$, $v_{70}/v_{30} = 0.95$, and flow is turbulent?

A: Using Eq. (22), we have

$$v_1 W_1^{1.78}\mu_1^{0.22} = v_2 W_2^{1.78}\mu_2^{0.22}$$

$$4500^{1.78} \times 0.5^{0.22} \times 0.95 = W_2^{1.78}$$

or

$$W_2 = 4013 \text{ lb/hr}$$

3.21

Q: What is the flow in gpm if 1000 lb/hr of an oil of specific gravity (60/60°F) = 0.91 flows in a pipe at 60°F and at 168°F?

A: We need to know the density at 60°F and at 168°F.
    At 60°F:

Density = ρ = 0.91 × 62.4 = 56.78 lb/cu ft

$$v_{60} = \frac{1}{56.78} = 0.0176 \text{ cu ft/lb}$$

Hence at 60°F,

$$q = \frac{1000}{60 \times 56.78} = 0.293 \text{ cu ft/min (cfm)}$$
$$= 0.293 \times 7.48 = 2.2 \text{ gpm}$$

At 168°F, the specific volume of fuel oils increases with temperature:

$$v_t = v_{60}[1 + E(t - 60)] \tag{23}$$

where E is the coefficient of expansion as given in Table 3.8 [13]. For this fuel oil, E = 0.0004. Hence,

$$v_{168} = 0.0176 \times (1 + 0.0004 \times 108) = 0.01836 \text{ cu ft/lb}$$

Table 3.8 Expansion Factor for Fuel Oils

°API	E
14.9	0.00035
15–34.9	0.00040
35–50.9	0.00050
51–63.9	0.00060
64–78.9	0.00070
79–88.9	0.00080
89–93.9	0.00085
94–100	0.00090

Hence

$$q_{168} = 1000 \times \frac{0.01836}{60} = 0.306 \text{ cfm} = 0.306 \times 7.48$$

$$= 2.29 \text{ gpm}$$

3.22

Q: How is the pressure loss in natural gas lines determined? Determine the line size to limit the gas pressure drop to 20 psi when 20,000 scfh of natural gas of specific gravity 0.7 flows with a source pressure of 80 psig. The length of the pipeline is 150 ft.

A: The Spitzglass formula is widely used for compressible fluids [13]:

$$q = 3410 \times F \sqrt{\frac{P_1^2 - P_2^2}{sL}} \tag{24}$$

where

$$
\begin{aligned}
q &= \text{gas flow, scfh} \\
s &= \text{gas specific gravity} \\
P_1, P_2 &= \text{gas inlet and exit pressures, psia} \\
F &= \text{a function of pipe inner diameter (see Table 3.9)} \\
L &= \text{length of pipeline, ft}
\end{aligned}
$$

Substituting, we have

$$20,000 = 3410 \times F \sqrt{\frac{95^2 - 75^2}{0.7 \times 150}}$$

Hence $F = 1.03$. From Table 3.9 we see that d should be $1\frac{1}{4}$ in. Choosing the next higher standard F or d limits the pressure drop to desired values. Alternatively, if q, d, L, and P_1 are given, P_2 can be found.

3.23

Q: Determine the pressure loss in a rectangular duct 2 ft by 2.5 ft in cross section if 25,000 lb/hr of flue gases at 300°F flow through it. The equivalent length is 1000 ft.

Table 3.9 Standard Steel Pipe[a] Data (Black, Galvanized, Welded, and Seamless)

Nominal pipe size [in. (mm)]	Schedule[b]	Outside diameter [in. (mm)]	Inside diameter (in.)	Wall thickness (in.)	Functions of inside diameter, F^c (in.)
⅛ (6)	40	0.405 (10.2)	0.269	0.068	0.00989
¼ (8)	40	0.540 (13.6)	0.364	0.088	0.0242
⅜ (10)	40	0.675 (17.1)	0.493	0.091	0.0592
½ (15)	40	0.840 (21.4)	0.622	0.109	0.117
¾ (20)	40	1.050 (26.9)	0.824	0.113	0.265
1 (25)	40	1.315 (33.8)	1.049	0.113	0.533
1¼ (32)	40	1.660 (42.4)	1.380	0.140	1.17
1½ (40)	40	1.900 (48.4)	1.610	0.145	1.82
2 (50)	40	2.375 (60.2)	2.067	0.154	3.67
2½ (65)	40	2.875 (76.0)	2.469	0.203	6.02
3 (80)	40	3.500 (88.8)	3.068	0.216	11.0
4 (100)	40	4.500 (114.0)	4.026	0.237	22.9
5 (125)	40	5.563 (139.6)	5.047	0.258	41.9
6 (150)	40	6.625 (165.2)	6.065	0.280	68.0
8 (200)	40	8.625 (219.1)	7.981	0.322	138
8 (200)	30	8.625	8.071	0.277	142
10 (250)	40	10.75 (273.0)	10.020	0.365	247
10 (250)	30	10.75	10.136	0.307	254
12 (300)	40	12.75 (323.9)	11.938	0.406	382
12 (300)	30	12.75	12.090	0.330	395

[a]ASTM A53-68, standard pipe.

[b]Schedule numbers are approx. values of $\dfrac{1000 \times \text{maximum internal service pressure, psig}}{\text{allowable stress in material, psi}}$.

[c]$F = \sqrt{d(1 + 0.03d + 3.6/d)}$ for use in Spitzglass formula 5/23 for gas line pressure loss
Source: Adapted from Ref. 13.

A: The equivalent diameter of a rectangular duct is given by

$$d_i = 2 \times a \times \frac{b}{a + b} = 2 \times 2 \times \frac{2.5}{4.5} = 2.22 \text{ ft}$$
$$= 26.64 \text{ in.}$$

The friction factor f in turbulent flow region for flow in ducts and pipes is given by [11]

$$f = \frac{0.316}{Re^{0.25}} \tag{25}$$

We make use of the equivalent diameter calculated earlier [Eq. (14)] while computing Re:

$$Re = 15.2 \; \frac{W}{d_i \mu}$$

From Table 3.7 at 300°F, $\mu = 0.05$ lb/ft hr.

$$Re = 15.2 \times \frac{25,000}{26.64 \times 0.05} = 285,285$$

Hence

$$f = \frac{0.316}{285,285^{0.25}} = 0.014$$

For air or flue gases, pressure loss is generally expressed in in. WC and not in psi. The following equation gives ΔP_g [11]:

$$\Delta P_g = 93 \times 10^{-6} \times f \times W^2 \times v \times \frac{L_e}{d^5} \tag{26}$$

where d_i is in inches and the specific volume is $v = 1/\rho$.

$$\rho = \frac{40}{460 + 300} = 0.526 \text{ lb/cu ft}$$

Hence

$$v = \frac{1}{0.0526} = 19 \text{ cu ft/lb}$$

Substituting into Eq. (26), we have

$$\Delta P_g = 93 \times 10^{-6} \times 0.014 \times 25,000^2 \times 19$$

$$\times \frac{1000}{(26.64)^5} = 1.16 \text{ in. WC}$$

3.24a

Q: Determine the Reynolds number when 500,000 lb/hr of super-heated steam at 1600 psig and 750°F flows through a pipe of inner diameter 10 in.

A: The viscosity of superheated steam does not vary as much with pressure as it does with temperature (see Table 3.5).

$$\mu = 0.062 \text{ lb/ft hr}$$

Using Eq. (14), we have

$$\text{Re} = 15.2 \times \frac{W}{d_i\mu} = 15.2 \times \frac{500,000}{10 \times 0.062}$$

$$= 1.25 \times 10^7$$

3.24b

Q: Determine the Reynolds number when hot air flows over a tube bundle.

Air mass velocity $= 7000 \text{ lb/ft}^2 \text{ hr}$
Temperature of air film $= 800°F$
Tube size $= 2$ in. OD
Transverse pitch $= 4.0$ in.

A: The Reynolds number when gas or fluids flow over tube bundles is given by the expression

$$\text{Re} = \frac{Gd}{12\mu} \tag{27}$$

where

$G =$ fluid mass velocity, lb/ft^2 hr
$d =$ tube outer diameter, in.
$\mu =$ gas viscosity, lb/ft hr

At 800°F, the air viscosity from Table 3.7 is 0.08 lb/ft hr; thus

$$\text{Re} = 7000 \times \frac{2}{12 \times 0.08} = 14,580$$

3.25

Q: There are three tubes connected between two headers of a super-heater, and it is required to determine the flow in each parallel pass. Following are the details of each pass.

Tube no. (pass no.)	Inner diameter (in.)	Equivalent length (ft)
1	2.0	400
2	1.75	350
3	2.0	370

Total steam flow is 15,000 lb/hr, and average steam conditions are 800 psia and 750°F.

A: As the passes are connected between the same headers, the pressure drop in each will be the same. Also, the total steam flow will be equal to the sum of the flow in each. That is,

$$\Delta P_1 = \Delta P_2 = \Delta P_3$$

In other words, using the pressure drop correlation, we have

$$W_1^2 f_1 \times \frac{L_{e1}}{d_{i1}^5} = W_2^2 f_2 \times \frac{L_{e2}}{d_{i2}^5} = W_3^2 f_3 \times \frac{L_{e3}}{d_{i3}^5}$$

and

$$W_1 + W_2 + W_3 = \text{total flow}$$

The effect of variations in steam properties in the various tubes can be neglected, as its effect will not be very significant.

Substituting the data and using f from Table 3.6, we obtain

$$W_1 + W_2 + W_3 = 15,000$$

$$W_1^2 \times 0.0195 \times \frac{400}{2^5} = W_2^2 \times 0.02 \times \frac{350}{(1.75)^5}$$

$$= W_3^2 \times 0.0195 \times \frac{370}{2^5} = \text{a constant}$$

Simplifying and solving for flows, we have

$$W_1 = 5353 \text{ lb/hr}, \quad W_2 = 4054 \text{ lb/hr}, \quad W_3 = 5591 \text{ lb/hr}$$

This type of calculation is done to check if each pass receives adequate steam flow to cool it. Note that pass 2 had the least flow, and hence a metal temperature check must be performed. If the metal temperature is high, the tube length or tube sizes must be modified to ensure that the tubes are protected from overheating.

3.26

Q: How is the equivalent length of a piping system determined? 100 ft of a piping system has three globe valves, a check valve, and three 90° bends. If the line size is 2 in., determine the total equivalent length.

A: The total equivalent length is the sum of the developed length of the piping plus the equivalent lengths of valves, fittings, and bends. Table 3.10 gives the equivalent length of valves and fittings. A globe valve has 58.6 ft, a check valve has 17.2 ft, and

Table 3.10 Equivalent Length L_e for Valves and Fittings[a]

Pipe size (in.)	1[b]	2[b]	3[b]	4[b]
1	0.70	8.70	30.00	2.60
2	1.40	17.20	60.00	5.20
3	2.00	25.50	87.00	7.70
4	2.70	33.50	114.00	10.00
6	4.00	50.50	172.00	15.20
8	5.30	33.00	225.00	20.00
10	6.70	41.80	284.00	25.00
12	8.00	50.00	338.00	30.00
16	10.00	62.50	425.00	37.50
20	12.50	78.40	533.00	47.00

[a]$L_e = Kd_i/12f$, where d_i is the pipe inner diameter (in.) and K is the number of velocity heads (adapted from Crane Technical Paper 410). f is the Darcy friction factor.
[b]1, Gate valve, fully open; 2, swing check valve, fully open; 3, globe valve, fully open; 4, 90° elbow.

a 90° bend has 5.17 ft of equivalent length. The equivalent length of all valves and fittings is

$$3 \times 58.6 \times 17.2 + 3 \times 5.17 = 208.5 \text{ ft}$$

Hence the total equivalent length is $(100 + 208.5) = 308.5$ ft.

3.27

Q: Determine the pressure drop of flue gases and air flowing over a tube bundle under the following conditions:

Gas mass velocity = 7000 lb/ft^2 hr
Tube size = 2 in. OD
Transverse pitch = 4.0 in.
Longitudinal pitch = 3.6 in.
Arrangement: in-line
Average gas temperature = 800°F
Number of rows deep = 30

A: The following procedure may be used to determine gas pressure drop over tube bundles in in-line and staggered arrangements [11].

$$\Delta P_g = 9.3 \times 10^{-10} \times fG^2 \times \frac{N_H}{\rho_g} \qquad (28)$$

where

G = gas mass velocity, lb/ft^2 hr
ΔP_g = gas pressure drop, in. WC
f = friction factor
ρ_g = gas density, lb/cu ft
N_H = number of rows deep

where, for in-line arrangement for $S_T/d = 1.5$ to 4.0 and for $2000 < \text{Re} < 40{,}000$ [12],

$$f = \text{Re}^{-0.15} \times \left(0.044 + \frac{0.08 S_L/d}{(S_T/d - 1)^{0.43 + 1.13 d/S_L}} \right) \qquad (29)$$

where S_T is the transverse pitch and S_L is the longitudinal pitch, in.

For a staggered arrangement for $S_T/d = 1.5$ to 4.0,

$$f = Re^{-.16} \times \left(0.25 + \frac{0.1175}{(S_T/d - 1)^{1.08}}\right) \qquad (30)$$

In the absence of information on gas properties, use a molecular weight of 30 for flue gas. Then, from Chapter 1,

$$\rho_g = 30 \times \frac{492}{359 \times (460 + 800)} = 0.0326 \text{ lb/cu ft}$$

The viscosity is to be estimated at the gas film temperature; however, it can be computed at the average gas temperature, and the difference is not significant for Reynolds number computations.

From Table 3.7, $\mu = 0.08$ lb/ft hr. From Eq. (27),

$$Re = \frac{Gd}{12\mu} = \frac{7000 \times 2}{12 \times 0.08} = 14{,}580$$

From Eq. (29),

$$f = (14{,}580)^{-0.15} \left(0.044 + \frac{0.08 \times 2}{1}\right) = 0.0484$$

$$\Delta P_g = 9.3 \times 10^{-0} \times 0.0484 \times 7000^2 \times \frac{30}{0.0326}$$

$$= 2.03 \text{ in. WC}$$

Similarly, using Eq. (30) we can estimate ΔP_g for a staggered arrangement.

Note: The foregoing procedure may be used in the absence of field-tested data or correlation.

3.28

Q: Determine the gas pressure drop over a bundle of circumferentially finned tubes in an economizer when

Gas mass velocity of flue gas = 6000 lb/ft^2 hr

(The method of computing G for plain and finned tubes is discussed in Chapter 4.)

Average gas temperature = 800°F
Tube size = 2.0 in.
Transverse pitch S_T = 4.0 in.
Longitudinal pitch S_L = 3.6 in.
Number of rows deep = 10

A: The equation of Robinson and Briggs [11] may be used in the absence of site-proven data or correlation provided by the manufacturer for staggered arrangement:

$$\Delta P_g = \frac{1.58 \times 10^{-8} \times G^{1.684} \times d^{0.611} \times \mu^{0.316} \times (460 + t) \times N_H}{S_T^{0.412} \times S_L^{0.515} \times MW} \qquad (31)$$

where

G = gas mass velocity, lb/ft hr
MW = gas molecular weight
d = tube outer diameter, inc.
$F(t) = \mu^{0.316} \times (460 + t)$
S_T, S_L = transverse and longitudinal pitch, in.

$F(t)$ is given as a function of gas temperature in Table 3.11. Substituting into Eq. (31) gives us

$$\Delta P_g = 1.58 \times 10^{-8} \times 6000^{1.684} \times 2^{0.611} \times 556$$
$$\times \frac{10}{4^{0.412} \times 3.6^{0.515} \times 30}$$
$$= 3.0 \text{ in. WC}$$

3.29

Q: What is boiler circulation, and how is it determined?

A: The motive force driving the steam–water mixture through boiler tubes (water tube boilers) or over tubes (in fire tube boilers) is often the difference in density between the cooler water in the downcomer circuits and the steam–water mixture in the riser

Table 3.11 $F(t)$ Versus t for Air or Flue Gases

$t(F°)$	$F(t)$
200	251
400	348
600	450
800	556
1000	664
1200	776
1600	1003

tubes (Figure 3.4). A thermal head is developed because of this difference, which forces a certain amount of steam–water mixture through the system. This head overcomes several losses in the system such as

Friction loss in the downcomers
Friction loss and flow acceleration loss in the risers and connecting pipes to the drum
Gravity loss in the evaporator tubes and the riser system
Losses in the drum internals

Generally, the higher the drum operating pressure, the lower the difference between the densities of water and the steam–water mixture, and hence the lower the circulation rate.

Circulation ratio (CR) is defined as the ratio between the mass of the steam–water mixture flowing through the system and the mass of the steam generated. If CR = 15, then a boiler generating 10,000 lb/hr of steam would have 150,000 lb/hr of steam–water mixture flowing through the downcomers, risers, internals, etc. The quality of steam at the exit of the riser = 1/CR, or 0.067 if CR = 15. In other words, 6.7% would be the average wetness of steam in the mixture. Low-pressure systems have an average CR ranging from 10 to 40. If there are several parallel circuits for the steam–water mixture, each would have a different resistance to flow, and hence CR would vary from circuit to circuit. For natural circulation systems, CR is usually arrived at by trial and error or by iterative calculation, which first assumes

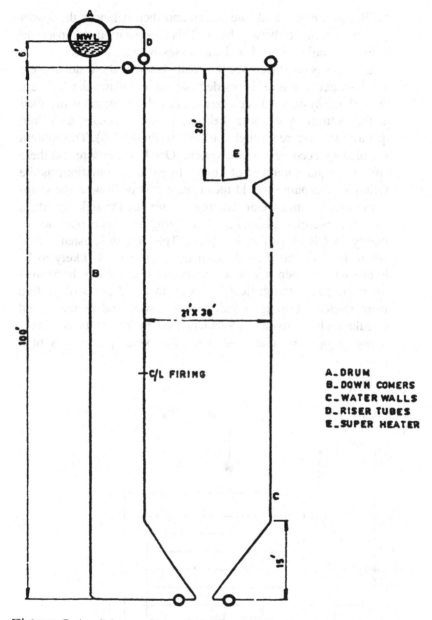

Figure 3.4 Scheme of natural circulation boiler showing furnace, drum, riser, and downcomer circuits.

a CR and computes all the losses and then balances the losses with the available thermal head. This computation is carried on until the available head and the losses balance.

Sometimes the difference in density between the water and the steam–water mixture is inadequate to circulate the mixture through the system. In such cases, a circulation pump is installed at the bottom of the steam drum, which circulates a desired quantity of mixture through the system (Figure 3.5). This system is called a *forced circulation* system. One has to ensure that there are an adequate number of pumps to ensure circulation, as the failure of the pump would mean starvation of flow in the evaporator tubes. Since we are forcing the mixture through the tubes, the CR is preselected, and the circulating pump is chosen accordingly. A CR of 3 to 10 is typical. This system is usually used when the pressure drop through the evaporator is likely to be high such as when horizontal tubes are used. When horizontal tubes are used, the critical heat flux to avoid DNB (departure from nucleate boiling) conditions is lower, and hence forced circulation helps to ensure adequate flow inside the tubes. Circulating pumps are also used when the boiler pressure is high

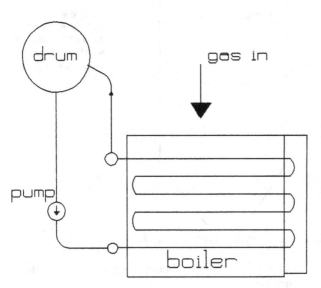

Figure 3.5   Scheme of forced circulation boiler.

owing to the lower difference in density between the water and the steam–water mixture.

3.30

Q: What is the main purpose of determining CR?

A: Determination of CR is not the end in itself. The CR value is used to determine whether a given circuit in the boiler has all the conditions necessary to avoid DNB (departure from nucleate boiling) problems. For each pressure and quality (or CR) there is a particular heat flux beyond which the type of boiling may change from nucleate, which is preferred, to film boiling, which is to be avoided, as this can cause the tube wall temperatures to rise significantly, resulting in tube failure. DNB occurs at heat fluxes of 100,000 to 400,000 Btu/ft² hr depending on size and orientation of tubes, pressure, mass velocity, quality, and roughness of tubes. DNB occurs at a much lower heat flux in a horizontal tube than in an equivalent vertical tube because the steam bubble formation and release occurs more freely and rapidly in vertical tubes than in horizontal tubes where there is a possibility of bubbles adhering to the top of the tube and causing overheating. More information on DNB and circulation can be found in references cited in Refs. 11 and 14.

Note that the heat flux in finned tubes is much higher than in bare tubes owing to the large ratio of external to internal surface area; this aspect is also discussed elsewhere. Hence one has to be careful in designing boilers with extended surfaces to ensure that the heat flux in the finned tubes does not reach critical levels or cause DNB. That is why boilers with very high gas inlet temperatures are designed with a few rows of bare tubes followed by a few rows of low-fin-density tubes and then high-fin-density tubes. As the gas cools, the heat flux decreases.

3.31a

Q: Describe the procedure for analyzing the circulation system for the water tube boiler furnace shown in Figure 3.4.

A: First, the thermal data such as energy absorbed, steam genera-
      tion, pressure, and geometry of downcomers, evaporator tubes,
      and risers should be known. These are obtained from an analysis
      of furnace performance (see example in Chapter 4). The circula-
      tion ratio (CR) is assumed; then the flow through the system is
      computed, followed by estimation of various pressure losses;
      Thom's method is used for evaluating two-phase flow losses
      [15,16].

      The losses can be estimated as follows. ΔP_f, the friction loss in
      two-phase flow (evaporators/risers), is given by

$$\Delta P_f = 4 \times 10^{-10} \times v_f \times \frac{fL}{d_i} \times G_i^2 \times r_3 \qquad (32)$$

      The factor r_3 is shown in Figure 3.6. G_i is the tube-side mass

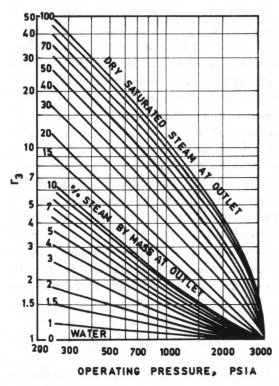

Figure 3.6   Thom's two-phase multiplication factor for friction loss
[11,15,16].

velocity in lb/ft^2 hr. The friction factor used is that of Fanning, which is 0.25 the Moody friction factor.

ΔP_g, the gravity loss in the heated riser/evaporator, is given by

$$\Delta P_g = 6.95 \times 10^{-3} \times L \times \frac{r_4}{v_f} \tag{33}$$

where r_4 is obtained from Figure 3.7.

ΔP_a, the acceleration loss, which is significant at lower pressures and at high mass velocities, is given by

$$\Delta P_a = 1.664 \times 10^{-11} \times v_f \times G_i^2 \times r_2 \tag{34}$$

Figure 3.8 gives r_2.

Single-phase pressure losses such as losses in downcomers, are obtained from

$$\Delta P = 12 \times f \times L_e \times \rho \times \frac{V^2}{2g \times d_i}$$

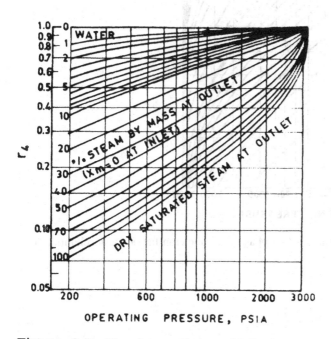

Figure 3.7   Thom's two-phase multiplication factor for gravity loss [11,15,16].

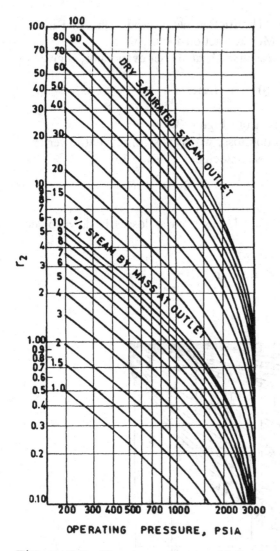

Figure 3.8 Thom's two-phase multiplication factor for acceleration loss [11,15,16].

or

$$\Delta P = 3.36 \times 10^{-6} \times fL_e v \ \frac{W^2}{d_i^5}$$

where

W = flow per tube, lb/hr
V = fluid velocity, fps
f = Moody's friction factor
L_e = effective or equivalent length of piping, ft
v = specific volume of the fluid, cu ft/lb

The unheated riser losses can be obtained from

$$\Delta P_f = f \times \frac{12L_e}{d_i} \times G_i^2 \times \frac{v_f r_f}{2g \times 144} \tag{35}$$

r_f is given in Figure 3.9.

The equivalent lengths have to be obtained after considering the bends, elbows, etc., in the piping; see Tables 3.10 and 3.12.

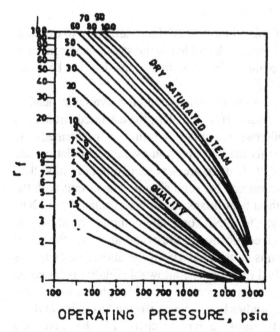

OPERATING PRESSURE, psia

Figure 3.9 Two-phase friction factor for unheated tubes [11,15,16].

Table 3.12 L_e/d_i Ratios for Fitting Turbulent Flow

Fitting	L_e/d_i
45° elbows	15
90° elbows, standard radius	32
90° elbows, medium radius	26
90° elbows, long sweep	20
180° close-return bends	75
180° medium-radius return bends	50
Tee (used as elbow, entering run)	60
Tee (used as elbow, entering branch)	90
Gate valves, open	7
Gate valves, one-quarter closed	40
Gate valves, half-closed	200
Gate valves, three-quarters closed	800
Gate valves, open	300
Angle valves, open	170

A heat balance is first done around the steam drum to estimate the amount of liquid heat to be added to the steam–water mixture before the start of boiling. The mixture is considered to be water until boiling starts.

Once all of the losses are computed, the available head is compared with the losses. If they match, the assumed circulation rate is correct; otherwise another iteration is performed. As mentioned before, this method gives an average circulation rate for a particular circuit. If there are several parallel circuits, then the CR must be determined for each circuit. The circuit with the lowest CR and highest heat fluxes should be evaluated for DNB.

In order to analyze for DNB, one may compute the allowable steam quality at a given location in the evaporator with the actual quality. The system is considered safe if the allowable quality is higher than the actual quality. The allowable quality is based on the heat flux, pressure, mass velocity, and roughness and orientation of the tubes. Studies have been performed to arrive at these values. Figure 3.10 shows a typical chart [14] that gives the

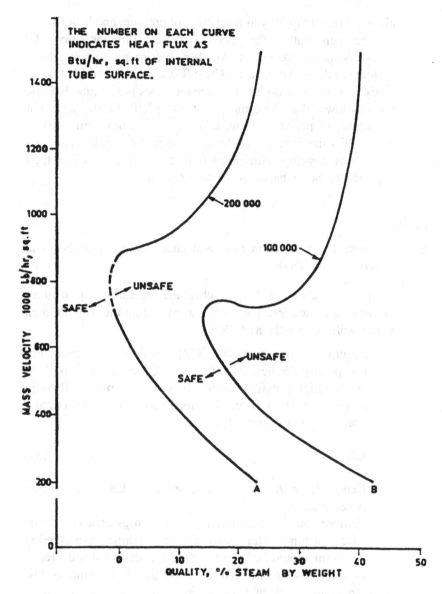

THE NUMBER ON EACH CURVE
INDICATES HEAT FLUX AS
Btu/hr, sq.ft OF INTERNAL
TUBE SURFACE.

Figure 3.10   Allowable quality for nucleate boiling at 2700 psia, as a function of mass velocity and heat flux inside tubes. (Courtesy of Babcock and Wilcox.)

allowable steam quality as a function of pressure and heat flux. It can be seen that as the pressure or heat flux increases, the allowable quality decreases. Another criterion for ensuring that a system is safe is that the actual heat flux on the steam side (inside tubes in water tube boilers and outside tubes in fire tube boilers) must be lower than the critical heat flux (CHF) for the particular conditions of pressure, flow, tube size, roughness, orientation, etc. CHF values are available in the literature; boiler manufacturers have developed their own CHF correlations based on their experience. See Chapter 4 for an example.

3.31b

Q: Compute the circulation ratio and check the system shown in Figure 3.4 for DNB.

A: Figure 3.4 shows a boiler schematic operating on natural-circulation principles. The basis for estimating the flow through water walls is briefly as follows.

1. Assume a circulation ratio (CR) based on experience. For low-pressure boilers ($<$ 1000 psia), CR could be from 20 to 50. For high-pressure boilers (1000 to 2700 psia), CR could range from 9 to 5. The following expression relates circulation ratio and dryness fraction, x:

$$CR = \frac{1}{x} \tag{37}$$

Hence, flow through the evaporator = CR $\times$ the steam generated.

2. Furnace thermal performance data such as efficiency, furnace exit temperature, and feedwater temperature entering the drum should be known before the start of this exercise, in addition to details such as the location of the drum, bends, size, and length of various circuits.

3. Mixture enthalpy entering downcomers is calculated as follows through an energy balance at the drum.

$$h_{fw} + CR \times h_e = h_g + CR \times h_m \tag{36}$$

4. As the flow enters the water walls, it gets heated, and boiling starts after a particular distance from the bottom of the furnace. This distance is called boiling height, and it increases as the subcooling increases. It is calculated as follows.

$$L_b = L \times CR \times W_s \ \frac{h_f - h_m}{Q}$$

Beyond the boiling height, the two-phase flow situation begins.

5. Friction loss in various circuits such as downcomers, connecting headers, water-wall tubes (single-phase, two-phase losses), riser pipes, and drums are calculated. Gravity losses, ΔP_g, are estimated along the acceleration losses, ΔP_a, in a boiling regime. The head available in the downcomer is calculated and equated with the losses. If they balance, the assumed CR is correct; otherwise, a revised trial is made until they balance. Flow through the water-wall tubes is thus estimated.

6. Checks for DNB are made. Actual quality distribution along furnace height is known. Based on the heat-flux distribution (Figure 3.11), the allowable quality along the furnace height may be found. If the allowable quality exceeds actual quality, the design is satisfactory; otherwise, burnout possibilities exist, and efforts must be made to improve the flow through water-wall tubes.

EXAMPLE

A coal-fired boiler has a furnace configuration as shown in Figure 3.4. Following are the parameters obtained after performing preliminary thermal design:

Steam generated	600,000 lb/hr
Pressure at drum	2700 psia
Feedwater temperature entering drum from economizer	570°F
Furnace absorption	320×10^6 BTU/hr
Number and size of downcomers	4, 12-in. ID

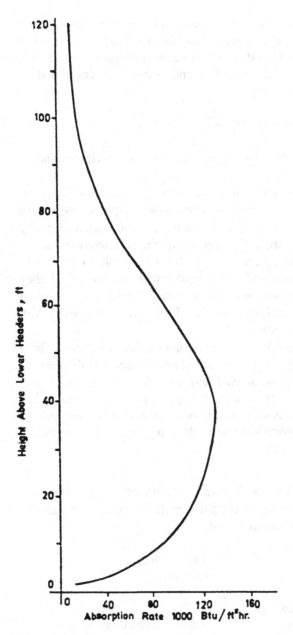

Figure 3.11   Typical heat absorption rates along furnace height.

Number and size of water-wall	
tubes	416, 2½-in. OD × 0.197 in. thick
Number and size of riser tubes	15, 6-in. ID
Drum ID	54 in.
Furnace projected area	8400 ft^2

Since it is difficult to estimate flow through parallel paths, let us assume that flow in each tube or circuit of downcomers, water walls, and risers may be near the average flow values. However, computer programs may be developed that take care of different circuits. Since the manual methods gives a good idea of the solution procedure (though approximate), it is described below.

METHOD

Let circulation ratio CR = 8. Then, $x = 0.125$. From the steam tables,

$t_{sat} = 680°F$
$h_g = 1069.7$ Btu/lb
$h_f = 753.7$ Btu/lb
$v_f = 0.0303$ cu ft/lb
$v_g = 0.112$ cu ft/lb
$h_{fw} = 568$ Btu/lb

Enthalpy of steam leaving water walls,

$h_e = 0.125 × 1069.7 + 0.875 × 753.7 = 793.2$ Btu/lb

Heat balance around the drum gives

Steam flow = 600,000 lb/hr
Water-wall, downcomer flow = 8 × 600,000 = 4,800,000 lb/hr

$600,000 × 568 + 8 × 600,000 × 793.2$
$= 600,000 × 1069.7 + 8 × 600,000 h_m$

Hence, $h_m = 731$ Btu/lb.
From the steam tables,

$v_m = 0.0286$ cu ft/lb
$v_e = 0.125 × 0.112 + 0.875 × 0.0303 = 0.0405$ cu ft/lb

a. ΔP_g = head available = $106/(0.0286 \times 144)$ = 25.7 psi.

b. ΔP_{dc} = losses in downcomer circuit.

The downcomer has one 90° bend and one entrance and exit loss. Using an approximate equivalent length of $7d_i$,

$$L_e = 104 + 16 + (7 \times 12) = 204 \text{ ft}$$

The value f_i from Table 3.6 is around 0.013.

$$V_{dc} = \frac{8 \times 600,000 \times 0.0286 \times 576}{3600 \times \pi \times 144 \times 4} = 12.1 \text{ fps}$$

$$\Delta P_{dc} = \frac{0.013 \times 204 \times (12.1)^2 \times 12}{2 \times 32 \times 12 \times 0.0286 \times 144} = 1.47 \text{ psi}$$

c. Estimate boiling height:

$$L_b = 100 \times 8 \times 600,000 \times \frac{753.7 - 731}{320 \times 10^6} = 31 \text{ ft}$$

Hence, up to a height of 31 ft, preheating of water occurs. Boiling occurs over a length of $100 - 31 = 69$ ft only.

d. Gravity loss in boiling height:

$$V_m, \text{ mean specific volume} = \frac{0.0286 + 0.0303}{2}$$

$$= 0.02945 \text{ cu ft/lb}$$

$$\Delta P_g = \frac{31}{0.02945 \times 144} = 7.3 \text{ psi}$$

e. Friction loss in boiling height. Compute velocity through water-wall tubes: $d_i = 2.1$ in.

$$V_w = \frac{8 \times 600,000 \times 576 \times 0.02945}{416 \times \pi \times (2.1)^2 \times 3600} = 3.93 \text{ fps}$$

From Table 3.6, $f_i = 0.019$.

One exit loss, one 135° bend, and one 45° bend can be considered for computing an equivalent length. L_e works out to about 45 ft.

$$\Delta P_w = \frac{0.019 \times 45 \times (3.93)^2 \times 12}{2 \times 32 \times 2.1 \times 0.02945 \times 144} = 0.28 \text{ psi}$$

f. Computing losses in two-phase flow, from Figures 3.6 to 3.8, for $x = 12.5\%$ and $P = 2700$ psi,

$$r_2 = 0.22, \qquad r_3 = 1.15, \qquad r_4 = 0.85$$

For computing two-phase losses:

$$\Delta P_a = 1.664 \times 10^{-11} \times v_f r_2 \, G_i^2$$

$$G_i = \frac{8 \times 600{,}000 \times 576}{416 \times \pi \times (2.1)^2} = 480{,}000 \text{ lb/ft}^2 \text{ hr}$$

$$\Delta P_a = 1.664 \times 10^{-11} \times 0.0303 \times (4.8 \times 10^5)^2$$
$$\times \, 0.22 = 0.026 \text{ psi}$$

Friction loss,

$$\Delta P_f = 4 \times 10^{-10} \times 0.0303 \times \frac{0.0019}{4} \times 69$$

$$\times \, (4.8 \times 10^5)^2 \times \frac{1.15}{2.1} = 0.5 \text{ psi}$$

Gravity loss,

$$\Delta P_g = \frac{6.944 \times 10^{-3} \times 69 \times 0.85}{0.0303} = 13.4 \text{ psi}$$

Total two-phase loss $= 0.026 + 0.5 + 13.4 = 13.926$ psi, or 14.0 psi

g. Riser circuit losses. Use Thom's method for two-phase unheated tubes. Let the total equivalent length, considering bends and inlet and exit losses, be 50 ft.

$$r_f = 1.4 \text{ (Figure 3.9)}, \qquad f_i = 0.015 \text{ from Table 3.6}$$

$$G_i = \frac{576 \times 8 \times 600{,}000}{\pi \times 36 \times 15} = 1.63 \times 10^6 \text{ lb/ft}^2 \text{ hr}$$

$$\Delta P_f = 0.015 \times \frac{50 \times 12}{6} \times \frac{(1.63 \times 10^6)^2}{2 \times 32 \times 3600^2}$$

$$\times \, \frac{1.4}{144} \times 0.0303 = 1.41 \text{ psi}$$

Note that in estimating pressure drop by Thom's method for heated tubes, the Darcy friction factor was used. For unheated tubes, Moody's friction factor could be used. Void fraction α' from Figure 3.12 = 0.36. From Eq. 38.

$$\Delta P_g = [\rho_f (1-\alpha') + \rho_g \alpha'] \frac{L}{144} \tag{38}$$

$$\Delta P_g = \left[\left(\frac{1}{0.0303} \times 0.64 \right) + \left(\frac{1}{0.112} \times 0.36 \right) \right]$$
$$\times \frac{5}{144} = 0.85 \text{ psi}$$

Total losses in riser circuit = 1.41 + 0.85 = 2.26 psi.

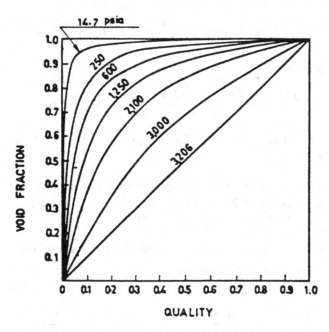

Figure 3.12   Void fraction as a function of quality and pressure for steam [11,16].

h. Losses in drum. This is a negligible value; use 0.2 psi. (Generally the supplier of the drums should furnish this figure.)

Total losses = b + d + e + f + g + h

   = 1.47 + 7.3 + 0.28 + 14.0 + 2.26

   + 0.2 = 25.51 psi

Available head = a = 25.70 psi

Hence, as these two match, an assumed circulation ratio of 8 is reasonable. This is only an average value for the entire system. If one is interested in a detailed analysis, the circuits should be separated according to heat loadings, and a rigorous computer analysis balancing flows and pressure drop in each circuit can be carried out.

ANALYSIS FOR DNB

Typical furnace absorption profiles for the actual fuel fired are desirable for DNB analysis. Since these data are generally based on field tests, for the problem at hand let us use Figure 3.11, which gives typical absorption profiles for a boiler.

Average heat flux = $320 \times 10^6/8400 = 38{,}095$ Btu/ft^2 hr

There is a variation at any plan cross section of a boiler furnace between the maximum heat flux and the average heat flux, based on the burner location, burners in operation, excess air used, etc. This ratio between maximum and average could be 20 to 30%. Let us use 25%.

Again, the absorption profile along furnace height shows a peak at some distance above the burner where maximum heat release has occurred. It decreases as the products of combustion leave the furnace. The average for the entire profile may be found, and the ratio of actual to average heat flux should be computed. For the sake of illustration, use the following ratios of actual to average heat flux at locations mentioned.

Distance from bottom (ft)	Ratio of actual to average heat flux
40	1.4
56	1.6
70	1.0
80	0.9
100	0.4

We must determine the maximum inside heat flux at each of the locations and correct it for flux inside the tubes to check for DNB. Hence, considering the tube OD to ID ratio of 1.19 and the 25% nonuniformity at each furnace elevation, we have the following local maximum inside heat flux at the locations mentioned

$$\left(q_i \text{ is taken as } q_p \times \frac{d}{d_i} \right)$$

Location (ft)	q_i (Btu/ft^2 hr)
40	$1.4 \times 38{,}095 \times 1.25 \times 1.19 = 79{,}335$
56	90,440
70	56,525
80	50,872
100	22,600

It is desirable to obtain allowable quality of steam at each of these locations and check to be sure actual quality does not exceed it.

DNB tests based on particular tube profiles, roughness, and water quality as used in the operation give the most realistic data for checking furnace tube burnout. Correlations, though available in the literature, may give a completely wrong picture since they are based on tube size, heating pattern, water quality, and tube roughness that may not tally with actual operating conditions. Correlations, however, give the trend, which could be useful. For the sake of illustrating our example, let us use Figure 3.10. This gives a good estimate only, as extrapolation must be carried out for the low heat flux in our case. We see the following trend at $G_i = 480,000$ lb/ft^2 hr and 2700 psia:

Location (ft)	Allowable quality (%)
40	25
56	22
70	30
80	34
100	42

Figure 3.13 shows the actual quality (assuming linear variation, perhaps in reality quadratic) vs. allowable quality. It shows that a large safety margin exists; hence, the design is safe. This exercise should be carried out at all loads (and for all circuits) before coming to a conclusion.

3.32a

Q: How is the circulation system analyzed in fire tube boilers?

A: The procedure is similar to that followed for water tube boilers in that the CR is assumed and the various losses are computed. If the losses associated with the assumed CR and the resulting mass flow balance the available head, then the assumed CR is correct; otherwise another iteration is done. Since fire tube boilers in general use horizontal tubes, the allowable heat flux to avoid DNB is lower than when vertical tubes are used. With gas streams containing hydrogen and steam as in hydrogen plant waste heat boilers, the tube-side and hence the overall heat transfer coefficient and heat flux will be rather high compared to flue gas stream from combustion of fossil fuels. Typical allowable heat fluxes for horizontal tubes range from 100,000 to 150,000 Btu/ft^2 hr.

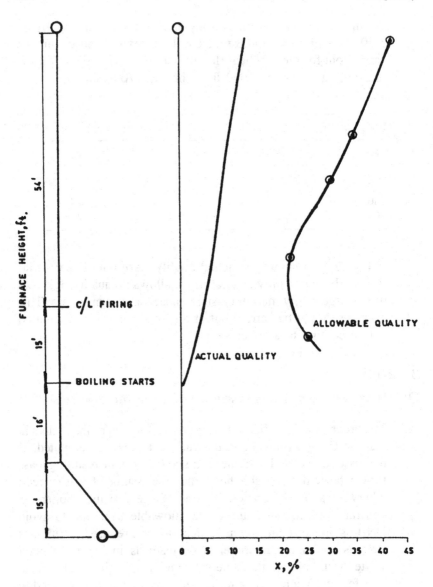

Figure 3.13   Actual quality vs. allowable quality along furnace height.

3.32b

Q: Perform the circulation calculations for the system shown in Figure 3.14 with the following data:

Steam flow = 20,000 lb/hr,   steam pressure = 400 psig

Assume that saturated water enters the drum.

A: From steam tables, v_f = 0.194 and v_g = 1.12 cu ft/lb Assume there are two downcomers of size 4 in schedule 40 (d_i = 4.026 in.) and two risers of size 8 in schedule 40 (d_i = 7.981 in.). The total developed length of each downcomer is 22.5 ft, and each has two 90° bends; the riser pipes have a total developed length of 5 ft. Exchanger diameter is 6 ft, and the center distance between the exchanger and the steam drum is 8 ft.

1. Assume CR = 15; then

Mixture volume = 0.067 × 1.12 + 0.933 × 0.0194

= 0.0931

The head available due to the column of saturated water is 11/(0.0194/144) = 3.94 psi, where 11 ft is the height of the water column.

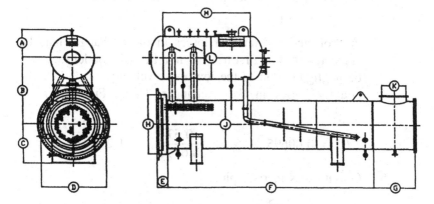

Figure 3.14   Circulation scheme in fire tube boiler.

2. Losses in downcomers:

   a. Water velocity $= 0.05 \times \left(15 \times \dfrac{20,000}{2}\right)$

   $\times \dfrac{0.0194}{(4.026)^2} = 9$ fps

   Inlet plus exist losses $= 1.5$ velocity head

   $= 1.5 \times 9 \times \dfrac{9}{2 \times 32 \times 144 \times 0.0194}$

   $= 0.68$ psi

   b. Total developed length $= 22.5 + 2 \times 10 = 42.5$ ft, where 10 ft is the equivalent length of a 90° bend from Table 3.10.

   $\Delta P_f = 3.36 \times 0.0165 \times \left(15 \times \dfrac{20}{2}\right)^2 \times 42.5$

   $\times \dfrac{0.0194}{(4.026)^5} = 0.98$ psi

   where 0.0165 is the friction factor. Equation (13) was used for pressure drop of single-phase flow.

   Total downcomer losses $= 0.68 + 0.98$

   $= 1.66$ psi

3. Friction and acceleration losses in the exchanger may be neglected for this first trial, as in a fire tube boiler they will be negligible due to the low mass velocity.

4. Gravity losses in the exchanger: Using Figure 3.7, r_4 $= 0.57$.

   $\Delta P_g = 0.00695 \times 6 \times \dfrac{0.57}{0.0194} = 1.22$ psi

5. Gravity loss in riser pipe:

   $\Delta P_g = \dfrac{5}{0.0931/144} = 0.37$ psi

6. Friction loss in riser:

$$\text{Velocity} = 0.05 \times \left(15 \times \frac{20}{2}\right) \times \frac{0.0931}{(7.981^2)} = 11 \text{ fps}$$

Inlet plus exit losses = 1.5 × velocity head

$$= \frac{1.5 \times 11 \times 11}{2 \times 32 \times 0.0931 \times 144} = 0.21 \text{ psi}$$

$$\text{Friction loss} = 3.36 \times 0.014 \times \left(15 \times \frac{20}{2}\right)^2$$

$$\times 5 \times \frac{0.0931}{(7.981)^5} = 0.02 \text{ psi}$$

where 5 ft is the developed length of the riser.

Let the losses in drum internals = 0.5 psi. This can vary depending on the type of internals used. Then

$$\text{Total losses} = 1.66 + 1.22 + 0.37 + 0.21 + 0.02$$
$$+ 0.50 = 3.98 \text{ psi}$$

This is close to the available head; hence CR = 15 is the circulation ratio for this system. The calculations can be fine tuned with actual dimensions after the layout is done. One can compute the heat flux and compare it with the allowable heat flux to check if the circulation rate is adequate. Usually circulation is not a problem in this type of boiler, as the heat flux is low, on the order of 20,000 to 30,000 Btu/ft^2 hr, while the allowable flux could be 100,000 to 150,000 Btu/ft^2 hr. See Chapter 4 for correlations for CHF (critical heat flux).

3.33

Q: How is the flow in steam blowoff lines determined?

A: Whenever steam flows to the atmosphere from a high-pressure vessel, the flow reaches critical flow conditions, and beyond a certain pressure further lowering of pressure does not increase the steam discharge. The flow is given by the equation [17]

$$W = 1891 \times Y \times d^2 \times \left(\frac{\Delta P}{K/\upsilon} \right)^{0.5} \tag{39}$$

The value of ΔP to be chosen depends on K, the system resistance, where

$$K = 12 \times \frac{fL_e}{d}$$

where

L_e = total equivalent length of all downstream piping including valves, fittings, ft
f = Darcy friction factor
d = pipe inner diameter, in.
Y = expansion factor (see Table 3.13)
υ = specific volume of steam before expansion, cu ft/lb
ΔP = pressure drop, lower of actual upstream pressure − downstream pressure or that obtained from Table 3.13

Table 3.13 Limiting Factors For Sonic Velocity $k = 1.3$

K	$\dfrac{\Delta P}{P'_1}$	Y
1.2	0.525	0.612
1.5	0.550	0.631
2.0	0.593	0.635
3	0.642	0.658
4	0.678	0.670
6	0.722	0.685
8	0.750	0.698
10	0.773	0.705
15	0.807	0.718
20	0.831	0.718
40	0.877	0.718
100	0.920	0.718

EXAMPLE

Determine the flow of saturated steam from a vessel at 170 psia to atmosphere if the total equivalent system resistance $K = 10$ and pipe inner diameter $= 2.067$ in.

Solution.   Specific volume of steam at 170 psia $= 2.674$ ft^3/lb. Actual $\Delta P = 170 - 14.7 = 155.3$ psia. From Table 3.13, for $K = 10$, $\Delta P/P_1 = 0.773$, or $\Delta P = 170 \times 0.773 = 131.5$ psia. Hence, use $\Delta P = 131.5$ psia. Also from Table 3.13 for $K = 10$, $Y = 0.705$. Hence

$$W = 1891.0 \times 0.705 \times (2.067)^2 \times \left(\frac{131.5}{10/2.674} \right)^{0.5}$$

$$= 12,630 \text{ lb/hr}$$

3.34

Q: How is the flow through boiler blowdown lines determined?

A: Sizing of blowdown or drain lines is very important in boiler or process plant operations.

Effect of friction can be easily studied.

The problem of estimating the discharge rates from a boiler drum or vessel to atmosphere or to a vessel at low pressures involves two-phase flow calculations and is a lengthy procedure [18].

Presented below is a simplified approach to the problem that can save considerable time for engineers who are involved in sizing or estimating discharge rates from boiler drums, vessels, or similar applications involving water.

Several advantages are claimed for these charts, including the following.

No reference to steam tables is required.

No trial-and-error procedure is involved.

Effect of friction can be easily studied.

Obtaining pipe size to discharge a desired rate of fluid, the reverse problem, is simple.

THEORY

The basic Bernoulli's equation can be written as follows for flow in a piping system:

$$10^4 \, v \, dP + \frac{V^2}{2g} \, dk + \frac{v}{g} \, dv + dH = 0 \qquad (40)$$

Substituting mass flow rate $m = V/v$:

$$\frac{m^2}{2g} \left(dk + 2 \frac{dv}{v} \right) = -10^4 \times \frac{dP}{v} - \frac{dH}{v^2} \qquad (41)$$

Integrating between conditions 1 and 2:

$$\frac{m^2}{2g} \left[k + 2\ln \frac{v_2}{v_1} \right] = -10^4$$

$$\int_1^2 \frac{dP}{v} - \int_1^2 \frac{dH}{v^2} \qquad (42a)$$

$$m = \left[\frac{2g}{k + 2\ln (v_2/v_1)} \times \left(-10^4 \right. \right.$$

$$\left. \left. \int_1^2 \frac{dP}{v} - \int_1^2 \frac{dH}{v^2} \right) \right]^{1/2} \qquad (42b)$$

where

$K = fl/d$, the equivalent pipe resistance

When the pressure of the vessel to which the blowdown pipe is connected is decreased, the flow rate increases until critical pressure is reached at the end of the pipe. Reducing the vessel pressure below critical pressure does not increase the flow rate.

If the vessel pressure is less than the critical pressure, critical flow conditions are reached and sonic flow results.

From thermodynamics, the sonic velocity can be shown to be:

$$V_c = \sqrt{- v^2 \, g \left(\frac{dP}{dv} \right)_s \times 10^4} \qquad (43)$$

and

$$m_c = 100 \sqrt{- g \left(\frac{dP}{dv} \right)_s} \qquad (44)$$

The term $(dP/dv)_s$ refers to the change in pressure to volume ratio at critical flow conditions at constant entropy.

Hence, in order to estimate m_c, Eqs. (42) and (44) have to be solved. This is an iterative procedure. For the sake of simplicity, the term involving the height differences will be neglected. For high-pressure systems the error in neglecting this term is marginal, on the order of 5%.

The problem is, then, given K and P_s, to estimate P_c and m. This is a trial-and-error procedure, and the steps are outlined below, followed by an example. Figure 3.16 and 3.17 are two charts that can be used for quick sizing purposes.

1. Assume a value for P_c.
2. Calculate $(dP/dv)_s$ at P_c for constant-entropy conditions. The volume change corresponding to 2 to 3% of P_c can be calculated, and then $(dP/dv)_s$ can be obtained.
3. Calculate m_c using Eq. (44).
4. Solve Eq. (42b) for m_c.

The term $[-10^{-4} \int_1^2 dP/v]$ is computed as follows using Simpson's rule:

$$-10^4 \int_1^2 \frac{dP}{v} = -10^4 \int_1^2 \rho \, dP$$

$$= \frac{P_s - P_c}{6} \times (\rho_s + 4\rho_m + \rho_c)$$

where

ρ_m = density at a mean pressure of $(P_s + P_c)/2$.

The densities are computed as isenthalpic conditions. The term $2 \ln (v_2/v_1) = 2 \ln (\rho_s/\rho_c)$ is then found.

Then m is computed using Eq. (42b). If m computed using Eqs. (42b) and (44) tally, the assumed P_c and the resultant m_c are correct. Otherwise P_c has to be changed, and all steps have to be repeated until m and m_c agree.

EXAMPLE

A boiler drum blowdown line is connected to a tank set at 8 atm. Drum pressure is 100 atm, and the resistance K of the blowdown line is 80. Estimate the critical mass flow rate m_c and the critical pressure P_c.

The procedure will be detailed for an assumed pressure P_c of 40 atm.

For steam table $P_s = 100$ atm, $s = 0.7983$ (kcal/kg °C); $h_l = 334$ kcal/kg, $v_l = 0.001445$ m³/kg or $\rho = 692$ kg/m³.

Let $\rho_c = 40$ atm; then $h_l = 258.2$ kcal/kg, $h_v = 669$ kcal/kg, $S_l = 0.6649$, $S_v = 1.4513$, $v_l = 0.001249$, $v_v = 0.05078$.

Hence:

$$x = \frac{S - S_l}{S_v - S_l} = \frac{0.7983 - 0.6649}{1.4513 - 0.6649}$$

$$= 0.1696$$

$$v = v_l + x(v_v - v_l)$$

$$= 0.001249 + 0.1696 \times (0.05078 - 0.001249)$$

$$= 0.009651 \text{ m}^3/\text{kg}$$

Again, compute v at 41 atm (2.5% more than P_c). Using steps similar to those described above, $v = 0.0093$ m³/kg.

Hence,

$$m_c = 100 \sqrt{- g\left(\frac{dP}{dv}\right)_s}$$

$$= 100 \sqrt{\frac{9.8 \times 1}{0.00965 - 0.0093}} = 16,733 \text{ kg/m}^2 \text{ sec}$$

Compute the densities as

$$\rho_s = \frac{1}{0.001445} = 692 \text{ kg/m}^3$$

The dryness fraction at 40 atm at isenthalphic condition is

$$x = \frac{334 - 258.2}{669 - 258.2} = 0.1845$$

$$v_c = 0.001249 + 0.1845 \times (0.05078 - 0.001249)$$

$$= 0.010387 \text{ m}^3/\text{kg}$$

$$\rho_c = 96.3 \text{ kg/m}^3$$

Similarly, at $P_m = (100 + 40)/2 = 70$ atm,

$v_m = 0.03785$ m^3/kg   or $\rho_m = 264$ kg/m^3

$$-10^4 \int_1^2 \frac{dP}{v} = \frac{100 - 40}{6}$$

$\times (692 + 4 \times 264 + 96.3)$

$$= 184 \times 10^6 \times 2 \times \ln\left(\frac{v_2}{v_1}\right) \, .$$

$$= 2 \times \ln\left(\frac{\rho_s}{\rho_c}\right) = 4.6$$

Substituting the various quantities into Eq. (42b),

$$m = \sqrt{\frac{2 \times 9.8}{80 + 4.6} \times 184 \times 10^6}$$
$$= 6530 \text{ kg/m}^2 \text{ sec}$$

The two values m and m_c do not agree. Hence we have to repeat the calculations for another P_c.

This has been done for $P_c = 30$ and 15, and the results are presented in Figure 3.15.

At about 19 atm, the two curves intersect, and the mass flow rate is about 7000 kg/m^2 sec.

However, one may do the calculations at this pressure and check.

USE OF CHARTS

As seen above, the procedure is lengthy and tedious, and trial and error is involved. Also, reference to steam tables makes it cumbersome. Hence with various K values and initial pressure P_s, a calculator was used to solve for P_c and m, and the results are presented in Figures 3.16 and 3.17.

3.35

Q: What is the effect of stack height on friction loss, draft?

A: Whenever hot flue gases flow in a vertical stack, a natural draft is created owing to the difference in density between the low-

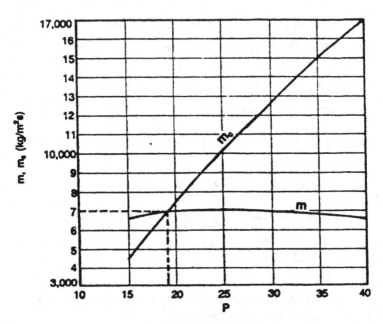

Figure 3.15   Calculation results.

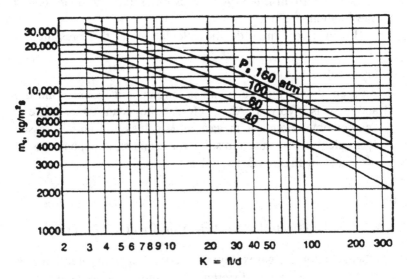

Figure 3.16   Solving for *m*.

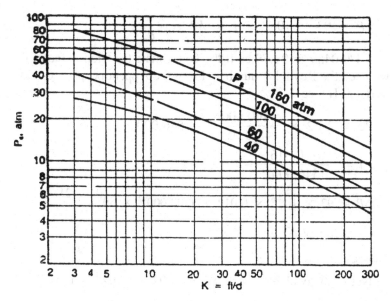

Figure 3.17   Solving for P.

density flue gases and ambient air, which has a higher density. However, due to the friction losses in the stack, this available draft is reduced.

EXAMPLE

If 100,000 lb/hr of flue gases at 400°F flow in a 48-in. ID stack of 50 ft height, determine the net stack effect. Ambient air temperature is 70°F.

Solution.   Density of flue gases (see Q1.02) at 400°F = 39.5/860 = 0.0459 lb/cu ft. Density of air at 70°F = 40/530 = 0.0755 lb/cu ft. Hence

Total draft available = $(0.0755 - 0.0459) \times 50 = 1.48$ lb/ft^2

$$= (0.0755 - 0.0459) \times 50 \times \frac{12}{62.4} = 0.285 \text{ in. WC}$$

(The factor 62.4 is density of water, and 12 converts ft. to in.)

Let us see how much the friction loss per unit length is. From Eq. (26),

$$\Delta P = 93 \times 10^{-6} \times f \times W^2 \times \frac{v}{d^5}$$

$v = 1/0.0459 = 21.79$ cu ft/lb. To estimate the friction factor f, we need the Reynolds number. From the Appendix, $\mu = 0.058$ lb/ft hr. Hence

$$\text{Re} = 15.2 \times \frac{100,000}{48/0.058} = 546,000.$$

$$f = \frac{0.316}{(546,000)^{0.25}} = 0.012$$

$$\Delta P = 93 \times 10^{-6} \times 0.012 \times (100,000)^2 \times 50 \times \frac{21.79}{48^5}$$

$$= 0.048 \text{ in. WC}$$

Hence

Net draft available $= 0.285 - 0.048 = 0.237$ in. WC

NOMENCLATURE

A	Area of orifice, in.^{2}
C	A constant depending on ratio of gas specific heats
CR	Circulation ratio
C_d	Discharge coefficient
C_v	Control valve coefficient
d	Tube or pipe outer diameter, in.
d_o, d_i	Orifice diameter and pipe or duct inner diameter, in.
E	Expansion factor for fuel oils
f	Friction factor
G	Gas mass velocity, lb/ft^2 hr
h	Differential pressure across flow meter, in. WC
h_e	Enthalpy of mixture at exit, Btu/lb
h_f, h_g, h_m, h_{fw}	Enthalpy of saturated liquid, saturated steam, mixture, and feedwater, respectively, Btu/lb
K	System resistance
K_m	Valve recovery coefficient

K_{sh}	Superheat correction factor
L	Length of pipe, ft
L_e	Equivalent length, ft
M	Constant used in Q3.25
m_c	mass flow at critical condition, kg/m^2 sec
MW	Molecular weight of gas or vapor
N_H	Number of rows deep in a tube bundle
P_a	Accumulated inlet pressure, psia
P_b	Back pressure, psig
P_s	Set pressure, psig
P_v	Vapor pressure, psia
P_1, P_2	Inlet and exit pressures, psia
ΔP	Pressure drop, psi
ΔP_g	Gas pressure drop, in. WC
$\Delta P_a, \Delta P_f, \Delta P_g$	Acceleration loss, friction loss, and loss due to gravity, psi
q	Fluid flow, gpm
Re	Reynolds number
r_2, r_3, r_4, r_f	Factors used in two-phase pressure drop calculation
S	Entropy
s	Specific gravity of fluid
S_T, S_L	Transverse and longitudinal pitch, in.
t, T	Fluid temperature, °F or °R
t_s	Saturation temperature, °F
v	Specific volume of fluid, cu ft/lb
V_c	Critical velocity, m/s
V	Fluid velocity, ft/sec
v_f, v_g, v_m	Specific volume of saturated liquid, steam, and mixture, cu ft/lb
W	Flow, lb/hr
x	Steam quality, fraction
y	Volume fraction of gas
Y	Expansion factor
β	d_o/d_i ratio
μ	Fluid viscosity, lb/ft hr
ρ	Density of fluid, lb/cu ft; subscript g stands for gas
μ	Specific volume of fluid, m^3/kg

REFERENCES

1. V. Ganapathy, Determining flowmeter sizes, *Plant Engineering*, Sept. 18, 1980, p. 127.
2. *Chemical Engineers' Handbook*, 5th ed., McGraw-Hill, New York, 1974, pp. 5–7.
3. V. Ganapathy, Converting pilot tube readings, *Plant Engineering*, June 24, 1982, p. 61.
4. ASME, *Boiler and Pressure Vessel Code*, Sec. 1, New York, 1980, pp. 59, 67.
5. Crosby Valve Catalog 402, Crosby, Wrentham, Mass., 1968, p. 27.
6. ASME, *Boiler and Pressure Vessel Code*, Sec. 8, New York, 1980, Appendix 11, p. 455.
7. V. Ganapathy, Control valve coefficients, *Plant Engineering*, Aug. 20, 1981, p. 80.
8. V. Ganapathy, Nomogram estimates control valve coefficients, *Power Engineering*, December 1978, p. 60.
9. F. D. Jury, *Fundamentals of valve sizing for liquids*, Fisher Technical Monograph 30, Fisher Controls Co., Marshalltown, Iowa, 1974, p. 2.
10. Masoneilan, *Handbook for Control Valve Sizing*, 6th ed., Norwood, Mass., 1977, p. 3.
11. V. Ganapathy, *Applied Heat Transfer*, PennWell Books, Tulsa, Okla., 1982, pp. 500–530.
12. V. Ganapathy, Chart speeds estimates of gas pressure drop, *Oil and Gas Journal*, Feb. 4, 1980, p. 71.
13. *North American Combustion Handbook*, 2nd ed., North American Mfg. Co., Cleveland, Ohio, 1978, pp. 20–25.
14. Babcock and Wilcox, *Steam: Its Generation and Use*, 38th ed.
15. J. R. S. Thom, Prediction of pressure drop during forced circulation boiling of water, *International Journal of Heat Transfer*, no. 7, 1964.
16. W. Roshenow and J. P. Hartnett, *Handbook of Heat Transfer*, McGraw-Hill, New York, 1972.
17. Crane Company Technical Paper 410.
18. F. J. Moody, Maximum two-phase vessel blowdown from pipes, *Transactions of ASME, Journal of Heat Transfer*, August 1966, p. 285.

4

Heat Transfer Equipment Design and Performance

4.01: Estimating surface area of heat transfer equipment; overall heat transfer coefficient; approximating overall heat transfer-coefficient in water tube boilers, fire tube boilers, and air heaters; log-mean-temperature difference

4.02: Estimating tube-side heat transfer coefficient; simplified expression for estimating tube-side coefficient

4.03: Estimating tube-side coefficient for air, flue gas, water, and steam

4.04: Estimating heat transfer coefficient outside tubes

4.05: Estimating convective heat transfer coefficient outside tubes using Grimson's correlations

4.06: Effect of in-line vs. staggered arrangement

4.07a: Evaluating nonluminous radiation heat transfer using Hottel's charts

4.07b: Nonluminous radiation using equations

4.08a: Predicting heat transfer in boiler furnaces

4.08b: Design of radiant section for heat recovery application

4.09:   Evaluating distribution of radiation to tube banks

4.10:   Sizing fire tube boilers

4.11:   Effect of gas velocity, tube size on fire tube boiler size

4.12:   Computing heat flux, tube wall temperatures

4.13:   Effect of scale formation on tube wall temperature and boiler performance

4.14:   Design of water tube boilers

4.15a: Predicting off-design performance

4.15b: Logic for off-design performance evaluation for water tube boilers

4.16:   Estimating metal temperature in a boiler superheater tube; thermal resistances in heat transfer; calculating heat flux

4.17:   Predicting performance of fire tube and water tube boilers

4.18:   Why finned tubes are used and their design aspects

4.19a: Heat transfer and pressure drop in finned tubes using ESCOA correlations

4.62:   Effect of gas pressure on heat transfer

4.63:   Converting gas analysis from weight to volume basis

4.01

Q: How is the surface area of heat transfer equipment determined?
      What terms can be neglected while evaluating the overall heat
      transfer coefficient in boilers, economizers, and superheaters?

A: The energy transferred in heat transfer equipment, Q, is given by
      the basic equation

$$Q = U \times A \times \Delta T \tag{1}$$

Also,

$$W_h \times \Delta h_h = W_c \, \Delta h_c \tag{2}$$

where

$\quad A = $ surface area, ft^2
$\quad W = $ fluid flow, lb/hr
$\quad \Delta h = $ change in enthalpy (subscripts h and c stand for hot
$\qquad\quad$ and cold)
$\quad \Delta T = $ corrected log-mean temperature difference, °F
$\quad U = $ overall heat transfer coefficient, Btu/ft^2 hr °F

For extended surfaces, U can be obtained from [1]

$$\frac{1}{U} = \frac{A_t}{h_i A_i} + \mathrm{ff}_i \times \frac{A_t}{A_i} + \mathrm{ff}_o + \frac{A_t}{A_w} \times \frac{d}{24 K_m}$$
$$\times \ln\left(\frac{d}{d_i}\right) + \frac{1}{\eta h_o} \tag{3}$$

where

$\quad A_t = $ surface area of finned tube, ft^2/ft
$\quad A_i = $ tube inner surface area $= \pi d_i/12$, ft^2/ft
$\quad A_w = $ average wall surface area $= \pi(d + d_i)/24$, ft^2/ft
$\quad K_m = $ thermal conductivity of the tube wall, Btu/ft hr
$\qquad\quad$ °F
$\quad d, d_i = $ tube outer and inner diameter, in.

ff_i, ff_o = fouling factors inside and outside the tubes, ft^2 hr °F/Btu

h_i, h_o = tube-side and gas-side coefficients, Btu/ft^2 hr °F

η = f in effectiveness

If bare tubes are used instead of finned tubes, $A_t = \pi d/12$. Equation (3) can be simplified to

$$\frac{1}{U} = \frac{d}{h_i d_i} + \frac{1}{h_o} + \frac{d}{24 K_m} \times \ln \frac{d}{d_i}$$

$$+ ff_i \times \frac{d}{d_i} + ff_o \qquad (4)$$

where h_o is the outside coefficient.

Now let us take the various cases.

WATER TUBE BOILERS, ECONOMIZERS, AND SUPERHEATERS

The gas-side heat transfer coefficient h_o is significant; the other terms can be neglected. In a typical bare-tube economizer, for example, $h_i = 1500$ Btu/ft^2 hr °F, ff_i and $ff_o = 0.001$ ft^2 hr °F/Btu, and $h_o = 12$ Btu/ft^2 hr °F. $d = 2.0$ in., $d_i = 1.5$ in., and $K_m = 25$ Btu/ft hr °F.

Substituting in Eq. (4) yields

$$\frac{1}{U} = \frac{2.0}{1500 \times 1.5} + \frac{1}{12} + \frac{2.0}{24 \times 25} \times \ln \frac{2}{1.5}$$

$$+ 0.001 \times \frac{2.0}{1.5} + 0.001$$

$$= 0.0874$$

Hence,

$U = 11.44$ Btu/ft^2 hr °F

Thus we see that the overall coefficient is close to the gas-side coefficient, which is the highest thermal resistance. The metal thermal resistance and the tube-side resistance are not high enough to change the resistance distribution much.

However, in a liquid-to-liquid heat exchanger, all the resistances will be of the same order, and hence none of the resistances can be neglected.

Even if finned tubes were used in the case above, with A_t/A_i = 9 substituting in Eq. (3), U = 9.3 Btu/ft^2 hr °F, which is close to h_o. Thus, while trying to figure U for economizers, water tube boilers, or gas-to-liquid heat exchangers, U may be written as

$$U = (0.8 \text{ to } 0.9) \times h_o \tag{5}$$

FIRE TUBE BOILERS, GAS COOLERS, AND HEAT EXCHANGERS WITH GAS FLOW INSIDE TUBES WITH LIQUID OR STEAM–WATER MIXTURE ON THE OUTSIDE

h_o is large, on the order of 1000 to 1500 Btu/ft^2 hr °F, while h_i will be about 10 to 12 Btu/ft^2 hr °F. Again, using Eq. (4), it can be shown that

$$U \approx h_i \times \frac{d_i}{d} \tag{6}$$

All the other thermal resistances can be seen to be very small, and U approaches the tube-side coefficient h_i.

GAS-TO-GAS HEAT EXCHANGERS (EXAMPLE: AIR HEATER IN BOILER PLANT)

In gas-to-gas heat transfer equipment, both h_i and h_o are small and comparable, while the other coefficients are high.

Assuming that h_o = 10 and h_i = 15, and using the tube configuration above,

$$\frac{1}{U} = \frac{2.0}{15 \times 1.5} + \frac{1}{10} + 0.001 + 9.6 \times 10^{-4}$$
$$+ 0.001 \times \frac{2}{1.5} = 0.1922$$

or

$$U = 5.2 \text{ Btu/ft}^2 \text{ hr °F}$$

Simplifying Eq. (4), neglecting the metal resistance term and fouling, we obtain

$$U = h_o \times \frac{h_i d_i/d}{h_o + h_i d_i/d} \tag{7}$$

Thus both h_o and h_i contribute to U.

ΔT, the corrected log-mean-temperature difference, can be estimated from

$$\Delta T = F_T \times \frac{\Delta T_{max} - \Delta T_{min}}{\ln(\Delta T_{max}/\Delta T_{min})}$$

where F_T is the correction factor for flow arrangement. For counterflow cases, $F_T = 1.0$. For other types of flow, textbooks may be referred to for F_T. It varies from 0.6 to 0.95 [2]. ΔT_{max} and ΔT_{min} are the maximum and minimum terminal differences.

In a heat exchanger the hotter fluid enters at 1000°F and leaves at 400°F, while the colder fluid enters at 250°F and leaves at 450°F. Assuming counterflow, we have

$$\Delta T_{max} = 1000 - 450 = 550°F,$$
$$\Delta T_{min} = 400 - 250 = 150°F$$

Then

$$\Delta T = \frac{550 - 150}{\ln(550/150)} = 307°F$$

In boiler economizers and superheaters, F_T could be taken as 1. In tubular air heaters, F_T could vary from 0.8 to 0.9. If accurate values are needed, published charts can be consulted [1,2].

4.02

Q: How is the tube-side heat transfer coefficient h_i estimated?

A: The widely used expression for h_i is [1]

$$Nu = 0.023 \times Re^{0.8} \times Pr^{0.4} \tag{8}$$

where the Nusselt number is

$$Nu = \frac{h_i \times d_i}{12k} \tag{9}$$

the Reynolds number is

$$Re = 15.2 \times \frac{w \times d_i}{\mu} \tag{10}$$

where w is the flow in the tube in lb/hr, and the Prandtl number is

$$\text{Pr} = \frac{\mu \times C_p}{k} \tag{11}$$

where

μ = viscosity, lb/ft hr
C_p = specific heat, Btu/lb °F
k = thermal conductivity, Btu/ft hr °F

all estimated at the fluid bulk temperature.

Substituting Eqs. (9) through (11) into Eq. (8) and simplifying, we have

$$h_i = 2.44 \times \frac{w^{0.8}k^{0.6}C_p^{0.4}}{d_i^{1.8}\mu^{0.4}} = 2.44 \times \frac{w^{0.8}C}{d_i^{1.8}} \tag{12}$$

where C is a factor given by

$$C = \frac{k^{0.6}C_p^{0.4}}{\mu^{0.4}}$$

C is available in the form of charts for various fluids [1] as a function of temperature. For air and flue gases, C may be taken from Table 4.1.

For hot water flowing inside tubes, Eq. (8) has been simplified and can be written as below [3], for $t < 300°F$:

$$h_i = (150 + 1.55t) \times \frac{V^{0.8}}{d_i^{0.2}} \tag{13}$$

Table 4.1   Factor C for Air and Flue Gases

Temp. (°F)	C
200	0.162
400	0.172
600	0.180
800	0.187
1000	0.194
1200	0.205

where

V = velocity, ft/s
t = water temperature, °F

For very viscous fluids, Eq. (8) has to be corrected by the term involving viscosities at tube wall temperature and at bulk temperature [1].

4.03a

Q: Estimate h_i when 200 lb/hr of air at 800°F and at atmospheric pressure flows in a tube of inner diameter 1.75 in.

A: Using Table 4.1 and Eq. (12), we have C = 0.187.

$$h_i = 2.44 \times 200^{0.8} \times \frac{0.187}{1.75^{1.8}} = 11.55 \text{ Btu/ft}^2 \text{ hr °F}$$

where

w = flow, lb/hr
d_i = inner diameter, in.

For gases at high pressures, Ref. 1 gives the C values.

4.03b

Q: In an economizer, 50,000 lb/hr of water at an average temperature of 250°F flows in a pipe of inner diameter 2.9 in. Estimate h_i.

A: Let us use Eq. (13). first the velocity has to be calculated. From Q1.07a, $V = 0.05(wv/d_i^2)$. v, the specific volume of hot water at 250°F, is 0.017 cu ft/lb. Then,

$$V = 0.05 \times 50,000 \times \frac{0.017}{2.9^2} = 5.05 \text{ ft/sec}$$

Hence, from Eq. (13),

$$h_i = (150 + 1.55 \times 250) \times \frac{5.05^{0.8}}{2.9^{0.2}} = 1586 \text{ Btu/ft}^2 \text{ hr °F}$$

4.03c

Q: Estimate the heat transfer coefficient when 5000 lb/hr of super-heated steam flows inside a tube of inner diameter 1.78 in. in a boiler superheater. The steam pressure is 1000 psia, and the temperature is 800°F. Quick estimates of heat transfer coefficients for saturated and superheated steam in tubes or pipes can be made using Eq. (12) with factor C given in Table 4.2.

A: Using Eq. (12) and a C value of 0.345, we have

$$h_i = 2.44 \times 5000^{0.8} \times \frac{0.345}{1.78^{1.8}} = 271 \text{ Btu/ft}^2 \text{ hr °F}$$

If the steam were saturated, C would be 0.49. Then we would have

$$h_i = 2.44 \times 5000^{0.8} \times \frac{0.49}{1.78^{1.8}} = 385 \text{ Btu/ft}^2 \text{ hr °F}$$

4.04

Q: How is the outside gas heat transfer coefficient h_o in boilers, air heaters, economizers, and superheaters determined?

Table 4.2   Factor C for Steam

Pressure (psia):	100	500	1000	2000
Sat.:	0.244	0.417	0.490	0.900
Temp. (°F)				
400	0.271	—	—	—
500	0.273	0.360	—	—
600	0.281	0.322	0.413	—
700	0.291	0.316	0.358	0.520
800	0.305	0.320	0.345	0.420
900	0.317	0.327	0.347	0.394
1000	0.325	0.340	0.353	0.386

A: The outside gas heat transfer coefficient h_o is the sum of the convective heat transfer coefficient h_c and nonluminous heat transfer coefficient h_N.

$$h_o = h_c + h_N \tag{14}$$

For finned tubes, h_o should be corrected for fin effectiveness. h_N is usually small if the gas temperature is less than 800°F and can be neglected.

ESTIMATING h_C FOR BARE TUBES

A conservative estimate of h_c for flow of fluids over bare tubes in in-line and staggered arrangements is given by [1]

$$\text{Nu} = 0.33 \times \text{Re}^{0.6} \times \text{Pr}^{0.33} \tag{15}$$

Substituting, we have the Reynolds, Nusselt, and Prandtl numbers

$$\text{Re} = \frac{G \times d}{12} \tag{16}$$

$$\text{Nu} = \frac{h_c \times d}{12k} \tag{17}$$

and

$$\text{Pr} = \frac{\mu \times C_p}{k} \tag{18}$$

where

G = gas mass velocity, lb/ft² hr
d = tube outer diameter, in.
μ = gas viscosity, lb/ft hr
k = gas thermal conductivity, Btu/ft hr °F
C_p = gas specific heat, Btu/lb °F

All the gas properties above are to be evaluated at the gas film temperature. Substituting Eqs. (16) to (18) into Eq. (15) and simplifying, we have

$$h_c = 0.9 \times G^{0.6} \times \frac{F}{d^{0.4}} \tag{19}$$

where

$$F = k^{0.67} \times \frac{C_p^{0.33}}{\mu^{0.27}} \tag{20}$$

Factor F has been computed for air and flue gases, and a good estimate is given in Table 4.3.

The gas mass velocity G is given by

$$G = 12 \times \frac{W_g}{N_w L(S_T - d)} \tag{21}$$

where

N_w = number of tubes wide
S_T = transverse pitch, in.
L = tube length, ft
W_g = gas flow, lb/hr

For quick estimates, gas film temperature t_f can be taken as the average of gas and fluid temperature inside the tubes.

EXAMPLE

Determine the gas-side convective heat transfer coefficient for a bare-tube superheater tube of diameter 2.0 in. with the following parameters:

Gas flow = 150,000 lb/hr
Gas temperature = 900°F
Average steam temperature = 500°F
Number of tubes wide = 12

Table 4.3   F Factor for Air and Flue Gases

Temp. (°F)	F
200	0.094
400	0.103
600	0.110
800	0.116
1000	0.123
1200	0.130

Length of the tubes = 10.5 ft
Transverse pitch = 4.0 in.
Longitudinal pitch = 3.5 in. (staggered)

Solution.   Estimate G. From Eq. (21),

$$G = 12 \times \frac{150,000}{12 \times 10.5 \times (4 - 2)} = 7142 \text{ lb/ft}^2 \text{ hr}$$

Using Table 4.3, at a film temperature of 700°F, $F = 0.113$.
Hence,

$$h_c = 0.9 \times 7142^{0.6} \times \frac{0.113}{2^{0.4}} = 15.8 \text{ Btu/ft}^2 \text{ hr °F}$$

As the gas temperature is not high, the h_N value will be low, so

$$U \approx h_o \approx h_c = 15.8 \text{ Btu/ft}^2 \text{ hr °F}$$

(Film temperature may be taken as the average of gas and steam temperature for preliminary estimates. If an accurate estimate is required, temperature drops across the various thermal resistances as discussed in Q4.16a must be determined.)

Convective heat transfer coefficient obtained by the above method or Grimson's method can be modified to include the effect of angle of attack a of the gas flow over the tubes. The correction factor F_n is 1 for perpendicular flow and decreases as shown in Table 4.4 for other angles [1].

If, for example, $h_c = 15$ and the angle of attack is 60°, then $h_c = 0.94 \times 15 = 14.1 \text{ Btu/ft}^2 \text{ hr °F}$.

4.05

Q: How is the convective heat transfer coefficient for air and flue gases determined using Grimson's correlation?

Table 4.4   Correction Factor for Angle of Attack

a, deg	90	80	70	60	50	40	30	20	10
F_n	1.0	1.0	0.98	0.94	0.88	0.78	0.67	0.52	0.42

A: Grimson's correlation, which is widely used for estimating h_c
[1], is

$$Nu = B \times Re^N \tag{22}$$

Coefficient B and power N are given in Table 4.5.

EXAMPLE

150,000 lb/hr of flue gases having an analysis CO_2 = 12, H_2O
= 12, N_2 = 70, and O_2 = 6, all in % volume, flows over a tube
bundle having 2-in. OD tubes at 4-in. square pitch. Tubes per
row = 18; length = 10 ft. Determine h_c if the fluid temperature
is 353°F and average gas temperature is 700 °F. The Appendix
tables give the properties of gases.

At a film temperature of 0.5 × (353 + 700) = 526°F, C_p =
0.2695, μ = 0.0642, and k = 0.02344. Then mass velocity G is

$$G = 12 \times \frac{150,000}{18 \times 10 \times (4 - 2)} = 5000 \text{ lb/ft}^2 \text{ hr}$$

From Table 4.5, for $S_T/d = S_L/d = 2$, B = 0.229 and N =
0.632, so

$$Re = \frac{5000 \times 2}{12 \times 0.0642} = 12,980$$

Table 4.5   Grimson's Values of B and N

S_L/d	$S_T/d = 1.25$		$S_T/d = 1.5$		$S_T/d = 2$		$S_T/d = 3$	
	B	N	B	N	B	N	B	N
Staggered								
1.25	0.518	0.556	0.505	0.554	0.519	0.556	0.522	0.562
1.50	0.451	0.568	0.460	0.562	0.452	0.568	0.488	0.568
2.0	0.404	0.572	0.416	0.568	0.482	0.556	0.449	0.570
3.0	0.310	0.592	0.356	0.580	0.44	0.562	0.421	0.574
In-line								
1.25	0.348	0.592	0.275	0.608	0.100	0.704	0.0633	0.752
1.50	0.367	0.586	0.250	0.620	0.101	0.702	0.0678	0.744
2.0	0.418	0.570	0.299	0.602	0.229	0.632	0.198	0.648
3.0	0.290	0.601	0.357	0.584	0.374	0.581	0.286	0.608

$$\text{Nu} = 0.229 \times 12,980^{0.632} = 91 = \frac{h_c \times 2}{12 \times 0.02344}$$

or $h_c = 12.8$ Btu/ft^2 hr °F

4.06

Q: How does an in-line arrangement compare with a staggered arrangement for bare tubes?

A: In-line arrangement is preferred to staggered with bare tubes because the gas-side heat transfer coefficient is nearly the same while the gas pressure drop is much higher. To illustrate, let us work out an example.

EXAMPLE

150,000 lb/hr of flue gases at 900°F is cooled to 500°F in a boiler generating 125 psig steam. Neglecting the effect of the non-luminous heat transfer coefficient, study the effect of using longitudinal pitches of 3, 4, and 6 in. Tube diameter is 2 in., and transverse pitch is 4 in. Boiler duty is 16 MM Btu/hr. Tubes per row = 18. Effective length = 10 ft. Gas analysis (% vol) is CO_2 = 12, H_2O = 12, N_2 = 70, O_2 = 6. Log-mean temperature difference (ΔT) = 319°F.

Using Grimson's correlation, h_c was computed for the various configurations. Overall heat transfer coefficient was then arrived at and used to determine surface areas. Then the number of rows deep for each case and the gas pressure drop were evaluated for each case. The results are shown in Table 4.6.

1. The staggered arrangement does not have a significant increase in h_c over the in-line arrangement when the longitudinal pitch-to-diameter ratio exceeds 2. It appears attractive below a ratio of 1.5. However, this ratio is not widely used in the industry because of low ligament efficiency, the possibility of bridging of particulates, cleanability, access for welding, etc.

2. The gas pressure drop is much higher for the staggered arrangement. For example, when we have a $S_T/d = S_L/d$

Table 4.6 Study of In-line Versus Staggered Arrangements

	$S_L/d = 1.5$		$S_L/d = 2.0$		$S_L/d = 3.0$	
	In-line	Staggered	In-line	Staggered	In-line	Staggered
h_c, Grimson	10.96	13.79	12.81	13.13	12.91	12.69
h_c, Fishenden	11.90	14.35	13.20	13.56	13.30	13.20
Friction factor	0.0396	0.0807	0.04928	0.0807	0.0686	0.0807
h_c, average	11.50	14.07	13	13.35	13.10	12.95
No. of rows	49	40	43	42	43	43
ΔP_o, in WC	1.32	2.12	1.44	2.29	2.00	2.33

= 2, the staggered arrangement has nearly the same number of rows but the gas pressure drop is 40 to 50% higher.

In conclusion, in-line arrangements are widely used for bare tubes. For finned tubes, see Q4.22.

4.07a

Q: How is nonluminous radiation heat transfer coefficient evaluated?

A: In engineering heat transfer equipment as boilers, fired heaters, and process steam superheaters where gases at high temperatures transfer energy to fluid inside tubes, nonluminous heat transfer plays a significant role. During combustion of fossil fuels such as coal oil, or gas—triatomic gases—for example, water vapor, carbon dioxide, and sulfur dioxide are formed, which contribute to radiation. The emissivity pattern of these gases has been studied by Hottel, and charts are available to predict gas emissivity if gas temperature, partial pressure of gases, and beam length are known.

Net interchange of radiation between gases and surroundings (e.g., a wall or tube bundle or a cavity) can be written as

$$\frac{Q}{A} = \sigma \times (\epsilon_g T_g^4 - \alpha_g T_o^4) \qquad (22)$$

where

ϵ_g = emissivity of gases at T_g
α_g = absorptivity at T_o
T_g = absolute temperature of gas, °R
T_o = absolute temperature of tube surface, °R

ϵ_g is given by

$$\epsilon_g = \epsilon_c + \eta \times \epsilon_w - \Delta\epsilon \tag{23}$$

α_g is calculated similarly at T_o. η is the correction factor for the water pressure, and $\Delta\epsilon$ is the decrease in emissivity due to the presence of water vapor and carbon dioxide.

Although it is desirable to calculate heat flux by (22), it is tedious to estimate α_g at temperature T_o. Considering the fact that T_o^4 will be much smaller than T_g^4, with a very small loss of accuracy, we can use the following simplified equation, which lends itself to further manipulations.

$$\frac{Q}{A} = \sigma\epsilon_g(T_g^4 - T_o^4) = h_N(T_g - T_o) \tag{24}$$

Nonluminous heat transfer coefficient h_N can be written as

$$h_N = \sigma\epsilon_g \frac{T_g^4 - T_o^4}{T_g - T_o} \tag{25}$$

To estimate h_N, partial pressures of triatomic gases and beam length L are required. L is a characteristic dimension that depends on the shape of the enclosure. For a bundle of tubes interchanging radiation with gases, it can be shown that

$$L = 1.08 \times \frac{S_T S_L - 0.785d^2}{d} \tag{26a}$$

L is taken approximately as 3.4 to 3.6 times the volume of the space divided by the surface area of the heat-receiving surface. For a cavity of dimensions a, b and c,

$$L = \frac{3.4 \times abc}{2(ab + bc + ca)} = \frac{1.7}{1/a + 1/b + 1/c} \tag{26b}$$

In the case of fire tube boilers, $L = d_i$.

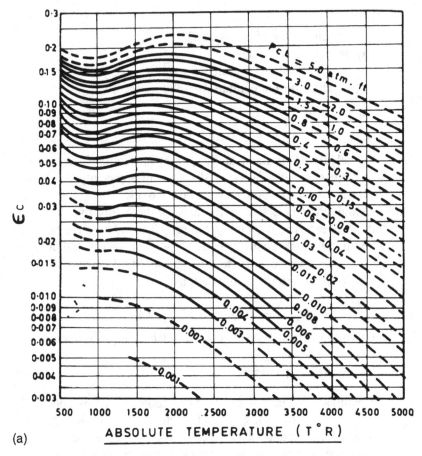

(a)

ABSOLUTE TEMPERATURE ($T°R$)

Figure 4.1   (a) Emissivity of carbon dioxide.
              (b) Emissivity of water vapor.
              (c) Correction factor for emissivity of water vapor.
              (d) Correction term due to presence of water vapor and
                  carbon dioxide. (From. Ref. 1.)

ϵ_g can be estimated using Figure 4.1, which gives ϵ_c, ϵ_w, η, and $\Delta\epsilon$. For purposes of engineering estimates, radiation effects of SO_2 can be taken as similar to those of CO_2. Hence, partial pressures of CO_2 and SO_2 can be added and Figure 4.1 used to get ϵ_c.

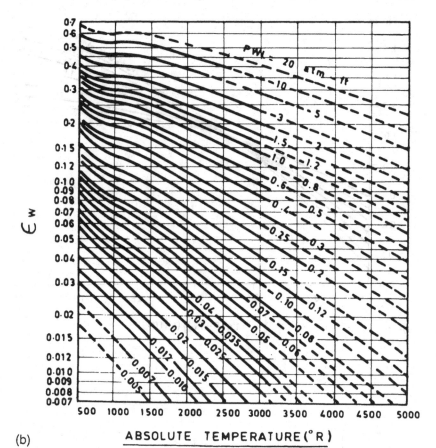

(b)   **ABSOLUTE TEMPERATURE (°R)**

EXAMPLE 1

Determine the beam length L if $S_T = 5$ in., $S_L = 3.5$ in., and $d = 2$ in.

Solution

$$L = 1.08 \times \frac{5 \times 3.5 - 0.785 \times 4}{2} = 7.8 \text{ in.}$$

EXAMPLE 2

In a fired heater firing a waste gas, CO_2 in flue gases $= 12\%$ and $H_2O = 16\%$. The gases flow over a bank of tubes in the

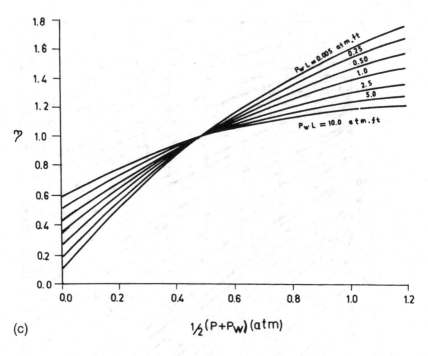

(c)

$\frac{1}{2}(P+P_W)$ (atm)

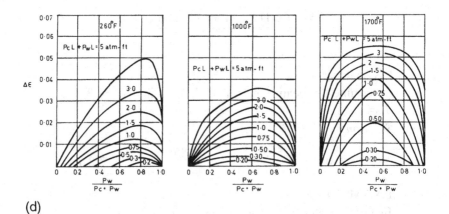

(d)

Figure 4.1 Continued

convective section where tubes are arranged as in Example 1 (hence $L = 7.8$). Determine h_N if $t_g = 1650°F$ and $t_o = 600°F$.

Solution.

$$P_cL = 0.12 \times \frac{7.8}{12} = 0.078 \text{ atm ft}$$

$$P_wL = 0.16 \times \frac{7.8}{12} = 0.104 \text{ atm ft}$$

In Figure 4.1a at $T_g = (1650 + 460) = 2110°R$ and $P_cL = 0.078$. $\epsilon_c = 0.065$. In Figure 4.1b, at $T_g = 2110°R$ and $P_wL = 0.104$, $\epsilon_w = 0.05$. In Figure 4.1c, corresponding to $(P + P_w)/2 = 1.16/2 = 0.58$ and $P_wL = 0.104$, $\eta = 1.1$. In Figure 4.1d, corresponding to $P_w/(P_c + P_w) = 0.16/0.28$ and $(P_c + P_w) L = 0.182$, $\Delta\epsilon = 0.002$. Hence,

$$\epsilon_g = 0.065 + (1.1 \times 0.05) - 0.002 = 0.118$$

Using Eq. (25) with the Boltzmann constant $\sigma = 0.173 \times 10^{-8}$,

$$h_N = 0.173 \times 10^{-8} \times 0.118 \times \frac{2110^4 - 1060^4}{2110 - 1060}$$

$$= 3.6 \text{ Btu/ft}^2 \text{ hr °F}$$

Thus, h_N can be evaluated for gases.

4.07b

Q: Can gas emissivity be estimated using equations?

A: Gas emissivity can be obtained as follows. h_N is given by

$$h_N = \sigma\epsilon_g \frac{T_g^4 - T_o^4}{T_g - T_o} \tag{25}$$

where

$$\sigma = \text{Stefan-Boltzmann constant} = 0.173 \times 10^{-8},$$

T_g and T_o = gas and tube outer wall temperature, °R

ϵ_g, gas emissivity, is obtained from Hottel's charts or from the expression [1]

$$\epsilon_g = 0.9 \times (1 - e^{-KL}) \tag{27a}$$

$$K = \frac{(0.8 + 1.6p_w) \times (1 - 0.38T_g/1000)}{\sqrt{(p_c + p_w)L}}$$
$$\times (p_c + p_w) \tag{27b}$$

T_g is in K. L is the beam length in meters, and p_c and p_w are the partial pressures of carbon dioxide and water vapor in atm. L, the beam length, can be estimated for a tube bundle by

$$L = 1.08 \frac{S_T \times S_L - 0.785d^2}{d} \tag{26a}$$

S_T and S_L are the transverse pitch and longitudinal pitch. Methods of estimating p_c and p_w are given in Chapter 1.

EXAMPLE
In a boiler superheater with bare tubes, the average gas temperature is 1600°F and the tube metal temperature is 700°F. Tube size is 2.0 in., and transverse pitch S_T = longitudinal pitch S_L = 4.0 in. Partial pressure of water vapor and carbon dioxide are p_w = 0.12, p_c = 0.16. Determine the nonluminous heat transfer coefficient.

From Eq. (26a), the beam length L is calculated.

$$L = 1.08 \times \frac{4 \times 4 - 0.785 \times 2 \times 2}{2} = 6.9 \text{ in.}$$

$$= 0.176 \text{ m}$$

Using Eq. (27b) with $T_g = (1600 - 32)/1.8 + 273 = 1114$ K, we obtain

$$\text{Factor } K = \frac{(0.8 + 1.6 \times 0.12) \times (1 - 0.38 \times 1.114)}{\sqrt{0.28 \times 0.176}}$$

$$\times 0.28 = 0.721$$

From Eq. (27a),

$$\epsilon_g = 0.9 \times [1 - \exp(-0.721 \times 0.176)] = 0.107$$

Then, from Eq. (25),

$$h_N = 0.173 \times 0.107 \times 10^{-8} \times \frac{2060^4 - 1160^4}{1600 - 700}$$

$$= 3.33 \text{ Btu/ft}^2 \text{ hr } {}^\circ\text{F}$$

4.08a

Q: How is heat transfer in a boiler furnace evaluated?

A: Furnace heat transfer is a complex phenomenon, and a single formula or correlation cannot be prescribed for sizing furnaces of all types. Basically, it is an energy balance between two fluids— gas and a steam–water mixture. Heat transfer in a boiler furnace is predominantly radiation, partly due to the luminous part of the flame and partly due to nonluminous gases. A general, approximate expression can be written for furnace absorption using an energy approach:

$$Q_F = A_p \epsilon_w \, \epsilon_f \, \sigma \, (T_g^4 - T_o^4)$$
$$= W_f \times \text{LHV} - W_g h_e \tag{28}$$

Gas temperature (T_g) is defined in many ways; some authors define it as the exit-gas temperature itself. Some put it as the mean of theoretical flame temperature and t_e. However, plant experience shows that a better agreement between measured and calculated values prevails when $t_g = t_e + 300$ to 400 °F [1].

The emissivity of gaseous flame is evaluated as follows [1]:

$$\epsilon_f = \beta(1 - e^{-KPL}) \tag{29}$$

β characterizes flame-filling volumes.

$\beta = 1.0$ for nonluminous flames
  $= 0.75$ for luminous sooty flames of liquid fuels
  $= 0.65$ for luminous and semiluminous flames of solid fuels
$L =$ beam length, m

K = attenuation factor, which depends on fuel type and presence of ash and its concentration. For non-luminous flame it is

$$K = \frac{0.8 + 1.6p_w}{\sqrt{(p_c + p_w)L}} \times \left(\frac{1 - 0.38T_e}{1000}\right) \times (p_c + p_w) \quad (27b)$$

For semiluminous flame, the ash particle size and concentration enter into the calculation:

$$K = \frac{0.8 + 1.6p_w}{\sqrt{(p_c + p_w)L}} \times \left(\frac{1 - 0.38T_e}{1000}\right) \times (p_c + p_w)$$
$$+ 7 \mu \times \left(\frac{1}{d_m^2 \times T_e^2}\right)^{1/3} \quad (27c)$$

where

d_m is the mean effective diameter of ash particles, in micrometers

d_m = 13 for coals ground in ball mills

= 16 for coals ground in medium- and high-speed mills

= 20 for combustion of coals milled in hammer mills

μ = ash concentration in g/Nm^3

T_e = furnace exit temperature, K

For luminous oil or gas flame,

$$K = \frac{1.6T_e}{1000} - 0.5 \quad (27d)$$

p_w and p_c are partial pressures of water vapor and carbon dioxide in the flue gas.

The above equations only give a trend. A wide variation could exist due to the basic combustion phenomenon itself. Again, the flame does not fill the furnace fully. Unfilled portions are subjected to gas radiation only, the emissivity of which is far below (0.15 to 0.30) that of the flame. Hence, ϵ_f decreases. Godridge reports that in a pulverized coal-fired boiler, emissivity varied as follows with respect to location [3]:

Excess air	15%	25%
Furnace exit	0.6	0.5
Middle	0.7	0.6

Also, furnace tubes coated with ferric oxide have emissivities, ϵ_w, of the order of 0.8, depending on whether a slag layer covers them. Soot blowing changes ϵ_w considerably. Thus, only an estimate of ϵ_f and ϵ_w can be obtained, which varies with type of unit, fuel, and operation regimes.

To illustrate these concepts, a few examples are worked out. The purpose is only to show the effect of variables like excess air and heat release rates on furnace absorption and furnace exit-gas temperature.

EXAMPLE 1

Determine the approximate furnace exit-gas temperature of a boiler when net heat input is about 2000×10^6 Btu/hr, of which 1750×10^6 Btu/hr is due to fuel and the rest is due to air. HHV and LHV of coals fired are 10,000 and 9000 Btu/lb, respectively, and a furnace heat release rate of 80,000 Btu/ft^2 hr (projected area basis) has been used. The values ϵ_w and ϵ_f may be taken as 0.6 and 0.5, respectively; 25% is the excess air used. Water-wall outer temperature is 600°F. Ash content in coal is 10%.

Solution

$$\frac{Q}{A_P} = 80,000 = W_f \times \frac{LHV}{A_P}$$

From combustion calculation methods discussed in Chapter 1, using 1 MM Btu fired basis, we have the following ratio of flue gas to fuel:

$$\frac{W_g}{W_f} = \frac{760 \times 1.24 \times 10^4}{10^6} + 1 - (10/100)$$

$$= 10.4 \text{ lb/lb}$$

$$Q = A_P \, \epsilon_w \, \epsilon_f \, \sigma \times (T_g^4 - T_o^4) = W_f \times LHV - W_g h_e$$

Dividing throughout by W_f,

$$\frac{A_p}{W_f} \; \epsilon_w \; \epsilon_f \; \sigma(T_g^4 - T_o^4) = \text{LHV} - \frac{W_g}{W_f} \, h_e$$

$A_p/W_f = \text{LHV}/80{,}000 = 0.1125$

Assume $t_e = 1900°\text{F}$.

$C_{pm} = 0.3 \; \text{Btu/lb} \; °\text{F}$

$t_g = (1900 + 300) = 2200°\text{F} = 2660°\text{R}$

Let us see if the assumed t_e is correct. Substituting for A_p/W_f, ϵ_w, ϵ_f, σ, T_g, T_e in the above equation, we have

$\text{LHS} = 0.1125 \times 0.6 \times 0.5 \times 0.173 \times (26.6^4$
$\quad\quad - 10.6^4) = 2850$

$\text{RHS} = (9000 - 10.4 \times 1900 \times 0.3) = 3072$

Since these do not tally, try $t_e = 1920 \, °\text{F}$. Neglect the effect of variation in C_{pm}:

$\text{LHS} = 0.1125 \times 0.6 \times 0.5 \times (26.8^4 - 10.6^4)$
$\quad\quad \times 0.173 = 2938$

$\text{RHS} = 9000 - 1920 \times 0.3 \times 10.4 = 3009$

Since these agree closely, furnace exit-gas temperature is around 1920°F. Note that the effect of external radiation to superheaters has been neglected in the energy balance. This may give rise to an error of 1.5 to 2.5% in t_e, but its omission greatly simplifies the calculation procedure. Also, losses occurring in the furnace were omitted to simplify the procedure. The error introduced is quite low.

EXAMPLE 2

It is desired to use a heat loading of 100,000 Btu/ft^2 hr in the furnace in Example 1. Other factors such as excess air and emissivities remain unaltered. Estimate the furnace exit-gas temperature.

Solution:

$$\frac{Q}{A_p} = 100{,}000 = W_f \times \frac{\text{LHV}}{A_p} \, , \; \frac{A_p}{W_f}$$

$$= \frac{LHV}{100,000} = 0.09$$

$$\frac{w_g}{w_f} = 10.4, \qquad t_e = 2000 \text{ °F}; \qquad t_g = 2300°F$$

$$C_{pm} = 0.3 \text{ Btu/lb °F}; \qquad T_g = 2300 + 460 = 2760°R$$

$$LHS = 0.09 \times 0.6 \times 0.5 \times 0.173 \times (27.6^4$$
$$- 10.6^4) = 2664$$
$$RHS = (9000 - 10.4) \times 2000 \times 0.3 = 2760$$

From this it is seen that t_e will be higher than assumed. Let

$$t_e = 2030°F, \qquad Tg = 2790°R$$

Then

$$LHS = 0.09 \times 0.6 \times 0.5 \times 0.173 \times [(27.9)^4$$
$$- (10.6)^4] = 2771$$
$$RHS = 9000 - 10.4 \times 2030 \times 0.3 = 2667$$

Hence, t_e will lie between 2000 and 2030°F, perhaps 2015°F.

The exercise shows that the exit-gas temperature in any steam generator will increase as more heat input is given to it; that is, the higher the load of the boiler, the higher the exit-gas temperature. Example 3 shows the effect of excess air on t_e.

EXAMPLE 3

What will be the furnace exit-gas temperature when 40% excess air is used instead of 25%, heat loading remaining at about 100,000 Btu/ft² hr in the furnace mentioned in earlier examples?

Solution

$$\frac{Q}{A_p} = 100,000 = W_f \frac{LHV}{A_p}, \qquad \frac{A_p}{W_f} = 0.09$$

$$\frac{W_g}{W_f} = \frac{760 \times 1.4 \times 10^4}{10^6} + 0.9 = 11.54 \text{ lb/lb}$$

$$t_e = 1950°F, \qquad C_{pm} = 0.3 \text{ Btu/lb °F}, \qquad T_g = 1950$$
$$+ 300 + 460 = 2710°R$$

$$\text{LHS} = 0.09 \times 0.6 \times 0.5 \times 0.173 \times [(27.1)^4$$
$$- (10.6)^4] = 2460$$
$$\text{RHS} = 9000 - (11.54 \times 1950 \times 0.3) = 2249$$

These nearly tally; hence, t_e is about 1950 °F, compared to about 2,030 °F in Example 2. The effect of the higher excess air has been to lower t_e.

EXAMPLE 4

If $\epsilon_w \times \epsilon_f = 0.5$ instead of 0.3, what will be the effect on t_e when heat loading is 100,000 Btu/ft^2 hr and excess air is 40%?

Solution

Let

$$t_e = 1800°F; \qquad T_g = 1800 + 300 + 460 = 2560°R$$
$$\text{LHS} = 0.09 \times 0.5 \times 0.173 \times [(25.6)^4 - (10.6)^4]$$
$$= 3245$$
$$\text{RHS} = 9000 - (11.54 \times 1800 \times 0.3) = 2768$$

Try:

$$t_e = 1700°F; \qquad T_g = 2460°R$$

Then

$$\text{LHS} = 0.09 \times 0.5 \times 0.173 \times [(24.6)^4 - (10.6)^4]$$
$$= 2752$$
$$\text{RHS} = 9000 - (11.54 \times 1700 \times 0.3) = 3115$$

Try:

$$t_e = 1770°F; \qquad T_g = 2530°R.$$

Then

$$\text{LHS} = 3091; \qquad \text{RHS} = 2872$$

Hence, t_e will be around 1760 °F. This example shows that when surfaces are cleaner and capable of absorbing more radiation, t_e decreases.

In practice, furnace heat transfer is not evaluated as simply as shown above because of the inadequacy of accurate data on soot emissivity, particle size, distribution, flame size, excess air, presence and effect of ash particles, etc. Hence, designers develop data based on field tests. Estimating t_e is the starting point for the design of superheaters, reheaters, and economizers.

Some boiler furnaces are equipped with tilting tangential burners, while some furnaces have only front or rear nontiltable wall burners. The location of the burners affects t_e significantly. Hence, in these situations, correlations with practical site data would help in establishing furnace absorption and temperature profiles.

A promising technique for predicting furnace heat transfer performance is the zone method of analysis. It is assumed that the pattern of fluid flow, chemical heat release, and radiating gas concentration are known, and equations describing conservation of energy within the furnace are developed. The furnace is divided into many zones, and radiation exchange calculations are carried out.

4.08b

Q: How is heat transfer evaluated in unfired furnaces?

A: Radiant sections using partially or fully water-cooled membrane wall designs are used to cool gas streams at high gas temperatures (Figure 4.2). They generate saturated steam and may operate in parallel with convective evaporators if any. The design procedure is simple and may involve an iteration or two. The higher the partial pressures of triatomic gases, the higher will be the nonluminous radiation and hence the duty.

If a burner is used as in the radiant section of a furnace-fired HRSG, the emissivity of the flame must also be considered. As explained elsewhere [8], radiant sections are necessary to cool the gases to below the softening points of any eutectics present so as to avoid bridging or slagging at the convection section. They are also required to cool gases to a reasonable temperature at the superheater if it is used.

EXAMPLE

150,000 pph of flue gases at 1700°F has to be cooled to 1500°F in a radiant section of a waste heat boiler of cross section 8 ft ×

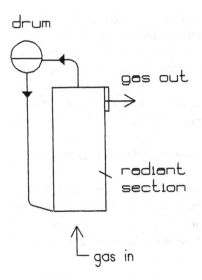

drum

gas out

radiant
section

gas in

Figure 4.2   Radiant furnace in a waste heat boiler.

10 ft. Saturated steam at 150 psig is generated by the gases. Determine the length. Gas analysis (% vol) is CO_2 = 12, H_2O = 12, N_2 = 70, and O_2 = 6.

Solution.   Let the length = 25 ft.

$$\text{Beam length} = 3.4 \times \frac{\text{volume}}{\text{surface}}$$

$$= \frac{3.4 \times 8 \times 10 \times 25}{8 \times 25 \times 2 + 10 \times 25 \times 2 + 2 \times 10 \times 8}$$

$$= 6.4 \text{ ft} = 2 \text{ m}$$

Average gas temperature = 1600°F = 1144 K. Using (27),

$$K = (0.8 + 1.6 \times 0.12)(1 - 0.38 \times 1.144)$$

$$\times \frac{0.24}{(0.24 \times 2)^{0.5}} = 0.194$$

$$\epsilon_g = 0.9 \times (1 - e^{-0.194 \times 2}) = 0.29$$

Let the average surface temperature of the radiant section be 400°F. Surface area for heat transfer = $(8 + 10) \times 2 \times 25 = 900$ ft.^{2}. Transferred energy is

$$Q_t = 0.173 \times 0.9 \times 0.29 \times [(20.6)^4 - (8.6)^4] \times 900$$
$$= 7.1 \text{ MM Btu/hr}$$

$$\text{Required duty} = 150{,}000 \times 0.99 \times 0.308 \times 200$$
$$= 9.15 \text{ MM Btu/hr}$$

Since there is a large discrepancy and the required duty is more, a larger surface area is required. As the furnace length does not significantly change the beam length and hence the gas emissivity, we can assume that the transferred duty is proportional to the surface area. Hence the required furnace length = $(9.15/7.1) \times 25 = 32$ ft.

4.09

Q: How is the distribution of external radiation to tube bundles evaluated? Discuss the effect of tube spacing.

A: Tube banks are exposed to direct or external radiation from flames, cavities, etc., in boilers. Depending on the tube pitch, the energy absorbed by each row of tubes varies, with the first row facing the radiation zone receiving the maximum energy. It is necessary to compute the energy absorbed by each row, particularly in superheaters as the contribution of the radiation can result in high tube wall temperatures.

The following formula predicts the radiation to the tubes [8].

$$a = 3.14 \times \frac{d}{2S} - \frac{d}{S} \times \left[\sin^{-1} \left(\frac{d}{S} \right) \right. $$
$$\left. + \sqrt{\left(\frac{S}{d} \right)^2 - 1} - \frac{S}{d} \right] \tag{30}$$

where a is the fraction of energy absorbed by the first row. The second row would then absorb $(1-a) \times a$; the third row, $\{1-[a+(1-a)a]\}a$; and so on.

EXAMPLE

1 MM Btu/hr of energy from a cavity is radiated to a superheater tube bank that has 2-in. OD tubes at a pitch of 8 in. If there are six rows, estimate the distribution of energy to each row.

Solution.   Substituting $d = 2$, $S = 8$ into Eq. (30), we have

$$a = 3.14 \times \left(\frac{2/8}{2}\right) - \frac{2}{8}\left[\sin^{-1}\left(\frac{2}{8}\right)\right.$$

$$\left. + \sqrt{4 \times 4 - 1} - \frac{8}{2}\right]$$

$$= 0.3925 - 0.25\,(0.2526 + \sqrt{15} - 4) = 0.361$$

Hence the first row absorbs 0.361 MM Btu/hr.

The second row would receive $(1 - 0.361) \times 0.361 = 0.231$ or 0.231 MM Btu/hr.

The third row receives $[1 - (0.361 + 0.231)]\,0.361 = 0.147$ MM Btu/hr

The fourth row, $[1 - (0.361 + 0.231 + 0.147)] \times 0.361 = 0.094$ MM Btu/Hr, and so on.

It can be seen that the first row receives the maximum energy and the amount lessens as the number of rows increases. For a tube pitch S of 4 in., $a = 0.6575$. The first row receives 0.6575 MM Btu/hr; the second, 0.225 M Btu/hr; and the third, 0.077 MM Btu/hr. Hence if the tube pitch is small, a large amount of energy is absorbed within the first two to three rows, resulting in high heat flux in those tubes and consequently high tube wall temperatures. Hence it is better to use a wide pitch when the external radiation is large so that the radiation is spread over more tubes and the intensity is not concentrated within two or three tubes. Screen tubes in boilers and fired heaters perform this function.

4.10

Q: Determine the size of a fire tube waste heat boiler required to cool 100,000 lb/hr of flue gases from 1500°F to 500°F. Gas analysis is (% vol) $CO_2 = 12$, $H_2O = 12$, $N_2 = 70$, and $O_2 = 6$; gas pressure is 5 in. WC. Steam pressure is 150 psig, and feedwater enters at 220°F. Tubes used are 2 × 1.77 in. (outer and inner diameter); fouling factors are gas-side ff Btu/ft hr °F = 0.002 and steam-side ff = 0.001 ft²°F/Btu. Tube metal thermal conductivity = 25 Btu/ft hr F. Steam-side boiling heat transfer coefficient = 2000 Btu/ft²°F. Assume that heat losses and margin = 2% and blowdown = 5%.

A: Use Eq. (4) to compute the overall heat transfer coefficient, and then arrive at the size from (1).

$$\frac{1}{U} = \frac{d_o}{d_i h_i} + ff_o + ff_i \frac{d_o}{d_i} + d_o \frac{\ln(d_o/d_i)}{24K_m}$$

$$+ \frac{1}{h_o} \qquad (4)$$

h_i, the tube-side coefficient, is actually the sum of a convective portion h_c plus a nonluminous coefficient h_n. h_c is obtained from Q4.04:

$$h_c = 2.44 \times w^{0.8} \times \frac{C}{d_i^{1.8}}$$

At the average gas temperature of 1000°F, the gas properties can be shown to be C_p = 0.287 Btu/lb°F, μ = 0.084 lb/ft hr, and k = 0.0322 Btu/ft hr°F. Hence,

$$C = \left(\frac{0.287}{0.084}\right)^{0.4} \times (0.0322)^{0.6} = 0.208.$$

Boiler duty Q = 100,000 × 0.98 × 0.287 × (1500 − 500)
$$= 28.13 \times 10^6 \text{ Btu/hr}$$

Enthalpies of saturated steam, saturated water, and feedwater from steam tables are 1195.5, 338, and 188 Btu/lb, respectively. The enthalpy absorbed by steam is then (1195.5 − 188) + 0.05 × (338 − 188) = 1015 Btu/lb, where 0.05 is the blowdown factor corresponding to 5% blowdown.
Hence

$$\text{Steam generation} = \frac{28.13 \times 10^6}{1015} = 27,710 \text{ lb/hr}$$

In order to compute h_i, the flow per tube w is required. Typically w ranges from 100 to 200 lb/hr for a 2-in. tube. Let us start with 600 tubes; hence $w = 100,000/600 = 167$ lb/hr.

$$h_c = 2.44 \times 0.208 \times \frac{167^{0.8}}{(1.77)^{1.8}} = 10.9 \text{ Btu/ft}^2 \text{ hr } °F$$

The nonluminous coefficient is usually small in fire tube boilers as the beam length corresponds to the tube inner diameter. However, the procedure used in Q4.07 can also be used here. Let us assume that it is 0.45 Btu/ft^2 hr °F. Hence

$$h_i = 10.90 + 0.45 = 11.35 \text{ Btu/ft}^2 \text{ hr } °F$$

Let us compute U. Since it is based on tube outside surface, let us call it U_o.

$$\frac{1}{U_o} = \frac{2/1.77}{11.35} + 0.001 + 0.002 \times \frac{2}{1.77}$$

$$+ \ln\left(\frac{2}{1.77}\right) \times \frac{2}{24 \times 25} + 0.0005$$

$$= 0.10 + 0.001 + 0.00226 + 0.00041 + 0.0005$$

$$= 0.10417$$

Hence, $U_o = 9.6$ Btu/ft^2 hr °F.

The various resistances in ft^2 hr °F/Btu are

Gas-side heat transfer	0.10
Gas-side fouling	0.00226
Metal resistance	0.00041
Steam-side fouling	0.001
Steam-side heat transfer	0.0005

If U is computed on the basis of tube inner surface area, then U_i is given by the expression

$A_i \times U_i = A_o \times U_o$. Hence

$$U_i = 9.6 \times \frac{2}{1.77} = 10.85 \text{ Btu/ft}^2 \text{ hr } °F$$

Log-mean temperature difference is

$$\Delta T = \frac{(1500 - 366) - (500 - 366)}{\ln[(1500 - 366)/(500 - 366)]} = 468°F$$

Hence

$$A_o = \frac{28.13 \times 10^6}{468 \times 9.6} = 6261 \text{ ft}^2 = 3.14 \times 2 \times 600$$

$$\times \frac{L}{12}$$

so required length L of the tubes $= 19.93$ ft. Use 20 ft. Then

$$A_o = 3.14 \times 2 \times 600 \times \frac{20}{12} = 6280 \text{ ft}^2,$$

$$A_i = 5558 \text{ ft}^2$$

Let us compute the gas pressure drop using Eq. (12) of Chapter 3.

$$\Delta P_g = 93 \times 10^{-6} \times w^2 \times f \times L_e \times \frac{v}{d_i^5}$$

Friction factor f depends on tube inner diameter and can be taken as 0.02. The equivalent length L_e can be approximated by $L + 5d_i$ to include the tube inlet and exit losses.

Specific volume v obtained as 1/density, or $v = 1/\rho$. Gas density at the average gas temperature of 1000°F is $\rho_g = 39/1460 = 0.0267$ lb/cu ft. Therefore,

$$\Delta P_g = 93 \times 10^{-6} \times 167^2 \times 0.02$$

$$\times \frac{20 + 5 \times 1.77}{0.0267 \times (1.77)^5} = 3.23 \text{ in. WC}$$

This is only one design. Several variables such as tube size and mass flow could be changed to arrive at several options that could be reviewed for optimum operating and installed costs.

4.11

Q: What is the effect of tube size and gas velocity on boiler size? Is surface area the sole criterion for boiler selection?

A: Surface area should not be used as the sole criterion for selecting or purchasing boilers, as tube size and gas velocity affect this variable.

Shown in Table 4.7 are the design options for the same boiler duty using different gas velocities and tube sizes; the procedure described in Q4.10 was used to arrive at these options. The purpose behind this example is to bring out the fact that surface area can vary by as much as 50% for the same duty.

1. As the gas velocity increases, the surface area required decreases, which is obvious.
2. The smaller the tubes, the higher the heat transfer coefficient for the same gas velocity, which also decreases the surface area.
3. For the same gas pressure drop, the tube length is smaller if the tube size is smaller. This fact helps when we try to fit a boiler into a small space.
4. For the same tube size, increasing the gas velocity results in a longer boiler, a greater gas pressure drop, but smaller surface area.

In the case of water tube boilers, more variables such as tube spacing and in-line or staggered arrangement, in addition to gas

Table 4.7 Effect of Tube Size and Gas Velocity

1. Size	1.75×1.521			2.0×1.77			2.5×2.238		
2. Velocity, fps	98	123	163	98	123	162	98	123	162
3. Tubes	1000	800	600	750	600	450	470	375	280
4. L, ft	15.75	16.75	18	18.75	20	21.5	24.75	26.0	28.50
5. S_i, ft^2	6269	5333	4298	6513	5558	4480	6812	5710	4673
6. U_i	19.47	11.08	13.70	9.07	10.68	13.19	8.73	10.29	12.72
7. ΔP_i, in. WC	2.05	3.34	6.23	1.97	3.20	6.00	1.95	3.16	6.00
8. P, kW	15.6	25.4	47.4	15.0	24.4	45.6	14.9	24.0	45.6

velocity and tube size can affect surface area. This is discussed elsewhere.

4.12

Q: How is the tube wall temperature in fire tube boilers evaluated? Discuss the importance of heat flux.

A: To compute the tube wall temperatures, heat flux must be known.

$$q_o = \text{heat flux outside tubes} = U_o \times (t_g - t_i) \text{ Btu/ft}^2 \text{ hr}$$

Similarly, q_i (heat flux inside the tube) would be $U_i \times (t_g - t_i)$. However, heat flux outside the tubes is relevant in fire tube boilers as boiling occurs outside the tubes, whereas in water tube boilers, the heat flux inside the tubes would be relevant. A high heat flux can result in a condition called DNB (departure from nucleate boiling), which will result in overheating of tubes. It is preferable to keep the actual maximum heat flux below the critical heat flux, which varies from 150,000 to 250,000 Btu/ft^2 hr depending on steam quality, pressure, and tube condition [1].

An electrical analogy can be used in determining the tube wall temperatures. Heat flux is analogous to current, electrical resistance to thermal resistance, and voltage drop to temperature drop. Using the example worked in Q4.10, we have that at average gas conditions the product of current (heat flux) and resistance (thermal resistance) gives the voltage drop (temperature drop):

$$q_o = \text{heat flux} = 9.6 \times (1000 - 366) = 6086 \text{ Btu/ft}^2 \text{ hr}$$

Temperature drop across gas film = $6086 \times 0.1 = 609°F$

Temperature drop across gas-side fouling = 6086×0.00226 = $14°F$

Temperature drop across tube wall = $6086 \times 0.00041 = 3°F$

Temperature drop across steam-side fouling = 6086×0.001 = $6°F$

Temperature drop across steam film = $6085 \times 0.0005 = 3°F$

Hence,

Average inside tube wall temperature $= 1000 - 609$
$- 14 = 377°F$

Outside tube wall temperature $= 377 - 3 = 374°F$.

The same results are obtained working from the steam side.

Outside tube wall temperature $= 366 + 6 + 3 = 375°F$

One can also compute the maximum tube wall temperature by obtaining the heat flux at the hot gas inlet end.

4.13

Q: What is the effect of scale formation on tube wall temperatures?

A: If nonsoluble salts such as calcium or magnesium salts or silica are present in the feedwater, they can deposit in a thin layer on tube surfaces during evaporation, thereby resulting in higher tube wall temperatures.

Table 4.8 lists the thermal conductivity k of a few scales. Outside fouling factor ff_o can be obtained if the scale information is available.

Table 4.8   Thermal Conductivities of Scale
Materials

Material	Thermal conductivity [(Btu/ft^2 hr °F)/in.]
Analcite	8.8
Calcium phosphate	25
Calcium sulfate	16
Magnesium phosphate	15
Magnetic iron oxide	20
Silicate scale (porous)	0.6
Boiler steel	310
Firebrick	7
Insulating brick	0.7

$$\text{ff}_o = \frac{\text{thickness of scale}}{\text{conductivity}}$$

Let us use the same example as in Q4.10 and check the effect of ff_o on boiler duty and tube wall temperatures. Let a silicate scale of thickness 0.03 in. be formed. Then,

$$\text{ff}_o = \frac{0.03}{0.6} = 0.05 \text{ ft}^2 \text{ hr } °\text{F/Btu}$$

Assume that other resistances have not changed. (Because of different duty and gas temperature profile, the gas-side heat transfer coefficient will be slightly different. However, for the sake of illustration, we neglect this.) We have

$$\frac{1}{U_o} = 0.10 + 0.00226 + 0.00041 + 0.05 + 0.0005$$

$$= 0.15317$$

Hence, $U_o = 6.52$ Btu/ft^{2}hr°F

Heat flux $q_o = 6.52 \times (1000 - 366) = 4133$ Btu/ft^2 hr.

Temperature drop across outside steam film = $0.0005 \times 4133 = 2°$F.

Temperature drop across steam-side fouling layer or scale = $4133 \times 0.05 = 207°$F.

Temperature drop across tube wall = $4133 \times 0.00041 = 2°$F.

We see that *average* tube wall temperature has risen to $366 + 2 + 207 + 2 = 577°$F from an earlier value of about 375°F. Scale formation is a serious problem. Note that the heat flux is now lower, but that does not help. *At the front end, where the heat flux is higher, the tubes would be much hotter.*

Now let us check the effect on boiler duty. It can be shown [1, 8] that

$$\ln \frac{t_{g1} - t_{sat}}{t_{g2} - t_{sat}} = \frac{UA}{W_g \times C_p \times h_{lf}}$$

where h_{lf} is the heat loss factor. If 2% losses are assumed, then $h_{lf} = 0.98$.

We know that $U_o = 6.52$, $A_o = 6820$, $t_{g1} = 1500$, $t_{sat} = 366$. Hence,

$$\ln \frac{1500 - 366}{t_{g2} - 366} = \frac{6.52 \times 6280}{100,000 \times 0.98 \times 0.287}$$

$$= 1.456, \text{ or } \frac{1500 - 366}{t_{g2} - 366} = 4.29$$

Hence $t_{g2} = 630°F$ compared to 500°F earlier. The reason for t_{g2} going up is the lower U_o caused by scale formation.

Hence new duty $= 100,000 \times 0.98 \times 0.287 \times (1500 - 630) = 24.47 \times 10^6$ Btu/hr. The decrease in duty is $28.13 - 24.47 = 3.66$ MM Btu/hr. Even assuming a modest energy cost of \$3/MM Btu, the annual loss due to increased fouling is $3.66 \times 3 \times 8000 = \$87,800$. The steam production in turn gets reduced.

Plant engineers should check the performance of their heat transfer equipment periodically to see if the exit gas temperature rises for the same inlet gas flow and temperature. If it does, then it is likely due to fouling on either the gas or steam side, which can be checked. Fouling on the gas side affects only the duty and steam production, but fouling on the steam side increases the tube wall temperature in addition to reducing the duty and steam production.

To ensure that variations in exit gas temperature are not due to fouling but are due to changes in gas flow or temperature, one can use simulation methods. For example, if, for the same gas flow, the inlet gas temperature is 1800°F, we can expect the exit gas temperature to rise. Under clean conditions, this can be estimated using Eq. (35).

$$\frac{1500 - 366}{500 - 366} = \frac{1800 - 366}{t_{g2} - 366}, \quad \text{or } t_{g2} = 535°F$$

Now if, in operation, the exit gas temperature were 570 to 600°F, then fouling could be suspected; but if the gas temperature were only about 535°F, this would only be due to the increased gas inlet temperature. Similarly, one can consider the effect of gas flow and saturation temperature.

4.14

Q: How is the size of a water tube boiler determined?

A: The starting point in the design of an evaporator bundle (Figure 4.3) is the estimation of overall heat transfer coefficient. The cross-sectional data such as the number of tubes wide, spacing, and length of tubes are assumed. U is estimated on the basis of mass velocity. The surface area is then determined, followed by the number of rows deep. The gas pressure drop is then evaluated.

This is only a feasible design. Several options are possible depending on the size and operating cost, which is influenced by the gas pressure drop. Optimization is then resorted to, prefera-bly using a computer program.

EXAMPLE

150,000 pph of flue gas must be cooled from 900°F to 520°F in an evaporator generating saturated steam at 125 psig with 230°F feedwater. Blowdown is 5%. Gas analysis (% vol) is as follows: CO_2 = 12, H_2O = 12, N_2 = 70, and O_2 = 6. Heat loss from the casing = 1.0%. Fouling factors on steam and gas side = 0.001 ft^2 hr °F/Btu.

Solution.   Let us assume the following. Tube size = 2 × 0.105 in. (inner diameter = 1.77 in.); number wide = 18; tube length = 10 ft; Transverse and longitudinal pitch are 4 in. each. Aver-age gas temperature = 0.5 × (900 + 520) = 710°F = 650 K. Fluid temperature inside tubes = 353°F. Hence film temperature = 0.5 × (353 + 710) = 531°F. The gas properties from the Appendix are C_p = 0.270, μ = 0.0645, and k = 0.02345. The gas specific heat at the average gas temperature = 0.277.

Duty Q = 150,000 × 0.99 × 0.277 × (900 − 520)

= 15.60 MM Btu/hr

Steam enthalpy change = (1193 − 198) + 0.05 × (325

− 198) = 1001.4 Btu/lb

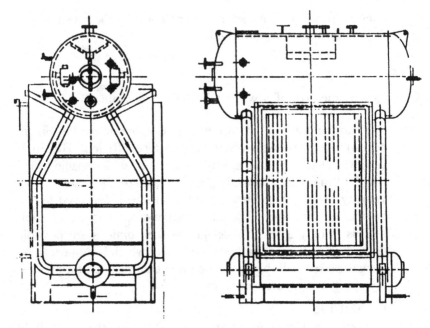

Figure 4.3 Boiler evaporator bundle.

Hence

$$\text{Steam generation} = \frac{15.60 \times 10^6}{1001.4} = 15,580 \text{ lb/hr}$$

$$G = 150,000 \times \frac{12}{18 \times 10 \times (4-2)} = 5000 \text{ lb/ft}^2 \text{ hr}$$

$$Re = 5000 \times \frac{2}{12 \times 0.0645} = 12920.$$

Using Grimson's equation, for a spacing of 4 in. in-line,

$$\text{Nu} = 0.229 \times (12,920)^{0.632} = 90.8 = h_c$$

$$\times \frac{2}{12 \times 0.02345}$$

Hence, $h_c = 12.78$ Btu/ft hr°F.

Let us compute the nonluminous coefficient. Partial pressures of CO_2 and H_2O = 0.12; beam length L = 1.08 × (4 × 4 − 0.785 × 4)/2 = 6.95 in. = 0.176 m;

$$K = (0.8 + 1.6 \times 0.12) \, (1 - 0.38 \times 0.650)$$
$$\times \frac{0.24}{(0.24 \times 0.176)^{0.5}} = 0.872,$$
$$\epsilon_g = 0.9 \times (1 - e^{-0.872 \times 0.176}) = 0.128$$

Assume that the surface temperature is 400°F. A reasonable assumption is 30 to 40°F above the average fluid temperature.

$$h_N = 0.173 \times 0.9 \times 0.128 \times \frac{11.7^4 - 8.6^4}{1170 - 860} = 0.85$$

(An additional factor of 0.9 was used to account for the emissivity of the surface.)

Using an h_i value of 2000 Btu/ft^2 hr°F, fouling factors of 0.001 for both inside and outside the tubes, and a tube metal conductivity of 25 Btu/ft hr°F

$$\frac{1}{U} = \frac{1}{0.85 + 12.78} + 0.001 + 0.001 \times \frac{2}{1.77}$$
$$+ \frac{1}{2000} \times \frac{2}{1.77} + 1 \times \ln \frac{(2/1.77)}{24 \times 25}$$
$$= 0.07326 + 0.001 + 0.0011 + 0.000565$$
$$+ 0.0004 = 0.07633$$

Hence, U = 13.1 Btu/ft^2 hr°F.

Log-mean temperature difference = [(900 − 353) − (520 − 353)]/ln (547/167) = 320°F

$$\text{Required surface area} = A = \frac{15,600,000}{320 \times 13.1} = 3721 \text{ ft}^2$$
$$= 3.14 \times \frac{2}{12} \times 18 \times 10 \times N_d$$

or N_d = 39.5; use 40. Provided surface area = 3768 ft^2.

Let us compute the gas pressure drop.

$$\text{Gas density} = \frac{MW}{359} \times \frac{492}{460 + t_g} = 29$$

$$\times \frac{492}{359 \times 1170}$$

$$= 0.034 \text{ lb/cu ft}$$

$$f = 12{,}950^{-0.15} \times (0.044 + 0.08 \times 2) = 0.0493$$

$$\Delta P_o = 9.3 \times 10^{-10} \times 5000^2 \times 40 \times \frac{0.0493}{0.034}$$

$$= 1.35 \text{ in. } WC$$

The average heat flux q (based on inner diameter) is

$$q = U \times (t_g - t_s) = 13.1 \times (710 - 353) \times \frac{2}{1.77}$$

$$= 5285 \text{ Btu/ft}^2 \text{ hr.}$$

The temperature drops across the various resistances are computed.

Drop across inside film = 5285/2000 = 2.7°F

Drop across inside fouling = 5285 × 0.001 = 5.3°F

$$\text{Drop across tube wall} = \left(0.0004 \times \frac{1.77}{2}\right) \times 5285$$

$$= 1.9°F$$

Hence the outer wall temperature = 353 + 2.7 + 5.3 + 1.9 = 363°F.

Note that this is only an average outer wall temperature. The heat flux and the tube wall temperature should be evaluated at the gas inlet, considering the nonuniformity in gas flow and temperature profile across the boiler cross section. Note also that the heat flux with bare tubes is low compared to that with finned tubes, as will be shown later.

Also, an evaporator section may have several combinations of tube spacings and fin configurations. This design would be used in clean gas applications when the inlet gas temperature is very high, where, due to heat flux and tube wall or fin tip temperature concerns, a few bare tubes would be used followed by tubes with increasing fin density. Also, in dirty gas applications, the tube spacing would be wide at the gas inlet because of slagging or bridging concerns. In these cases, an analysis of each sort of arrangement for heat transfer and pressure drop may be warranted.

4.15a

Q: How is the off-design performance of a boiler evaluated?

A: Let us see how the boiler in Q4.14 behaves when the gas inlet temperature and flow change while the steam parameters remain unchanged.

EXAMPLE

Predict the performance of the boiler in Q4.14 when the gas flow = 130,000 pph and inlet gas temperature = 1500°F. Gas analysis is unchanged.

Solution.   Performance calculations are more involved than design calculations, as we have a given surface area and the heat balance has to be arrived at through a trial-and-error procedure. The basic steps are

1. Assume a value for the exit gas temperature.
2. Calculate U.
3. Compute the assumed duty Q_a, $= W_g \times C_p \times (T_1 - T_2) \times h_{lf}$.
4. Calculate the log-mean temperature difference ΔT.
5. Compute the transferred duty $Q_t = U \times A \times \Delta T$.
6. If Q_a and Q_t do not agree, go back to step 1 and try another exit gas temperature.

For evaporators this is not a tedious calculations, but for economizers or superheaters, particularly when there is more than one stage, the calculations are time-consuming. Let the exit gas temperature = 700°F. The average film temperature = [0.5 × (1500 + 700) + 353] = 725°F. Gas properties at 725°F from the Appendix are C_p = 0.277, μ = 0.07256, and k = 0.027.

$$G = 130,000 \times \frac{12}{18 \times 10 \times (4 - 2)} = 4333 \text{ lb/ft}^2 \text{ hr}$$

$$Re = 4333 \times \frac{2}{12 \times 0.007256} = 9954$$

$$Nu = 0.229 \times 9954^{0.632} = 77 = h_c \times \frac{2}{12 \times 0.027}$$

Hence $h_c = 12.47$. Average gas temperature $= 0.5 \times (1500 + 700) = 1100°F = 866$ K

$$K = \frac{(0.8 + 1.6 \times 0.12)(1 - 0.38 \times 866)}{\sqrt{0.0.24 \times 0.176}} = .782$$

$$\epsilon_g = 0.9 \times (1 - e^{-0.782 \times 0.176}) = 0.1286.$$

$$h_n = 0.173 \times 0.9 \times 0.1286 \times \frac{[(15.4)^4 - (8.6)^4]}{1540 - 860} = 1.49$$

$$\frac{1}{U} = \frac{1}{12.47 + 1.49} + \frac{1}{2000} \times \frac{2}{1.77}$$

$$+ 0.001 \times \frac{2}{1.77} + 2 \frac{\ln(2/1.77)}{(24 \times 25)}$$

$$= 0.0747, \quad U = 13.38$$

The assumed duty $= 130,000 \times 0.99 \times (1500 - 700) = 29.8$ MM Btu/hr (C_p at average gas temperature $= 0.289$,)

Log-mean temperature difference

$$= \frac{(1500 - 353) - (700 - 353)}{\ln(1147/347)}$$

$$= 669°F$$

Transferred duty $Q_t = 3768 \times 13.38 \times 669 = 33.72$ MM Btu/hr. Since the discrepancy is large and the transferred duty is higher than the assumed duty, another trial is required. Try 650°F as exit gas temperature. There is no need to compute U again as the difference will be marginal.

Try assumed duty $Q_a = 130,000 \times 0.289 \times (1500 - 650) \times 0.99 = 31.61$ MM Btu/hr

$$\Delta T = 629; \quad U = 13.38; \quad A = 3768$$

Transferred duty $Q_t = 629 \times 13.38 \times 3768 = 31.71$ MM Btu/hr. This is close enough. Hence the new duty is 31.71 MM Btu/hr, and the corresponding steam flow $= (31.71/15.60) \times 15,580 = 31,670$ lb/hr.

Gas pressure drop is computed as before.

Average gas temperature $= 1075°F = 1535°R$

$$\rho = 29 \times \frac{492}{(359 \times 1534)} = 0.0259 \text{ lb/cu ft}$$

$$f = 9954^{-0.15} \times (0.044 + 0.08 \times 2) = 0.0512$$

$$\Delta P_q = 9.3 \times 10^{-10} \times 4333^2 \times 40 \times \frac{0.0512}{0.0259}$$

$$= 1.39 \text{ in. WC}$$

Tube wall temperature and heat flux can be computed as before.

4.15b

Q: Discuss the logic for determining the off-design performance of a water tube waste heat boiler with the configuration shown in Figure 4.4

A: In the design procedure one calculates the size of the various heating surfaces such as superheaters, evaporators, and economizers by the methods discussed earlier based on the equation $A = Q/(U \times \Delta T)$. In this situation, the duty Q, log-mean temperature difference ΔT, and overall heat transfer coefficient U are known or can be obtained easily for a given configuration.

In the off-design procedure, which is more involved, the purpose is to predict the performance of a given boiler under different conditions of gas flow, inlet gas temperature, and steam parameters. In these calculations several trial-and-error steps are required before arriving at the final heat balance and duty, as the surface area is now known. The procedure is discussed for a simple case, configuration 1 of Figure 4.4, which consists of a screen section, superheater, evaporator, and economizer.

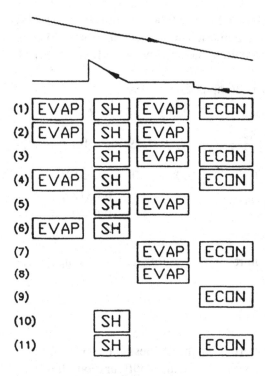

Key EVAP = Evaporator
     ECON = Economizer
      SH = Superheater

Figure 4.4   Configurations for water tube boiler.

1. Assume a steam flow W_s based on gas conditions.
2. Solve for the screen section, which is actually an evapora-
    tor, by using the methods discussed in Q4.15a.
3. Solve for the superheater section, either using the NTU
    method or by trial and error. Assume a value for the duty
    and compute the exit gas/steam temperatures and then ΔT.

    Assumed duty $Q_a = W_g \times C_p \times (T_{gi} - T_{go})$ $h_{lf} = W_s$
    $\times (h_{so} - h_{si})$

where

h_{so}, h_{si} = enthalpies of steam at exit and inlet
T_{gi}, T_{go} = gas inlet and exit temperatures.

Compute U. Then transferred duty is $Q_t = U \times A \times \Delta T$. If Q_a and Q_t are close, then the assumed duty and gas/steam temperatures are correct, and proceed to the next step; otherwise assume another duty and repeat step 3.

4. Solve for the evaporator section as in step 1. No trial and error is required, as the steam temperature is constant.

5. Solve for the economizer as in step 3. Assume a value for the duty and then compute exit gas/water temperatures, ΔT, and Q_t. Iteration proceeds until Q_a and Q_t match. The NTU method can also be used to avoid several iterations.

6. The entire HRSG duty is now obtained by adding all of the transferred duty of the four sections. The steam flow is corrected based on the actual total duty and enthalpy rise.

7. If the actual steam flow from step 6 equals that assumed in step 1, then the iterations are complete and the solution is over; if not, go back to step 1 with the revised steam flow.

The calculations become more complex if supplementary firing is added to generate a desired quantity of steam; the gas flow and analysis change as the firing temperature changes, and the calculations for U and the gas/steam temperature profile must take this into consideration. Again, if multipressure HRSGSs are involved, the calculations are even more complex and cannot be done without a computer.

4.16a

Q: Determine the tube metal temperature for the case of a superheater under the following conditions:

    Average gas temperature = 1200°F
    Average steam temperature = 620°F
    Outside gas heat transfer coefficient = 15 Btu/ft² hr °F
    Steam-side coefficient = 900 Btu/ft² hr °F

(Estimation of steam and gas heat transfer coefficients is discussed in Q4.03 and 4.04.)

    Tube size = 2 × 0.142 in. (2 in. OD and 0.142 in.
            thick)

Table 4.9 Thermal Conductivity of Metals, Btu/ft hr °F

Material	Temperature (°F)													
	200	300	400	500	600	700	800	900	1000	1100	1200	1300	1400	1500
Aluminum (annealed)														
Type 1100-0	126	124	123	122	121	120	118							
Type 3003-0	111	111	111	111	111	111	111							
Type 3004-0	97	98	99	100	102	103	104							
Type 6061-0	102	103	104	105	106	106	106							
Aluminum (tempered)														
Type 1100 (all tempers)	123	122	121	120	118	118	118							
Type 3003 (all tempers)	96	97	98	99	100	102	104							
Type 3004 (all tempers)	97	98	99	100	102	103	104							
Type 6061-T4 and T6	95	96	97	98	99	100	102							
Type 6063-T5 and T6	116	116	116	116	116	115	114							
Type 6063-T42	111	111	111	111	111	111	111							
Cast iron	31	31	30	29	28	27	26	25						
Carbon steel	30	29	28	27	26	25	24	23						
Carbon moly (½%) steel	29	28	27	26	25	25	24	23						

Material														
Chrome moly steels														
1% Cr, ½% Mo	27	27	26	25	24	24	23	21	21	21					
2¼% Cr, 1% Mo	25	24	23	23	22	22	21	21	20	20	20				
5% Cr, ½% Mo	21	21	21	20	20	20	20	19	19	19	19	17			
12% Cr	14	15	15	15	16	16	16	16	17	17	18				
Austenitic stainless steels														
18% Cr, 8% Ni	9.3	9.8	10	11	11	12	12	13	13	14	14	14	14	15	15
25% Cr, 20% Ni	7.8	8.4	8.9	9.5	10	11	11	12	12	13	13	14	14	15	15
Admiralty metal	70	75	79	84	89										
Naval brass	71	74	77	80	83										
Copper (electrolytic)	225	225	224	224	223										
Copper and nickel alloys														
90% Cu, 10% Ni	30	31	34	37	42	47	49	51	53						
80% Cu, 20% Ni	22	23	25	27	29	31	34	37	40						
70% Cu, 30% Ni	18	19	21	23	25	27	30	33	37						
30% Cu, 70% Ni (Monel)	15	16	16	16	17	18	18	19	20	20					
Nickel	38	36	33	31	29	28	28	29	31	33					
Nickel-chrome-iron	9.4	9.7	9.9	9.9	10	10	11	11	11	12	12	12	13	13	13
Titanium (gr B)	10.9	10.4	10.4	10.5	11	11	11	12	12	13	13				

Tube thermal conductivity = 21 Btu/ft hr °F (carbon steel)

(Thermal conductivity of metals can be looked up from Table 4.9.)

A: Since the average conditions are given and the average tube metal temperature is desired, we must have the parameters noted above under the most severe conditions of operation, such as the highest gas temperature, steam temperature, heat flux, and so on.

Let us use the concept of electrical analogy, in which the thermal and electrical resistances, heat flux, and current, temperature difference, and voltage are analogous. For the thermal resistance of the tube metal,

$$R_m = \frac{d}{24K_m} \ln\left(\frac{d}{d_i}\right) = \ln\left(\frac{2}{1.72}\right) \times \frac{2}{24 \times 21}$$

$$= 0.0006 \text{ ft}^2 \text{ hr } °F/Btu$$

Outside gas film resistance $R_o = \dfrac{1}{15} = 0.067 \text{ ft}^2 \text{ hr}$

°F/Btu

Inside film resistance $R_i = \dfrac{1}{900} = 0.0011 \text{ ft}^2 \text{ hr } °F/Btu$

Total resistance $R_t = 0.067 + 0.0006 + 0.0011$
$$= 0.0687 \text{ ft}^2 \text{ hr } °F/Btu$$

Hence

Heat flux $Q = \dfrac{1200 - 620}{0.0687} = 8443 \text{ Btu/ft}^2 \text{ hr}$

Temperature drop across the gas film = 8443 × 0.067 = 565°F

Temperature drop across the tube metal = 8443 × 0.0006 = 5°F

Temperature drop across steam film = 8443 × 0.0011 = 9.3°F

(Here we have applied the electrical analogy, where voltage drop is equal to the product of current and the resistance.) Hence,

Average tube

$$\text{metal temperature} = \frac{(1200 - 565) + (620 - 9.3)}{2}$$

$$= 632°F$$

We note that the tube metal temperature is close to the tube-side fluid temperature. This is because of the high tube-side coefficient compared to the gas heat transfer coefficient. This trend would prevail in equipment such as water tube boilers, superheaters, economizers, or any gas–liquid heat transfer equipment.

An approximate estimate of the tube metal temperature for bare tubes in a gas–liquid or gas–gas heat transfer device is

$$t_m = t_o - \frac{h_i}{h_i + h_o} \times (t_o - t_i) \tag{31}$$

where

h_i, h_o = heat transfer coefficients inside and outside the tubes, Btu/ft^2 hr °F

t_i, t_o = fluid temperatures inside and outside, °F

4.16b

Q: In a boiler air heater, h_o = 9, h_i = 12, t_i = 200°F, and t_o = 800°F. Estimate the average tube wall temperature t_m.

A: Using Eq. (31), we have

$$t_m = 800 - \frac{12}{12 + 9} \times (800 - 200) = 457°F$$

4.17

Q: How is the performance of fire tube and water tube boilers evaluated? Can we infer the extent of fouling from operational data? A water tube boiler waste heat boiler as shown in Figure 4.5 generates 10,000 lb/hr of saturated steam at 300 psia when

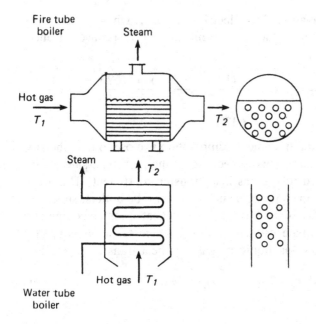

Figure 4.5   Sketch of fire tube and water tube boilers.

the gas flow is 75,000 lb/hr and gas temperatures in and out are 1000°F and 500°F. What should the steam generation and exit gas temperature be when 50,000 lb/hr of gas at 950°F enters the boiler?

A: It can be shown that in equipment with a phase change [1, 8],

$$\ln \frac{t_1 - t_{sat}}{t_2 - t_{sat}} = \frac{UA}{W_g C_p} \qquad (32)$$

where

t_1, t_2 = gas temperatures entering and leaving the boiler, °F

t_{sat} = saturation steam temperature, °F

A = surface area, ft²

U = overall heat transfer coefficient, Btu/ft² hr °F

W_g = total gas flow, lb/hr

C_p = instantaneous gas specific heat at the average temperature of $(t_1 + t_2)/2$ Btu/lb °F

For fire tube boilers, the overall heat transfer coefficient is dependent on the gas coefficient inside the tubes; that is, U is proportional to $W_g^{0.8}$. In a water tube boiler, U is proportional to $W_g^{0.6}$. Substituting these onto Eq. (32) gives us the following.

For fire tube boilers:

$$\ln \frac{t_1 - t_{sat}}{t_2 - t_{sat}} = \frac{K_1}{W_g^{0.2}} \tag{33}$$

For water tube boilers:

$$\ln \frac{t_1 - t_{sat}}{t_2 - t_{sat}} = \frac{K_2}{W_g^{0.4}} \tag{34}$$

As long as the fouling is not severe, Eqs. (33) and (34) predict the exit gas temperatures correctly. If t_2 is greater than predicted, we can infer that fouling has occurred. Also, if the gas pressure drop across the boiler is more than the calculated value (see Chapter 3 for pressure drop calculations), we can infer that fouling has taken place.

Calculate K_2 from Eq. (34). $t_{sat} = 417$ from the steam tables (see the Appendix).

$$K_2 = \ln \left(\frac{1000 - 417}{500 - 417} \right) \times (75,000)^{0.4} = 173$$

Let us predict the exit gas temperature when $W_g = 50,000$.

$$\ln \left(\frac{950 - 417}{t_2 - 417} \right) = \frac{(50,000)^{0.4}}{173} = 2.29$$

$$t_2 = 417 + \frac{950 - 417}{\exp(2.29)} = 471°F$$

Now the actual exit gas temperature is 520°F, which means that the fouling is severe.

The energy loss due to fouling is

$$Q = 50,000 \times 0.26 \times (520 - 471) = 0.63$$
$$\times 10^6 \text{ Btu/hr}$$

If energy costs \$3/MM Btu, the annual loss of energy due to fouling will be $3 \times 0.63 \times 8000 = \$15{,}120$ (assuming 8000 hours of operation a year).

4.18

Q: When and where are finned tubes used? What are their advantages over bare tubes?

A: Finned tubes are used extensively in boilers, superheaters, economizers, and heaters for recovering energy from clean gas streams such as gas turbine exhaust or flue gas from combustion of premium fossil fuels. If the particulate concentration in the gas stream is very low, finned tubes with a low fin density may be used. However, the choice of fin configuration, particularly in clean gas applications, is determined by several factors such as tube-side heat transfer coefficient, overall size, cost, and gas pressure drop, which affects the operating cost.

Solid and serrated fins (Figure 4.6) are used in boilers and heaters. Finned surfaces are attractive when the ratio between the heat transfer coefficients on the outside of the tubes to that inside is very small. In boiler evaporators or economizers, the tube-side

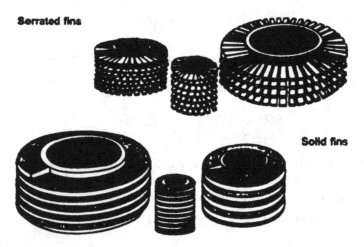

Figure 4.6   Solid and serrated fins.

coefficient could be in the range of 1500 to 3000 Btu/ft^2 hr °F, while the gas-side coefficient could be in the range of 10 to 20 Btu/ft^2 hr °F. A large fin density or a large ratio of external to internal surface area is justified in this case. As the ratio between the outside and inside coefficients decreases, the effectiveness of using a large ratio of external to internal surface areas decreases. For example, in superheaters or high-pressure air heaters, where the tube-side coefficient could be in the range of 30 to 300 Btu/ft^2 hr °F, it does not pay to use a large fin surface; in fact, it is counterproductive, as will be shown later. A moderate fin density such as two or three fins per inch would be adequate, while for economizers or evaporators, five or even six fins per inch may be justified if cleanliness permits.

The other important fact to be kept in mind is that more surface area does not necessarily mean more energy transfer. It is possible, through poor choice of fin configuration, to have more surface area and yet transfer less energy. One has to look at the product of surface area and overall heat transfer coefficient and not at surface area alone. The overall heat transfer coefficient is significantly reduced as we increase the fin surface or use more fins per inch.

Finned tubes offer several advantages over bare tubes such as a compact design that occupies less space, lower gas pressure drop, lower tube-side pressure drop due to the fewer rows of tubes, and a smaller overall weight and cost.

Solid fins offer slightly lower gas pressure drop over serrated fins, which have a higher heat transfer coefficient for the same fin density and configuration. Particulates, if present, are likely to accumulate on serrated finned tubes, which may be difficult to clean.

4.19a

Q: How are the heat transfer and pressure drop over finned tubes and tube and fin wall temperatures evaluated?

A: The widely used ESCOA correlations developed by ESCOA Corporation [9] will be used to evaluate the heat transfer and pressure drop over solid and serrated finned tubes in in-line and staggered arrangements. The basic equation for heat transfer coefficient with finned tubes is given by Eq. (3).

The calculation for tube-side coefficient h_i is discussed earlier. h_o consists of two parts, a nonluminous coefficient h_n, which is computed as discussed in Q4.07, and h_c, the convective heat transfer coefficient. Computation of h_c involves an elaborate procedure and the solving of several equations, as detailed below.

DETERMINATION OF h_c [9]

$$h_c = C_3 C_1 C_5 \left(\frac{d + 2h}{d} \right)^{0.5} \times \left(\frac{t_g + 460}{t_a + 460} \right)^{0.25}$$

$$\times\ GC_p \times \left(\frac{k}{\mu C_p} \right)^{0.67} \tag{35}$$

$$G = \frac{W_g}{[(S_T/12) - A_o]N_w L} \tag{36}$$

$$A_o = \frac{d}{12} + \frac{nbh}{6} \tag{37}$$

C_1, C_2, and C_3 are obtained from Table 4.10.

$$\text{Re} = \frac{Gd}{12} \tag{38}$$

$$s = \frac{1}{n} - b \tag{39}$$

FIN EFFICIENCY AND EFFECTIVENESS
For both solid and serrated fins; effectiveness η is

$$\eta = 1 - (1 - E) \times \frac{A_f}{A_t} \tag{40}$$

For solid fins,

$$A_f = \pi\, n \times \frac{(4dh + 4h^2 + 2bd + 4bh)}{24} \tag{41}$$

$$A_t = A_f + \pi\, \frac{d \times (1 - nb)}{12} \tag{42}$$

Table 4.10 Factors C_1 to C_6

Solids fins	Serrated fins
$C_1 = 0.25 \, Re^{-0.35}$	$C_1 = 0.25 \, Re^{-0.35}$
$C_2 = 0.07 + 8.0 \, Re^{-0.45}$	$C_2 = 0.07 + 8.0 \, Re^{-0.45}$

In-line

$C_3 = 0.20 + 0.65e^{-0.25h/s}$

$C_4 = 0.08[0.15S_T/d_o]^{-1.1(h/s)^{0.15}}$

$C_5 = 1.1 - (0.75 - 1.5e^{-0.70N_d})$

$C_6 = 1.6 - (0.75 - 1.5e^{-0.7N})e^{(-2.0S_L/S_T)^2}$

Serrated fins (In-line)

$C_3 = 0.35 + 0.5e^{-0.35h/s}$

$C_4 = 0.08(0.15S_T/d_o)^{1.1(h/s)^{0.20}}$

$C_5 = 1.1 - (0.75 - 1.5e^{-0.70N_d})e^{(-2.0S_L/S_T)}$

$C_6 = 1.6 - (0.75 - 1.5e^{-0.7N_d})e^{(-2.0S_L/S_T)^2}$

Staggered

$C_3 = 0.35 + 0.65e^{0.25h/s}$

$C_4 = 0.11 - (0.05S_T/D_o)^{-0.7\,(h/s)^{0.20}}$

$C_5 = 0.7 + (-0.70 - 0.8e^{-0.15N_d^2})e^{-1.0S_L/S_T}$

$C_6 = 1.1 + (1.8 - 2.1e^{(-0.15N_d^2)})e^{-2.0S_L/S_T} -$
$\quad (0.7 - 0.8e^{-0.15N_d^2})e^{-0.6S_L/S_T}$

Serrated fins (Staggered)

$C_3 = 0.55 + 0.45e^{-0.35h/s}$

$C_4 = 0.11 - (0.05S_T/d_o)^{-0.7(h/s)^{0.23}}$

$C_5 = 0.7 + (-0.70 - 0.8e^{(-0.15N_d^2)})e^{(-2.0S_L/S_T)}$

$C_6 = 1.1 + (1.8 - 2.1e^{(-0.15N_d^2)})e^{(-2.0S_L/S_T)} -$
$\quad (0.7 - 0.8e^{-0.15N_d^2})e^{-0.6S_L/S_T}$

$$E = 1/\{1 + 0.002292 \, m^2h^2 \, [(d+2h)/d]^{0.5}\} \tag{43}$$

where

$$m = (24 \, h_o/Kb)^{0.5} \tag{44}$$

For serrated fins,

$$A_f = \pi \, dn \, \frac{2h \, (ws + b) + bws}{12ws} \tag{45}$$

$$A_t = A_f + \pi \, d \, \frac{(1 - nb)}{12} \tag{46}$$

$$E = \frac{\tanh \, (mh)}{mh} \tag{47}$$

where

$$m = \left[\frac{24 \times h_o(b + ws)}{Kbws} \right]^{0.5} \tag{48}$$

Gas pressure drop ΔP_g is

$$\Delta P_g = (f + a) \, \frac{G^2N_d}{\rho_g \times 1.083 \times 10^9} \tag{49}$$

where

$$f = C_2C_4C_6 \times \left(\frac{d + 2h}{d} \right)^{0.5}$$

$$\text{for staggered arrangement} \tag{50}$$

$$= C_2C_4C_6 \times \frac{d + 2h}{d} \quad \text{for in-line arrangement} \tag{51}$$

$$a = \frac{(1 + B^2)}{4N_d} \times \frac{(t_{g2} - t_{g1})}{460 + t_g} \tag{52}$$

$$B = \left(\frac{\text{free gas area}}{\text{total area}} \right)^2 \tag{53}$$

C_2, C_4, C_6 are given in Table 4.10 for solid and serrated fins.

TUBE WALL AND FIN TIP TEMPERATURES

For solid fins the relation between tube wall and fin tip temperatures is given by

$$\frac{t_g - t_f}{t_g - t_b} = \frac{K_1(mr_e) \times I_0(mr_e) + I_1(mr_e) \times K_0(mr_e)}{K_1(mr_e) \times I_0(mr_0) + K_0(mr_0) \times I_1(mr_e)} \quad (54)$$

The various Bessel functional data are shown in Table 4.11 for serrated fins, treated as longitudinal fins:

$$\frac{t_g - t_f}{t_g - t_b} = \frac{1}{\cosh{(mb)}} \quad (55)$$

A good estimate of t_f can also be obtained for either type of fin as follows:

$$t_f = t_b + (t_g - t_b) \times (1.42 - 1.4 \times E) \quad (56)$$

t_b, the fin base temperature, is estimated as follows:

$$t_b = t_i + q \times (R_3 + R_4 + R_5) \quad (57)$$

where R_3, R_4, and R_5 are resistances to heat transfer of the inside film, fouling layer, and tube wall, respectively and heat flux q_o is given by

$$q_o = U_o (t_g - t_i) \quad (58)$$

The following example illustrates the use of the equations.

EXAMPLE

A steam superheater is designed for the following conditions.

Gas flow = 150,000 pph
Gas inlet temperature = 1000°F
Gas exit temperature = 861°F
Gas analysis (% vol): CO_2 = 12, H_2O = 12, N_2 = 70, O_2 = 6
Steam flow = 30,000 pph
Steam temperature in = 491°F (sat)
Steam exit temperature = 787°F
Steam pressure (exit) = 600 psig

Tubes used: 2 × 0.120 low alloy steel tubes; 18 tubes/row, 6 deep, in-line arrangement with 4-in. square pitch and nine streams. Tube inner diameter = 1.738 in.; outer diameter = 2 in.

Fins used: solid stainless steel, 2 fins/in., 0.5 in. high and 0.075 in thick. Fin thermal conductivity K = 15 Btu/ft hr. °F.

Table 4.11 I_0, I_1, K_0, and K_1 Values for Various Arguments

x	$I_0(x)$	$I_1(x)$	$K_0(x)$	$K_1(x)$
0	1.0	0	8	8
0.1	1.002	0.05	2.427	9.854
0.2	1.010	0.10	1.753	4.776
0.3	1.023	0.152	1.372	3.056
0.4	1.040	0.204	1.114	2.184
0.5	1.063	0.258	0.924	1.656
0.6	1.092	0.314	0.778	1.303
0.7	1.126	0.372	0.66	1.05
0.8	1.166	0.433	0.565	0.862
0.9	1.213	0.497	0.487	0.716
1.0	1.266	0.565	0.421	0.602
1.2	1.394	0.715	0.318	0.434
1.4	1.553	0.886	0.244	0.321
1.6	1.75	1.085	0.188	0.241
1.8	1.99	1.317	0.146	0.183
2.0	2.28	1.591	0.114	0.140
2.2	2.629	1.914	0.0893	0.108
2.4	3.049	2.298	0.0702	0.0837
2.6	3.553	2.755	0.554	0.0653
2.8	4.157	3.301	0.0438	0.0511
3.0	4.881	3.953	0.0347	0.0402
3.2	5.747	4.734	0.0276	0.0316
3.4	6.785	5.670	0.0220	0.0250
3.6	8.028	6.793	0.0175	0.0198
3.8	9.517	8.140	0.0140	0.0157
4.0	11.30	9.759	0.0112	0.0125
4.2	13.44	11.70	0.0089	0.0099
4.4	16.01	14.04	0.0071	0.0079
4.6	19.09	16.86	0.0057	0.0063
4.8	22.79	20.25	0.0046	0.0050
5.0	27.24	24.34	0.0037	0.0040

Determine the heat transfer coefficient and pressure drop.

Solution.

$$A_o = \frac{2}{12} + \frac{2 \times 0.5 \times 0.075}{6} = 0.17917 \text{ ft}^2/\text{ft}$$

$$G = \frac{150,000}{18 \times 10 \times [(4/12) - 0.17917]} = 5420 \text{ lb/ft}^2 \text{ hr}$$

The gas properties at the average gas temperature (from the Appendix) are

$$C_p = 0.2851, \qquad \mu = 0.08146, \qquad k = 0.03094$$

$$\text{Re} = \frac{5420 \times 2}{12 \times 0.08146} = 11,090$$

$$C_1 = 0.25 \times (11,090)^{-0.35} = 0.0096,$$

$$s = 1/2 - 0.075 = 0.425$$

$$C_3 = 0.2 + 0.65 \, e^{-0.25 \times 0.5/0.425} = 0.6843$$

$$C_5 = 1.1 - (0.75 - 1.5 \, e^{-0.7 \times 6}) \, (e^{-2 \times 4/4}) = 1.0015$$

Assume that the average fin temperature is 750°F. The average gas temperature = 930°F, and steam temperature = 640°F. The fin thermal conductivity K is assumed to be 15 Btu/ft hr °F. Then,

$$h_c = 0.0096 \times 0.6843 \times 1.0015 \times \left(\frac{3}{2}\right)^{0.5}$$

$$\times \left(\frac{930 + 460}{750 + 460}\right)^{0.25} \times 5420 \times 0.2851$$

$$\times \left(\frac{0.3094}{0.2851 \times 0.08146}\right)^{0.67} = 15.74$$

Using methods discussed in Q4.07, we find $h_n = 1.12$. The beam length for finned tubes is computed as $3.4 \times$ volume/surface areas. Hence

$$h_o = 15.74 + 1.12 = 16.86,$$

$$m = \left(\frac{24 \times 16.86}{15 \times 0.075}\right)^{0.5} = 19$$

$$E = 1/(1 + 0.002292 \times 19 \times 19 \times 0.5 \times 0.5 \times \sqrt{1.5})$$

$$= 0.80$$

$A_f = 3.14 \times 2$

$\times \dfrac{4 \times 2 \times 0.5 + 4 \times 0.5 \times 0.5 + 2 \times 0.075 \times 2 + 4 \times 0.075 \times 5}{24}$

$= 1.426$

$A_t = 1.426 + 3.14 \times 2 \times \dfrac{1 - 2 \times .075}{12} = 1.871$

Hence

$$\eta = 0.8 + (1 - 0.8) \times \dfrac{1.426}{1.871} = 0.848$$

Let us compute h_i for steam. $w = 30,000/9 = 3333$ lb/hr per tube. From Table 4.2, factor $C = 0.337$.

$$h_i = 2.44 \times 0.337 \times \dfrac{3333^{0.8}}{(1.738)^{1.8}} = 200 \text{ Btu/ft}^2 \text{ hr } °F$$

$$\dfrac{1}{U} = \dfrac{1}{16.85 \times 0.848} + 12$$

$$\times \dfrac{1.872}{200 \times 3.14 \times 1.738} + 0.001$$

$$+ 0.001 \times \dfrac{1.871 \times 12}{3.14 \times 1.738} + 24 \times \ln\left(\dfrac{2}{1.738}\right)$$

$$\times \dfrac{1.871}{24 \times 20 \times 3.14 \times 1.738} = 0.0699$$

$$+ 0.0211 + 0.001 + 0.0041 + 0.0024 = 0.0985$$

$U_o = 10.16 \text{ Btu/ft}^2 \text{ hr } °F$

Calculation of tube wall and fin tip temperature

Heat flux $q = 10.16 \times (930 - 640) = 2945 \text{ Btu/ft}^2 \text{ hr}$

$t_b = 640 + 2945 \times (0.0024 + 0.0041 + 0.0211)$

$\quad = 722 °F$

Using the elaborate Bessel functions,

$$mr_e = 19 \times \dfrac{1.5}{12} = 2.38 \text{ ft}, \qquad mr_o = 1.58 \text{ ft}$$

$$K_0(2.38) = 0.07, \qquad K_1(2.38) = 0.0837,$$

$$I_0 (2.38) = 3.048, \qquad I_1 (2.38) = 2.295$$
$$K_0 (1.58) = 0.186, \qquad I_0 (1.58) = 1.74$$

Hence,

$$\frac{930 - t_f}{930 - 722} = \frac{0.0837 \times 3.048 + 2.295 \times 0.07}{0.0837 \times 1.74 + 0.186 \times 2.295}$$
$$= 0.723; \qquad t_f = 780°F$$

Using the approximation

$$t_f = t_b + (1.42 - 1.4 \times 0.8) \times (930 - 722) = 785°F$$

Note that this is only an average base and fin tip temperature. For material selection purposes one should look at the maximum heat flux, which occurs, for instance, at the gas inlet in a counterflow arrangement, and also consider the nonuniformity or maldistribution in gas and steam flow. A computer program can be developed to compute the tube wall and fin tip temperatures at various points along the tube length and the results used to select appropriate materials.

It can be noted from the above that there are a few ways to reduce the fin tip temperature:

1. Increase fin thickness. This reduces the factor m and hence t_f.
2. Increase the thermal conductivity of the fin material. This may be difficult, as the thermal conductivity of carbon steels is higher than that of alloy steels, and carbon steels can withstand temperatures only up to 850°F, which alloy steels can withstand up to 1300°F depending on the alloy composition.
3. Reduce h_o or the gas-side coefficient by using a lower gas mass velocity.
4. Reduce fin height or density.
5. In designs where the gas inlet temperature is very high, use a combination of bare and finned rows. The first few rows could be bare, followed by tubes with a low fin density or height or increased thickness and then followed by tubes with higher fin density or height or smaller thickness to

obtain the desired boiler performance. A row-by-row analysis of the finned bundle is necessary, which requires the use of a computer program.

Computation of gas pressure drop

$$C_2 = 0.07 + 8 \times (11{,}090)^{-0.45} = 0.191$$

$$C_4 = 0.08 \times [0.15 \times 2]^{-1.11 \times (0.5/0.425)^{0.15}} = 0.3107$$

$$C_6 = 1, \quad f = 0.191 \times 0.3107 \times 1 \times \frac{3}{2} = 0.089$$

$$B^2 = \left(\frac{0.33 - 0.17917}{0.33} \right)^2 = 0.2089,$$

$$a = \frac{861 - 1000}{460 + 930} \times \frac{1 + 0.2089}{24} = -0.005$$

$$\Delta P_g = 0.084 \times 5420 \times 5420 \times \frac{6}{0.0288 \times 1.083 \times 10^9}$$

$$= 0.53 \text{ in. WC}$$

Computer solution of the above system of equations saves a lot of time. However, I have developed a chart (Figure 4.7) that can be used to obtain h_c (or h_g) and η values for serrated fins and in-line arrangement for various fin configurations and gas mass velocities for gas turbine exhaust gases at an average gas temperature of 600°F. Although a computer program is the best tool, the chart can be used to show trends and the effect of fin configuration on performance of finned surfaces. The use of the chart is explained later with an example. The following points should be noted.

1. From Figure 4.7, it can be seen that for a given mass velocity, the higher the fin density or height, the lower the gas-side coefficient or effectiveness, which results in lower U_o. The amount of energy transferred in heat transfer equipment depends on the product of the overall heat transfer coefficient and surface area and not on surface area alone. We will see later that one can have more surface area and yet transfer less duty due to poor choice of fin configuration.

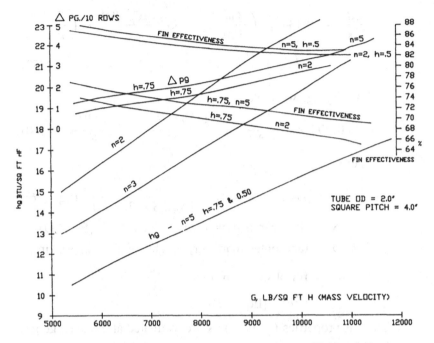

Figure 4.7 Chart of convective heat transfer coefficient and pressure drop versus fin geometry. [8, 10]

2. Higher fin density or height results in higher ΔP_g. Even after adjusting for the increased surface area per row, it can be shown that the higher the fin density or the greater the height, the higher the gas pressure drop will be for a given mass velocity.

4.19b

Q: Describe Briggs and Young's correlation.

A: Charts and equations provided by the manufacturer of finned tubes can be used to obtain h_c. In the absence of such data, the following equation of Briggs and Young for circular or helical finned tubes in staggered arrangement [4] can be used.

$$\frac{h_c d}{12k} = 0.134 \times \left(\frac{Gd}{12\mu}\right)^{0.681} \times \left(\frac{\mu C_p}{k}\right)^{0.33} \times \left(\frac{S}{h}\right)^{0.2}$$

$$\times \left(\frac{S}{b}\right)^{0.113} \tag{59}$$

Simplifying, we have

$$h_c = 0.295 \times \frac{G^{0.681}}{d^{0.319}} \times \frac{k^{0.67} C_p^{0.33}}{\mu^{0.351}} \times \frac{S^{0.313}}{h^{0.2} b^{0.113}} \tag{60}$$

where

$$G = \text{gas mass velocity} = \frac{W_g}{N_w L(S_T/12 - A_o)} \quad \text{[Eq. (36)]}$$

$$S = \text{fin clearance} = (1/n - b), \text{ in.} \quad \text{[Eq. (39)]}$$

$d, h, b = $ tube outer diameter, height, and thickness, in.

$$A_o = \text{fin obstruction area} = \frac{d}{12} + \frac{nbd}{6},$$

$$\text{ft}^2/\text{ft} \quad \text{[Eq. (37)]}$$

The gas properties C_p, μ, and k are evaluated at the average gas temperature.

The gas heat transfer coefficient b_c has to be corrected for the temperature distribution along the fin height by the fin efficiency

$$E = \frac{1}{1 + \dfrac{1}{3}\left(\dfrac{mh}{12}\right)^2 \sqrt{\dfrac{d + 2h}{d}}} \tag{61}$$

where

$$m = \sqrt{\frac{24h_c}{K_m b}} \quad \text{[Eq. (44)]}$$

K_m is the fin metal thermal conductivity, in Btu/ft hr °F.

In order to correct for the effect of finned area, a term called fin effectiveness is used. This term, η, is given by

$$\eta = 1 - (1 - E) \times \frac{A_f}{A_t} \quad \text{[Eq. 40)]}$$

where the finned area A_f and total area A_t are given by

$$A_f = \frac{\pi n}{24} \times (4dh + 4h^2 + 2bd + 4bh) \qquad \text{[Eq. (41)]}$$

$$A_t = A_f + \frac{\pi d}{12} \times (1 - nb) \qquad \text{[Eq. (42)]}$$

n is the fin density in fins/in. The factor

$$F = \frac{k^{0.67}C_p^{0.33}}{\mu^{0.35}} \qquad (62)$$

is given in Table 4.12.

The overall heat transfer coefficient with finned tubes, U, can be estimated as $U = 0.85\eta h_c$, neglecting the effect of non-luminous heat transfer coefficient.

EXAMPLE

Determine the gas-side heat transfer coefficient when 150,000 lb/hr of flue gases at an average temperature of 900°F flow over helically finned economizer tubes with the following parameters:

d = tube outer diameter = 2.0 in.
n = fins/in. = 3
h = fin height = 1 in.
b = fin thickness = 0.06 in.
L = effective length of tubes = 10.5 ft
N_w = number of tubes wide = 12
S_T = transverse pitch = 4.5 in. (staggered)

Table 4.12   Factor F for Finned Tubes

Temp. (°F)	F
200	0.0978
400	0.1250
600	0.1340
800	0.1439
1000	0.1473
1200	0.1540
1600	0.1650

Calculate A_o, A_f, and A_t. From Eq. (37),

$$A_o = \left(\frac{2}{12} + 3 \times 0.06 \times \frac{1}{6}\right) = 0.2 \text{ ft}^2/\text{ft}$$

From Eq. (41),

$$A_f = \left(\pi \times \frac{3}{24}\right) \times (4 \times 2 \times 1 + 4 \times 1 \times 1 + 2$$
$$\times 0.06 \times 2 + 4 \times 0.06) = 4.9 \text{ ft}^2/\text{ft}$$

From Eq. (42),

$$A_t = 4.9 + \pi \times \frac{2 \times (1 - 3 \times 0.06)}{12} = 5.33 \text{ ft}^2/\text{ft}$$

From Eq. (36),

$$G = \frac{150,000}{12 \times 10.5 \times (4.5/12 - 0.2)} = 6800 \text{ lb/ft}^2 \text{ hr}$$

Fin pitch $S = \dfrac{1}{3} - 0.06 = 0.27$

Using Eq. (62) with $F = 0.145$ from Table 4.12 gives us

$$h_c = 0.295 \times 6880^{0.681} \times 0.145$$
$$\times \frac{0.27^{0.313}}{2^{0.319} \times 1^{0.2} \times 0.06^{0.113}} = 12.74 \text{ Btu/ft}^2 \text{ hr } °F$$

Calculate fin efficiency from Eq. (61). Let metal thermal conductivity of fins (carbon steel) = 24 Btu/ft hr °F.

$$m = \sqrt{\frac{24 \times 12.74}{24 \times 0.06}} = 14.57$$

$$E = \frac{1}{1 + 0.33 \times (14.57 \times 1/12)^2 \times \sqrt{(2 + 2)/2}} = 0.6$$

Fin effectiveness $\eta = 1 - (1 - 0.6) \times \dfrac{4.9}{5.33} = 0.63$

Hence,

$$\eta h_c = 0.63 \times 12.74 = 8 \text{ Btu/ft}^2 \text{ hr } °F$$

K_m ranges from 23 to 27 Btu/ft hr °F for carbon steels, depending on temperature [1]. For alloy steels it is lower.

Table 4.13   Data for Gas Turbine Exhaust Gases[a]

Temp., °F	Sp. heat	Viscosity	Th. cond	F
200	0.2529	0.05172	0.0182	0.1152
400	0.2584	0.0612	0.02176	0.1238
600	0.2643	0.0702	0.02525	0.1316
800	0.2705	0.07885	0.02871	0.1392
1000	0.2767	0.0870	0.0321	0.1462

[a]% vol CO_2 = 3, H_2O = 7, N_2 = 75, O_2 = 15.

4.20

Q: Determine the overall heat transfer coefficient and pressure drop for a finned tube boiler for gas turbine exhaust under the following conditions:

Gas flow = 150,000 lb/hr (% vol CO_2 = 3, H_2O = 7, N_2 = 75, and O_2 = 15)
Gas inlet temperature = 1000°F
Exit gas temperature = 382°F
Duty = 150,000 × 0.2643 × 0.99 × (1000 − 382) = 24.25 MM Btu/hr (1% heat loss assumed); C_p = 0.2643 was taken from Table 4.13.
Steam pressure = 150 psig
Feedwater temperature = 240°F
Fouling factors ff_i, ff_o = 0.001 ft^2 hr °F/Btu
Boiler configuration: 18 tubes/row; square pitch = 4.0 in.; length = 10 ft; 18 tubes/row; serrated fins; 4 fins/in., 0.75 in. high, 0.05 in, thick; all-carbon steel (K_m = 25) surface area of finned tube = 5.35 ft^2/ft

Let us use Eq. (3) and Figure 4.7 to arrive at h_c and U_o. Note that h_c and h_g are used synonymously. The gas properties have to be computed first. Table 4.13 gives the properties along with the factor F used to compute h_g.

$$F = \left(\frac{C_p}{\mu} \right)^{0.33} \times k^{0.67}$$

To obtain h_g (or h_c), the gas mass velocity G must be computed. Using the equations given earlier,

$$A_o = \frac{2}{12} + 5 \times 0.75 \times \frac{0.06}{6} = 0.1979 \ \text{ft}^2/\text{ft}$$

$$G = \frac{150,000}{18 \times 10 \times (0.33 - 0.1979)} = 6308 \ \text{lb/ft}^2 \ \text{hr}$$

From Figure 4.7, h_g = 11.5; fin effectiveness = 0.745; gas pressure drop/10 rows = 1.5 in, WC. (Note that Figure 4.7 has been developed for gas turbine exhaust gases for an average gas temperature of 600°F and for serrated tubes; therefore, corrections for gas data or fin type should be done as required. The gas pressure drop is for 10 rows, and corrections should be made for actual number of rows.)

Let us assume that the tube-side coefficient h_i = 2000. The boiling heat transfer coefficient is very high compared to the gas-side coefficient and hence does not affect h_g.

$$\frac{A_t}{A_i} = \frac{5.35}{3.14 \times 1.77/12} = 11.55.$$

Then, substituting into Eq. (3), we have

$$\frac{1}{U_o} = \left(\frac{1}{11.5} \times 0.745 \right) + \frac{11.55}{2000}$$
$$+ \ 0.001 \times 11.55 + 0.001 + 2 \times 10.85$$
$$\times \ \frac{\ln(2/1.77)}{24/25} = 0.13946$$

Hence U_o = 7.17. Since the average gas temperature in our case is close to 600°F and the gas analysis is the same as that used for the chart, no correction is required. Otherwise it would be necessary to compute factor F and correct h_g. Also, because of the low gas temperature, the nonluminous heat transfer coefficient was neglected.

The log-mean temperature difference ΔT is

$$\Delta T = \left[\frac{(1000 - 366) - (382 - 366)}{\ln[(1000 - 366)/(382 - 366)]} \right]$$
$$= \ 168°F$$

$$\text{Surface area required} = \frac{24,250,000}{168 \times 7.17} = 20,140 \ \text{ft}^2$$

$20,140 = 5.35 \times 18 \times 1 \times N_d$, so N_d = number of rows deep = 21.

Hence gas pressure drop = $(21/10) \times 1.5 = 3.15$ in. WC. Thus the chart can be used to simplify the calculations. (h_c and h_g shown in the chart are the same.)

4.21

Q: How does a finned surface compare with a bare tube bundle for the same duty?

A: Let us try to design a boiler for the same application and duty using bare and finned tubes and compare the two designs. Q4.04 discusses the methodology for design of bare tubes.

Let us use the same cross section and tubes per row, length, tube size, and pitch as in Q4.20.

For the bare tube design,

$$G = 150,000 \times \frac{12}{18 \times 10 \times (4 - 2)} = 5000 \text{ lb/ft}^2 \text{ hr}$$

$$\text{Gas film temperature} = 0.5 \times \left(\frac{1000 + 382}{2} + 366 \right)$$

$$= 528°F$$

From the Appendix,

$$C_p = 0.262, \quad \mu = 0.675, \quad k = 0.024$$

$$F = 0.262^{0.33} \times \frac{0.024^{0.67}}{0.0675^{0.27}} = 0.1093$$

$$h_c = 0.9 \times 5000^{0.6} \times \frac{0.1093}{2^{0.4}} = 12.36$$

The nonluminous heat transfer coefficient is more significant for bare tubes than for finned tubes; h_N can be shown to be about 0.5 Btu/ft^2 hr °F.

Hence $h_o = h_c + h_N = 12.86$.

$$\frac{1}{U_o} = \frac{1}{12.86} + \frac{2}{1.77} \times 0.001 + 0.001$$

Table 4.14 Comparison of Bare Tube and Finned Tube Boilers[a]

	Bare tube	Finned tube
1. Gas flow, pph	150,000	
2. Inlet gas temp, °F	1000	
3. Exit gas temp, °F	382	
4. Duty, MM Btu/hr	24.25	
5. Steam pressure, psig	150	
6. Feedwater temp, °F	240	
7. Steam flow, pph	24,500	
8. Surface area, ft^2	11,670	20,140
9. U_0, Btu/ft^2 hr °F	12.86	7.17
10. Gas pressure drop, in. WC	4.5	3.15
11. Number of rows deep	124	21
12. Heat flux, Btu/ft^2 hr	9213	52,295
13. Tube wall temp, °F	385	484

[a]Number of tubes wide = 18; length = 10 ft; square pitch = 4.0 in.; finned tubes use four fins/in.; serrated fins, 0.75 in. high, 0.05 in. thick.

$$+ \; \frac{2/1.77}{2000} + 2 \; \frac{\ln(2/1.77)}{24/25} = 0.08085$$

$$U_o = 12.37 \text{ Btu/ft}^2 \text{ hr } °F$$

$$\text{Surface area required} = \frac{24,250,000}{(168 \times 12.37)} = 11,670 \text{ ft}^2$$

$$= 3.14 \times 2 \times 18 \times 10 \times \frac{N_d}{12} = 94.2 N_d;$$

$$N_d = 124$$

Gas pressure drop can be computed as in Chapter 3 and shown to equal 4.5 in. WC. Table 4.14 shows the results. The advantages of using finned tubes are obvious. The finned tube boiler is more compact, has lower gas pressure drop, and should also cost less, considering the labor involved with the smaller number of rows of tubes. It also weighs less.

4.22

Q: Which is the preferred arrangement for finned tubes, in-line or staggered?

A: Both in-line and staggered arrangements have been used with extended surfaces. The advantages of the staggered arrangement are higher overall heat transfer coefficients and smaller surface area; cost could be marginally lower depending on the configuration; gas pressure drop could be higher or lower depending on the gas mass velocity used. If cleaning lanes are required for soot blowing, an in-line arrangement is preferred.

   Both solid and serrated fins are used in the industry. Generally solid fins are used in applications where the deposition of solids is likely.

   The following example illustrates the effect of arrangement on boiler performance.

EXAMPLE

150,000 lb/h of turbine exhaust gases at 1000°F enter an evaporator of a waste heat boiler generating steam at 235 psig. Determine the performance using solid and serrated fins and in-line versus staggered arrangements. Tube size is 2 × 1.77 in. *Solution.*   Using the ESCOA correlations and the methodology discussed above for evaporator performance, the results shown in Table 4.15 were arrived at.

Table 4.15   Comparison Between Staggered and In-Line Designs for Nearly Same Duty and Pressure Drop[a]

	Serrated Fins		Solid Fins	
	In-line	Stagg.	In-line	Stagg.
Fin config.	5 × .75 × .05 × .157		2 × .75 × .05 × 0	
Tubes/row	18	20	18	20
No. of fins deep	20	16	20	16
Length	10	10	10	11
U_o	7.18	8.36	9.75	10.02
ΔP_g	3.19	3.62	1.72	1.42
Q	23.24	23.31	21.68	21.71
Surface	20524	18244	9802	9584

[a]Duty, MM Btu/hr; ΔP_g, in. WC; surface, ft^2; temperature, °F; U_o, Btu/ft^2 hr °F.

4.23

Q: How does the tube-side heat transfer coefficient or fouling factor
     affect the selection of fin configuration such as fin density,
     height, and thickness?

A: Fin density, height, and thickness affect the overall heat transfer
     coefficient as seen in Figure 4.7. However, the tube-side coeffi-
     cient also has an important bearing on the selection of fin con-
     figuration.
        A simple calculation can be done to show the effect of the tube-
     side coefficient on U_o. It was mentioned earlier that the higher
     the tube-side coefficient, the higher the ratio of external to
     internal surface area can be. In other words, it makes no sense to
     use the same fin configuration, say 5 fins/in. fin density, for a
     superheater as for an evaporator.
        Rewriting Eq. (3) based on tube-side area and neglecting other
     resistances,

$$\frac{1}{U_i} = \frac{1}{h_i} + \frac{A_i/A_t}{h_o\eta} \tag{63}$$

     Using the data from Figure 4.7, U_i values have been computed
     for different fin densities and for different h_i values for the
     configuration indicated in Table 4.16. The results are shown in

Table 4.16   Effect of h_i on $U_i{}^{a,b}$

h_i	20		100		1000	
n, fins/in.	2	5	2	5	2	5
G, lb/ft^2 hr	5591	6366	5591	6366	5591	6366
$A_t/(A_i\eta h_o)$	.01546	.00867	.01546	.00867	.01546	.00867
U_o	2.73	1.31	7.03	4.12	11.21	8.38
U_i	15.28	17.00	39.28	53.55	62.66	109
Ratio U_i		1.11		1.363		1.74
RatioΔP_g		1.6		1.3		1.02

[a]Calculations based on 2.0 × 0.105 tubes, 29 tubes/row, 6 ft long, 0.05 in. thick serrated fins; tubes
on 4.0 in. square pitch; fin height = 0.75 in.; gas flow = 150,000 pph; gas inlet temp = 1000°F.
[c]Surface area of 2 fins/in. tube = 2.59 ft^2/ft and for 5 fins/in. = 6.02 ft^2/ft.

Table 4.16. Also shown are the ratio of U_i values between the 5 and 2 fins/in. designs as well as their surface area.

The following conclusions can be drawn [10].

1. As the tube-side coefficient decreases, the ratio of U_i values (between 5 and 2 fins/in.) decreases. With $h_i = 20$, the U_i ratio is only 1.11. With an h_i of 2000, the U_i ratio is 1.74. What this means is that as h_i decreases, the benefit of increasing the external surface becomes less attractive. With 2.325 times the surface area we have only 1.11-fold improvement in U_i. With a higher h_i of 2000, the increase is better, 1.74.

2. A simple estimation of tube wall temperature can tell us that the higher the fin density, the higher the tube wall temperature will be. For the case of $h_i = 100$, with $n = 2$, $U_i = 39.28$, gas temperature $= 900°F$, and fluid temperature of 600°F,

$$\text{Heat flux } q_i = (900 - 600) \times 39.28 = 11,784$$
$$\text{Btu/ft}^2 \text{ hr}$$

The temperature drop across the tube-side film ($h_i = 100$) $= 11,784/100 = 118°F$. The wall temperature $= 600 + 118 = 718°F$.

With $n = 5$, $U_i = 53.55$, $q_i = 53.55 \times 300 = 16,065$ Btu/ft^2 hr. Tube wall temperature $= 600 + (16,065/100) = 761°F$. Note that we are comparing for the same height. The increase in wall temperature is 43°F.

3. The ratio of the gas pressure drop between the 5 and 2 fins/in. designs (after adjusting for the effect of U_i values and differences in surface area for the same energy transfer) increases as the tube-side coefficient reduces. It is 1.6 for $h_i = 20$ and 1.02 for $h_i = 2000$. That is, when h_i is smaller, it is prudent to use a smaller fin surface.

EFFECT OF FOULING FACTORS

The effects of inside and outside fouling factors ff$_i$ and ff$_o$ are shown in Tables 4.17 and 4.18. The following observations can be made.

Table 4.17   Effect of ff_i, Tube-Side Fouling Factor

Fins/in., n	2	2	5	5
U_o, clean	11.21	11.21	8.38	8.38
ff_i	0.001	0.01	0.001	0.01
U_o, dirty	10.54	6.89	7.56	4.01
U_o as %	100	65	100	53

[a]Tube-side coefficient = 2000.

Table 4.18   Effect of ff_o, Outside Fouling Factor

Fins/in., m	2	2	5	5
U_o, clean	11.21	11.21	8.38	8.38
ff_o	0.001	0.01	0.001	0.01
U_o, dirty	11.08	10.08	8.31	7.73
U_o as %	100	91	100	93

[a]Tube-side coefficient = 2000.

1. With a smaller fin density, the effect of ff_i is less. With 0.01 fouling and 2 fins/in., $U_o = 6.89$ compared with 10.54 with 0.001 fouling. The ratio is 0.65. With 5 fins/in., the corresponding values are 4.01 and 7.46, the ratio being 0.53. That means that with increased tube-side fouling, it makes sense to use a lower fin density or smaller ratio of external to internal surface area. The same conclusion was drawn with a smaller tube-side coefficient.

2. The effect of ff_o is less significant, as it is not enhanced by the ratio of external to internal surface area. A review of Eq. (1) tells us that the tube-side heat transfer coefficient or fouling factor is increased by the ratio of the external to internal surface area, and hence its effect is easily magnified.

4.24

Q: Compare the effect of tube-side fouling on bare, low, and high finned tubes.

A: Three boiler evaporators were designed using bare tubes, 2 fins/
in. and 5 fins/in., to cool 150,000 lb/hr of clean flue gases from
1000°F to 520°F. The effect of fouling factors of 0.001 and 0.01
on duty, tube wall temperatures, and steam production are shown
in Table 4.19. The following points may be observed [11].

1.   With bare tubes, the higher tube-side fouling results in the
lowest reduction in duty, from 19.65 to 18.65 MM Btu/hr,
with the exit gas temperature going up to 545°F from
520°F—see columns 1 and 2. With 2 fins/in., the exit gas
temperature increases to 604°F from 520°F, with the duty
reducing to 16.3 from 19.65 MM Btu/hr. The steam generation is about 3200 lb/hr lower. With 5 fins/in., the reduction
in duty and steam generation are the greatest.

2.   The heat flux increases with fin density. Therefore, with
high-temperature units one has to be concerned with DNB
(departure from nucleate boiling) conditions; however, heat
flux decreases because of fouling.

Table 4.19   Effect of Fouling Factors

Case	1	2	3	4	5	6
1. Gas temp in., °F	1000	1000	1000	1000	1000	1000
2. Exit temp, °F	520	545	520	604	520	646
3. Duty, MM Btu/hr	19.65	18.65	19.65	16.30	19.65	14.60
4. Steam flow, lb/hr	19,390	18,400	19,390	1,6110	19,390	14,400
5. ff_i, Ft^2 hr °F/Btu	0.001	.01	.001	.01	.001	.01
6. Heat flux Btu/ft^2 hr	9,314	8,162	35,360	23,080	55,790	30,260
7. Wall temp, °F	437	516	490	680	530	760
8. Fin temp, °F	—	—	730	840	725	861
9. A_t/A_i	1.13	1.13	5.6	5.6	12.3	12.3
10. Fins	bare	bare	(2 × 0.75 ×		(5 × 0.75 ×	
			0.05 × 0.157)		0.05 × 0.157)	
11. Tubes/row	20	20	20	20	20	20
12. No. deep	60	60	16	16	10	10
13. Length, ft	8	8	8	8	8	8
14. Surface area, ft^2	5024	5024	6642	6642	9122	9122
15. Gas Δp, in. WC	3.0	3.1	1.80	1.90	2.0	2.1

3. The tube wall temperature increases significantly with fin density. The same fouling factor results in a much higher tube wall temperature for finned tubes compared to bare tubes. The tube wall temperature increases from 530°F to 760°F with 5 fins/in., while it increases from 437°F to 516°F for bare tubes. The effect of fouling is more pronounced in tubes of high fin density, which means that high fin density tubes have to be kept cleaner than bare tubes. Demineralized water and good water treatment are recommended in such situations.

4.25

Q: How is the weight of solid and serrated fins determined?

A: The weight of fins is given by the formulas

$$W_f = 10.68 \times Fbn \times (d_o + h) \times (h + 0.03)$$
   for solid fins   (64a)

$$W_f = 10.68 \times Fbnd_o \times (h + 0.12)$$
   for serrated fins   (64b)

where

W_f = the fin weight, lb/ft (The segment of width does not affect the weight.)
b = fin thickness, in.
n = fin density, fins/in.
h = fin height, in.
d_o = tube outer diameter, in.

Factor F corrects for material of fins and is given in Table 4.20 [9].

The weight of the tubes has to be added to the fin weight to give the total weight of the finned tube. Tube weight per unit length is given by

$$W_t = 10.68 \times F \times d_m \times t_m$$ (65)

where

d_m = mean diameter of tube, in.
t_m = average wall thickness, in.

Table 4.20   Table of F Factors

Material	F
Carbon steel	1
Type 304, 316, 321 alloys	1.024
Type 409, 410, 430	0.978
Nickel 200	1.133
Inconel 600, 625	1.073
Incoloy 800	1.013
Incoloy 825	1.038
Hastelloy B	1.179

EXAMPLE

Determine the weight of solid carbon steel fins on a 2-in. OD tube if the fin density is 5 fins/in., height = 0.75 in., and thickness = 0.05 in. Average tube wall thickness is 0.120 in. *Solution.* F from Table 4.20 = 1. Using Eq. (64a), we have

$$W_f = 10.68 \times 1 \times 0.05 \times 5 \times (2 + 0.75) \times (0.75 + 0.03) = 5.725 \text{ lb/ft}$$

The tube weight has to be added to this. The tube weight is given by

$$W_t = 10.68 \times 1.94 \times 0.12 = 2.49 \text{ lb/ft}$$

Hence the total weight of the finned tube = 2.49 + 5.725 = 8.215 lb/ft.

4.26

Q: What is the effect of fin thickness and conductivity on boiler performance and tube and fin tip temperatures?

A: Table 4.21 gives the performance of a boiler evaporator using different fins.

2 × 0.120 carbon steel tubes; 26 tubes/row, 14 deep, 20 ft. long.

4 × 0.75 × 0.05 thick solid fins—surface area = 35,831 ft^2

4 × 0.75 × 0.102 thick solid fins—surface area = 36,426 ft^2

Table 4.21   Fin Configuration and Performance

Fin cond. (Btu/ft hr °F)	Fin thickness (in.)	Duty (MM Btu/hr)	Tube temp. (°F)	Fin temp. (°F)	U (Btu/ft^2 hr °F)
25	0.05	104	673	996	8.27
25	0.102	106.35	692	874	9.00
15	0.05	98.35	642	1164	6.78
15	0.102	103.48	670	990	7.98

In-line arrangement, 4 in. square pitch.

Gas flow = 430,000 lb/hr at 1400°F in; % vol CO_2 = 8.2, H_2O = 20.9, N_2 = 67.51, O_2 = 3.1

Steam pressure = 635 psig

Fouling factors = 0.001 ft^2 hr °F/Btu on both gas and steam.

It can be seen that

1. Due to the slightly larger surface area and higher heat transfer coefficient, more duty is transferred with higher fin thickness.
2. Overall heat transfer coefficient is increased owing to higher fin effectiveness for the same fin conductivity and greater fin thickness.
3. Lower fin conductivity reduces the fin effectiveness and the overall heat transfer coefficient U, and hence less duty is transferred.
4. Though fin tip temperature is reduced with greater fin thickness, owing to improved effectiveness, the tube wall temperature increases. This is due to the additional resistance imposed by the larger surface area.

4.27

Q: Is surface area an important criterion for evaluating different boiler designs?

A: The answer is yes if the person evaluating the designs is knowl-
edgeable in heat transfer–related aspects and no if the person
simply compares different designs looking only for surface area
information. As we have seen in the case of fire tube boilers
Q4.11, where, due to variations in tube size and gas velocity,
different designs with over 40 to 50% differences in surface areas
were seen for the same duty. In the case of water tube boilers
also, due to variations in tube size, pitch, and gas velocity, one
can have different surface areas for the same duty; hence one has
to be careful in evaluating boilers based only on surface areas.

   In the case of finned tube boilers, in addition to tube size,
pitch, and arrangement (staggered or in-line), one has to review
the fin configuration—the height, thickness, and fin density. The
higher the fin density or ratio of external to internal surface area,
the lower the overall heat transfer coefficient will be even though
the surface area can be 100 to 200% more. It is also possible to
transfer more duty with less surface area by proper selection of
fin geometry.

EXAMPLE

A superheater is to be designed for the conditions shown in Table
4.22. Study the different designs possible with varying fin configurations.

Table 4.22   Data for HRSG Superheater
Design

Gas flow, pph	200,000
Gas inlet temperature, °F	1200
Gas analysis, % by volume	
Carbon dioxide	7
Water	12
Nitrogen	75
Oxygen	6
Steam flow, pph	100,000
Entering steam temperature, °F	491
Leaving steam pressure, psig	600

Solution.   Using the methods discussed above, various designs were arrived at, with the results shown in Table 4.23 [10]. Several interesting observations can be made. In cases 1 and 2, the same energy of 14.14 MM Btu/hr is transferred; however, the surface area of case 2 is much higher because of the high fin density, which decreases *u*, the overall heat transfer coefficient. Also, the tube wall and fin tip temperatures are higher because of the large ratio of external to internal surface area.

Comparing cases 3 and 4, we see that case 3 transfers more energy with less surface area because of better fin selection. Thus it is not a good idea to select or evaluate designs based on surface area alone, as this can be misleading. In addition, excessive fin surface can lead to higher tube wall and fin tip temperatures, forcing one to use better materials and increasing the cost. Some purchasing managers believe incorrectly that if they can get more surface area for the same price, they are getting a good deal. Nothing could be further from the truth.

Table 4.23   Effect of Fin Geometry on Superheater Performance

	Case 1	Case 2	Case 3	Case 4
Duty, MM Btu/hr	14.14	14.18	17.43	17.39
Leaving steam temperature, °F	689	691	747	747
Gas pressure drop, in. WG	0.65	1.20	1.15	1.37
Leaving gas temperature, °F	951	950	893	892
Fins per in.	2	5	2.5	4
Fin height, in.	0.50	0.75	0.75	0.75
Fin thickness, in.	0.075	0.075	0.075	0.075
Surface area, ft^2	2471	5342	5077	6549
Max tube wall temperature, °F	836	908	905	931
Fin tip temperature, °F	949	1033	1064	1057
Overall heat transfer coefficient, Btu/ft^2 hr °F	11.79	5.50	8.04	6.23
Tube side pressure drop, psi	9.0	6.5	11.0	9.0
Number of rows deep	6	4	7	6

4.28

Q: How are tubular air heaters designed?

A: Let W_g and W_a be the gas and air quantities. Normally, flue gas flows inside the tubes while air flows across the tubes in cross-flow fashion, as shown in Figure 4.8. Carbon steel tubes of 1½ to 3.0 in. OD are generally used. Thickness ranges from 0.06 to 0.09 in. since high pressures are not involved. The tubes are arranged in in-line fashion and are connected to the tube sheets at the ends. More than one block may be used in series; in this case, air flows across the tube bundles with a few turns. Hence, while calculating log-mean temperature difference, we must consider correction factors F_T.

Flue gas velocity is in the range of 40 to 70 fps, while air-side mass velocities range from 4000 to 8000 lb/ft^2 hr. N_w and N_d, the number of tubes wide and deep, can be decided on the basis of duct dimensions leading to the air heater. In the case of a separate heater, we have the choice of N_w or N_d. In a boiler, for example, duct dimensions at the economizer section fix dimensions of the air heater also, since the air heater is located below the economizer.

To size the air heater, first determine the total number of tubes N_t: [1]

$$N_t = \frac{0.05\ W_g}{d_i^2\ \rho_g\ V_g} \tag{65}$$

S_T/d and S_L/d range from 1.25 to 2.0. For the gas-side heat transfer coefficient h_i, Eq. (12) is used:

$$h_i = 2.44 \times w^{0.8} \times \frac{C}{d_i^{1.8}}$$

Values of C are evaluated at average flue gas temperature.

Air-side heat transfer coefficient h_o is given by Eq. (19) (variation in h_o between staggered and in-line arrangement is small in the range of Reynolds number and pitches one comes across),

$$h_o = 0.9 \times G^{0.6} \times \frac{F}{d^{0.4}}$$

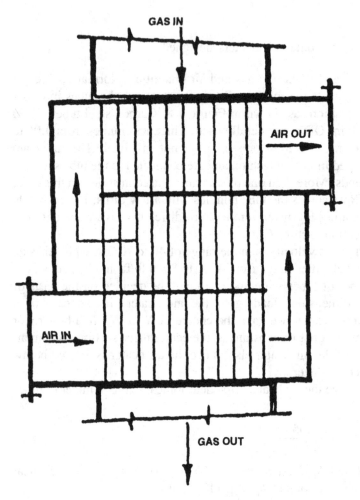

Figure 4.8 Tubular air heater.

The value h_o is calculated at air-film temperature.

Since the temperature drops across the gas and air films are nearly the same, unlike in an evaporator or superheater, film temperature is approximated as

$$t_f = (3 \, t_g + t_a)/4 \tag{66}$$

where t_g and t_a refer to the average of gas and air temperatures. Calculate U using

$$\frac{1}{U} = \left[\frac{1 \times d}{h_i d_i} + \frac{1}{h_o} \right] \tag{67}$$

Metal resistance is neglected. Air- and gas-side pressure drops can be computed by Eqs. (26) and (28) of Chapter 3, after surfacing is done:

$$\Delta P_g = 93 \times 10^{-6} \times f \, w^2 \, \frac{L + 5d_i}{\rho_g d_i^5}$$

$$\Delta P_{air} = 9.3 \times 10^{-10} \times f \times G^2 \, \frac{N_d}{\rho_{air}}$$

It is also good to check for partial-load performance to see if dew point corrosion problems are likely. Methods like air bypass or steam–air heating must be considered. Vibration of tube bundles also must be checked.

C and F are given in Table 4.24 for easy reference.

EXAMPLE

A quantity of 500,000 lb/hr of flue gas from a boiler is cooled from 700°F; 400,000 lb/hr of air at 80°F is heated to 400°F. Design a suitable tubular air heater. Carbon steel tubes of 2 in. OD and 0.087-in. thickness are available.

Solution. Assume that duct dimensions are not a limitation. Hence, the bundle arrangement is quite flexible. Choose S_T/d

Table 4.24 C and F Factors for Calculating h_i and h_o of Tubular Air

Temp. (°F)	C	F
200	0.162	0.094
400	0.172	0.103
600	0.18	0.110
800	0.187	0.116

$= 1.5$ and $S_L/d = 1.25$ in. in-line; use a maximum flue-gas velocity of 50 ft/sec.

From energy balance, assuming negligible losses and for a specific heat of 0.25 for gas and 0.24 for the air side,

$$Q = 500{,}000 \times 0.25 \times (700 - t) = 400{,}000 \times 0.24$$
$$\times (400 - 80) = 30.7 \times 10^6 \text{ Btu/hr}$$

Hence, the gas temperature leaving the air heater is 454°F. The average flue gas temperature is $(700 + 454)/2 = 577°F$. Let the molecular weight of the flue gas be 30. Then

$$\rho_g = \frac{30}{359} \times \frac{492}{460 + 577} = 0.0396 \text{ lb/cu ft}$$

From Eq. (65),

$$N_t = \frac{0.05 \times 500{,}000}{1.826^2 \times 0.0396 \times 50} = 3800$$

$$S_T = 3.0 \text{ in.}, \qquad S_L = 2.5 \text{ in.}$$

Let $N_w = 60$. Hence, the width of the air heater is

$$60 \times \frac{3.0}{12} = 15 \text{ ft}$$

$N_d = 63$ as $N_t = N_w \times N_d$, so

Depth $= 63 \times 2.5/12 = 13.2$ ft

At 577°F, from Table 4.24 we have $C = 0.178$:

$$h_i = 2.44 \times \left(\frac{500{,}000}{3780}\right)^{0.8} \times \frac{0.178}{(1.826)^{1.8}}$$

$$= 7.2 \text{ Btu/ft}^2 \text{ hr } °F$$

To estimate h_o, G is required. This requires an idea of L. We must assume a value for the length and check later to see if it is sufficient. Hence, it is a trial-and-error approach. Try $L = 15$ ft:

$$FGA = \frac{S_T - d}{12} \times N_w L = \frac{1}{12} \times 60 \times 15 = 75 \text{ ft}^2$$

$$G = 400{,}000/75 = 5333 \text{ lb/ft}^2 \text{ hr}$$

Average gas and air temperatures are

$$t_g = 577°F, \qquad t_a = 240°F,$$

$$t_f = \frac{3 \times 577 + 240}{4} = 492°F$$

From Table 4.24, F is 0.105. Then

$$h_o = 0.9 \times 5333^{0.6} \times 0.105/2^{0.4} = 12.3 \text{ Btu/ft}^2 \text{ hr }°F$$

$$U1 = \frac{1}{7.2} \times \frac{2.0}{1.826} + \frac{1}{12.3}$$

$$= 0.152 + 0.081 = 0.233$$

$$U = 4.3 \text{ Btu/ft}^2 \text{ hr }°F$$

We must calculate F_T, the correction factor for ΔT, for the case of one fluid mixed and other unmixed. From Figure 4.9 (single-pass crossflow),

$$R = \frac{700 - 454}{400 - 80} = 0.77$$

$$P = \frac{400 - 80}{700 - 80} = 0.516$$

$$F_T = 0.9$$

Therefore,

$$\Delta T = 0.9 \times \frac{(454 - 80) - (700 - 400)}{\ln (374/300)} = 302°F$$

$$A = \frac{Q}{U \times \Delta T} = \frac{30.7 \times 10^6}{4.3 \times 302} = 23,641 \text{ ft}^2$$

$$= \frac{\pi \times 2}{12} \times 3780 \, L$$

$$L = 11.95 \text{ ft}$$

Hence, the assumed L is not correct. Try $L = 11.0$ ft.

$$FGA = \frac{11}{15} \times 75 = 55 \text{ ft}^2$$

$$G = 7272 \text{ lb/ft}^2 \text{ hr}$$

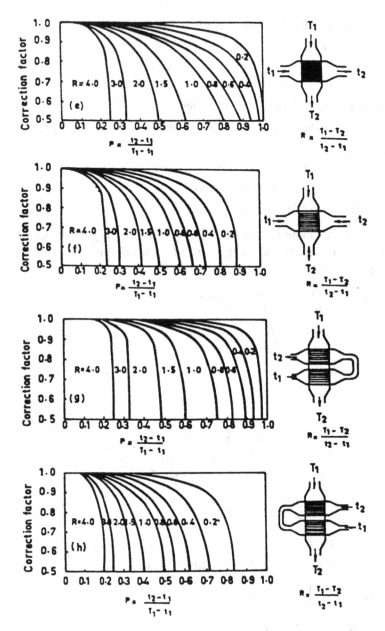

Figure 4.9   Crossflow correction factors for log-mean temperature difference. [1,2]

Taking ratios,

$$h_o = \left(\frac{7272}{5333} \right)^{0.6} \times 12.3 = 14.8 \text{ Btu/ft}^2 \text{ hr } °F$$

$$\frac{1}{U} = \frac{1 \times 2.0}{7.2 \times 1.826} + \frac{1}{14.8} = 0.152 + 0.067$$

$$= 0.219$$

$$U = 4.56 \text{ Btu/ft}^2 \text{ hr } °F$$

$$A = \frac{30.7 \times 10^6}{4.56 \times 302} = 22,293 \text{ ft}^2, \qquad L = 11.25 \text{ ft}$$

The calculated and assumed lengths are close to each other, and the design may be frozen. Check the metal temperature at the exit portion. Since the gas-side resistance and air-film resistances are 0.152 and 0.067, the metal temperature at the exit of the air heater can be calculated as follows. The drop across the gas film will be

$$\frac{0.152(454 - 80)}{0.152 + 0.067} = 260°F$$

Metal temperature will be $454 - 260 = 194°F$.

If the flue gas contains sulfur, dew point corrosion may occur at the exit. Since the air-side heat transfer coefficient is high, the drop across its film is low compared to the gas-side film drop. If we increase the flue gas heat transfer coefficient, the drop across its film will be low and the metal temperature will be higher.

4.29

Q: How is the off-design performance evaluated?

The air heater described in Q4.28 works at partial loads. W_g = 300,000 lb/hr, and flue gas enters the air heater at 620°F. W_a = 250,000 lb/hr, and the air temperature is 80°F. Check the exit gas temperatures of gas and air.

A: Assume the gas leaves the air heater at 400°F. Then

$$Q = 300{,}000 \times 0.25 \times (620 - 400) = 250{,}000$$
$$\times\ 0.24\ (t - 80) = 16.5 \times 10^6$$

Air temperature leaving $= 355°F$

To calculate h_i and h_o, see Table 4.24. At an average flue gas temperature of

$$\frac{620 + 400}{2} = 510°F, \qquad C = 0.175$$

And at a film temperature of $[3 \times 510 + (355 + 80)/2]/4$ $= 437°F$, $F = 0.104$

$$h_i = 2.44 \times \left(\frac{300{,}000}{3825}\right)^{0.8} \times \frac{0.175}{1.826^{1.8}}$$
$$= 4.75\ \text{Btu/ft}^2\ \text{hr °F}$$

$$G = \frac{250{,}000}{75} = 3333\ \text{lb/ft}^2\ \text{hr}$$

$$h_o = 0.9 \times \frac{(3333)^{0.6}}{2^{0.4}}$$
$$\times\ 0.104 = 9.22\ \text{Btu/ft}^2\ \text{hr °F}$$
$$\frac{1}{U} = \frac{2}{1.826 \times 4.73} + \frac{1}{9.22} = 0.238$$
$$U = 4.22\ \text{Btu/ft}^2\ \text{hr °F}$$

From Figure 4.9;

$$P = \frac{355 - 80}{620 - 80} = 0.51,$$

$$R = \frac{620 - 400}{355 - 80} = 0.8$$

$$F_T = 0.9$$

$$\Delta T = 0.9 \times \frac{(400 - 80) - (620 - 355)}{\ln\ (320/265)} = 262°F$$

Transferred $Q = 4.2 \times 262 \times 23{,}640 = 26 \times 10^6$ Btu/hr, and assumed $Q = 16.5 \times 10^6$ Btu/hr. They don't tally.

Since an air heater can transfer more energy, assume a higher air temperature, 390°F, at the exit:

$$Q = 250,000 \times 0.24 \, (390 - 80) = 18.6 \times 10^6$$
$$= 300,000 \times 0.25 \times (620 - t)$$

Then gas temperature leaving $= 372°F$

Assume U remains the same at 4.2 Btu/ft^2 hr °F. Then

$$R = 0.8, \qquad P = 0.574, \qquad F_T = 0.82,$$
$$\Delta T = 213°F$$

Transferred $Q = 4.2 \times 23,640 \times 213 = 21.1$
$$\times 10^6 \text{ Btu/hr}$$

Again, they don't tally. Next, try $Q = 20 \times 10^6$ Btu/hr.

Air temperature leaving $= 410°F$

Gas temperature leaving $= 353°F$

$$F_T = 0.75, \qquad \Delta T = 0.75 \times 242 = 182°F$$

Transferred $Q = 4.2 \times 23,640 \times 182 = 18 \times 10^6$ Btu/hr

Again, try an exit air temperature at 400°F. Then

$$Q = 250,000 \times 0.24 \times (400 - 80) = 19.2$$
$$\times 10^6 \text{ Btu/hr}$$

$$\text{Exit gas temperature} = 620 - \frac{19.2 \times 10^6}{300,000 \times 0.25}$$

$$= 364°F$$

$$R = 0.8, \qquad P = \frac{320}{540} = 0.593, \qquad F_T = 0.77$$

$$\Delta T = 0.77 \times \frac{284 - 220}{\ln \, (284/220)} = 193°F$$

Transferred $Q = 4.2 \times 193 \times 23,640 = 19.16 \times 10^6$ Btu/hr

$$Q = 19.2 \times 10^6 \text{ Btu/hr}$$

The gas leaves at 364°F against 454°F at full load.

Metal temperature can be computed as before. At lower loads, metal temperature is lower, and the air heater should be given some protection. This protection may take two forms: Bypass part of the air or use steam to heat the air entering the heater to

100 to 120°F. Either of these will increase the average metal temperature of the air heater. In the first case, the air-side heat transfer coefficient will fall. Because U decreases, the gas temperature leaving the air heater will increase and less Q will be transferred. Hence, metal temperature will increase. In the second case, since air temperature entering increases, protection of the metal is ensured. Again, the gas temperature differential at the exit will be higher, causing a higher exit gas temperature.

EXAMPLE

Solve the problem using the NTU method.

Solution. Often the NTU method is convenient when trial-and-error calculations of the type shown above are involved.

$$NTU = \frac{UA}{C_{min}} = \frac{4.2 \times 23,640}{250,000 \times 0.24} = 1.65$$

$$\frac{C_{mixed}}{C_{unmixed}} = \frac{250,000 \times 0.24}{300,000 \times 0.25} = 0.80$$

$$\epsilon = \text{effectiveness} = 1 - \exp\left\{ - \frac{C_{max}}{C_{min}} \right.$$

$$\left. [1 - \exp(-NTU \times C)] \right\} \tag{68}$$

$$= 1 - \exp\{ - 1.25$$
$$\times [1 - \exp(-1.65 \times 0.8)]\} = 0.59$$

$$\text{Effectiveness} = 0.59 = \frac{\text{air temperature rise}}{620 - 80}$$

Air temperature rise = 319°F

Air temperature leaving = 319 + 80 = 399°F

This compares well with the answer 400°F. When U does not change much, this method is very handy.

4.30

Q: Predict the exit gas and water temperatures and the energy transferred in an economizer under the following conditions:

t_{g1} = gas temperature in = 1000°F

t_{w1} = water temperature in = 250°F
A = surface area = 6000 ft^2
W_g = gas flow = 75,000 lb/hr
W_w = water flow = 67,000 lb/hr
U = overall heat transfer coefficient = 8 Btu/ft^2 hr °F
C_{pg} = gas specific heat = 0.265 Btu/lb °F
C_{pw} = water specific heat = 1 Btu/lb °F

A: Figure 4.10 shows the arrangement of an economizer. A trial-
and-error method is usually adopted to solve for the duty of any
heat transfer equipment if the surface area is known. This procedure is detailed in Q4.29. Alternatively, the *number of transfer
units* (NTU) *method* predicts the exit temperatures and duty. For
more on this theory, the reader is referred to any textbook on heat
transfer [2]. Basically, the duty Q is given by

$$Q = \epsilon C_{\min}(t_{g1} - t_{w1}) \tag{69}$$

where ϵ depends on the type of flow, whether counterflow,
parallel flow, or crossflow. In economizers, usually a counterflow arrangement is adopted. ϵ for this is given by

$$\epsilon = \frac{1 - \exp\left[-\text{NTU} \times (1 - C)\right]}{1 - C \exp\left[-\text{NTU} \times (1 - C)\right]} \tag{70}$$

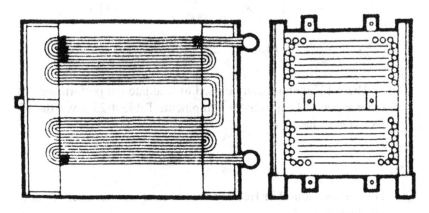

Figure 4.10   Economizer.

where

$$NTU = \frac{UA}{C_{min}} \quad \text{and} \quad C = \frac{(WC_p)_{min}}{(WC_p)_{max}}$$

$$(WC_p)_{min} = 75{,}000 \times 0.265 = 19{,}875$$

$$(WC_p)_{max} = 67{,}000 \times 1 = 67{,}000$$

$$C = \frac{19{,}875}{67{,}000} = 0.3$$

$$NTU = 8 \times \frac{6000}{19{,}875} = 2.42$$

Substituting into Eq. (70) yields

$$\epsilon = \frac{1 - \exp(-2.42 \times 0.7)}{1 - 0.3 \times \exp(-2.42 \times 0.7)} = 0.86$$

From Eq. (69),

$$Q = 0.86 \times 19{,}875 \times (1000 - 250) = 12.8$$
$$\times 10^6 \text{ Btu/hr}$$

Let us calculate the exit water and gas temperatures.

$$Q = W_w C_{pw}(t_{w2} - t_{w1}) = W_g C_{pg}(t_{g1} - t_{g2})$$

Hence,

$$t_{w2} = 250 + 12.8 \times \frac{10^6}{67{,}000 \times 1} = 441°F$$

$$t_{g2} = 1000 - 12.8 \times \frac{10^6}{75{,}000 \times 0.265} = 355°F$$

The NTU method can be used to evaluate the performance of other types of heat transfer equipment, Table 4.25 gives the effectiveness factor ϵ.

4.31

Q: How is the natural or free convection heat transfer coefficient in air determined?

Table 4.25 Effectiveness Factors

Exchanger type	Effectiveness
Parallel flow, single-pass	$\epsilon = \dfrac{1 - \exp[-NTU \times (1 + C)]}{1 + C}$
Counterflow, single-pass	$\epsilon = \dfrac{1 - \exp[-NTU \times (1 - C)]}{1 - C\exp[-NTU \times (1 - C)]}$
Shell-and-tube (one shell pass; 2, 4, 6, etc., tube passes)	$\epsilon_1 = 2\left[1 + C + \dfrac{1 + \exp[-NTU \times (1 + C^2)^{1/2}]}{1 - \exp[-NTU \times (1 + C^2)^{1/2}]}(1 + C^2)^{1/2}\right]^{-1}$
Shell-and-tube (n shell passes; $2n$, $4n$, $6n$, etc., tube passes)	$\epsilon_n = \left[\left(\dfrac{1 - \epsilon_1 C}{1 - \epsilon_1}\right)^n - 1\right]\left[\left(\dfrac{1 - \epsilon_1 C}{1 - \epsilon_1}\right)^n - C\right]^{-1}$
Crossflow, both streams unmixed	$\epsilon \approx 1 - \exp\{C \times NTU^{0.22}[\exp(-C \times NTU^{0.78}) - 1]\}$
Crossflow, both streams mixed	$\epsilon = NTU\left[\dfrac{NTU}{1 - \exp(-NTU)} + \dfrac{NTU \times C}{1 - \exp(-NTU \times C)} - 1\right]^{-1}$
Crossflow, stream C_{min} unmixed	$\epsilon = C\{1 - \exp[-C[1 - \exp(-NTU)]]\}$
Crossflow, stream C_{max} unmixed	$\epsilon = 1 - \exp\{-C[1 - \exp(-NTU \times C)]\}$

A: The situations of interest to steam plant engineers would be those involving heat transfer between pipes or tubes and air as when an insulated pipe runs across a room or outside it and heat transfer can take place with the atmosphere.

Simplified forms of these equations are the following [12].

1. Horizontal pipes in air:

$$h_c = 0.5 \times \left(\frac{\Delta T}{d_o}\right)^{0.25} \tag{71a}$$

where

ΔT = temperature difference between the hot surface and cold fluid, °F

d_o = tube outside diameter, in.

2. Long vertical pipes:

$$h_c = 0.4 \times \left(\frac{\Delta T}{d_o} \right)^{0.25} \tag{71b}$$

3. Vertical plates less than 2 ft high:

$$h_c = 0.28 \left(\frac{\Delta T}{z} \right)^{0.25} \tag{71c}$$

where

z = height, ft

4. Vertical plates more than 2 ft high:

$$h_c = 0.3 \times (\Delta T)^{0.25} \tag{71d}$$

5. Horizontal plates facing upward:

$$h_c = 0.38 \times (\Delta T)^{0.25} \tag{71e}$$

6. Horizontal plates facing downward:

$$h_c = 0.2 \times (\Delta T)^{0.25} \tag{71f}$$

EXAMPLE

Determine the heat transfer coefficient between a horizontal bare pipe of diameter 4.5 in. at 500°F and atmospheric air at 80°F. *Solution.*

$$h_c = 0.5 \times \left(\frac{500 - 80}{4.5} \right)^{0.25} = 1.55 \text{ Btu/ft}^2 \text{ hr °F}$$

Note that the above equations have been modified to include the effect of wind velocity in the insulation calculations; see Q4.51.

4.32

Q: How is the natural or free convection heat transfer coefficient between tube bundles and liquids determined?

A: One has to determine the free convection heat transfer coefficient when tube bundles such as desuperheater coils or drum preheat coils are immersed in boiler water in order to arrive at the overall heat transfer coefficient and then the surface area. Drum coil

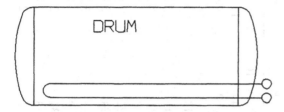

Figure 4.11   Exchanger inside boiler drum.

desuperheaters are used instead of spray desuperheaters when solids are not permitted to be injected into steam. The heat exchanger is used to cool superheated steam (Figure 4.11), which flows inside the tubes, while the cooler water is outside the tubes in the drum. Drum heating coils are used to keep boiler water hot for quick restart or to prevent freezing.

In this heat exchanger, steam condenses inside tubes while the cooler water is outside the tubes. The natural convection coefficient between the coil and drum water has to be determined to arrive at the overall heat transfer coefficient and then the size or surface area.

The equation that relates h_c with other parameters is [2]

$$\text{Nu} = 0.54 \left[\frac{d^3 \rho^2 g \beta \ \Delta T}{\mu^2} \times \frac{\mu C_p}{k} \right]^{0.25} \tag{72}$$

Simplifying the above we have

$$h_c = 144 \times \left(k^3 \times \frac{\rho^2 \beta \ \Delta T \ C_p}{\mu d_o} \right)^{0.25} \tag{73}$$

where

d_o = tube outer diameter, in.
k = fluid thermal conductivity, Btu/ft hr °F
C_p = fluid specific heat, Btu/lb °F
β = volumetric expansion coefficient, °R^{-1}
ΔT = temperature difference between tubes and liquid, °F
μ = viscosity of fluid, lb/ft hr
ρ = fluid density, lb/ft^3

Table 4.26 Properties of Saturated Water

t (°F)	C_p (Btu/lb °F)	ρ (lb ft³)	μ (lb/ft hr)	ν (ft²/hr)	k (Btu/hr ft °F)	α (ft²/hr)	β (°R⁻¹) $\times 10^{-3}$	N
32	1.009	62.42	4.33	0.0694	0.327	0.0052	0.03	13.37
40	1.005	62.42	3.75	0.0601	0.332	0.0053	0.045	11.36
50	1.002	62.38	3.17	0.0508	0.338	0.0054	0.070	9.41
60	1.000	62.34	2.71	0.0435	0.344	0.0055	0.10	7.88
70	0.998	62.27	2.37	0.0381	0.349	0.0056	0.13	6.78
80	0.998	62.17	2.08	0.0334	0.355	0.0057	0.15	5.85
90	0.997	62.11	1.85	0.0298	0.360	0.0058	0.18	5.13
100	0.997	61.99	1.65	0.0266	0.364	0.0059	0.20	4.52
110	0.997	61.84	1.49	0.0241	0.368	0.0060	0.22	4.04
120	0.997	61.73	1.36	0.0220	0.372	0.0060	0.24	3.65
130	0.998	61.54	1.24	0.0202	0.375	0.0061	0.27	3.30
140	0.998	61.39	1.14	0.0186	0.378	0.0062	0.29	3.01
150	0.999	61.20	1.04	0.0170	0.381	0.0063	0.31	2.72
160	1.000	61.01	0.97	0.0159	0.384	0.0063	0.33	2.53
170	1.001	60.79	0.90	0.0148	0.386	0.0064	0.35	2.33

180	1.002	60.57	0.84	0.0139	0.389	0.0064	0.37	2.16
190	1.003	60.35	0.79	0.0131	0.390	0.0065	0.39	2.03
200	1.004	60.13	0.74	0.0123	0.392	0.0065	0.41	1.90
210	1.005	59.88	0.69	0.0115	0.393	0.0065	0.43	1.76
220	1.007	59.63	0.65	0.0109	0.395	0.0066	0.45	1.66
230	1.009	59.38	0.62	0.0104	0.395	0.0066	0.47	1.58
240	1.011	59.10	0.59	0.0100	0.396	0.0066	0.48	1.51
250	1.013	58.82	0.56	0.0095	0.396	0.0066	0.50	1.43
260	1.015	58.51	0.53	0.0091	0.396	0.0067	0.51	1.36
270	1.017	58.24	0.50	0.0086	0.396	0.0067	0.53	1.28
280	1.020	57.94	0.48	0.0083	0.396	0.0067	0.55	1.24
290	1.023	57.64	0.46	0.0080	0.396	0.0067	0.56	1.19
300	1.026	57.31	0.45	0.0079	0.395	0.0067	0.58	1.17
350	1.044	55.59	0.38	0.0068	0.391	0.0067	0.62	1.01
400	1.067	53.65	0.33	0.0062	0.384	0.0068	0.72	0.91
450	1.095	51.55	0.29	0.0056	0.373	0.0066	0.93	0.85
500	1.130	49.02	0.26	0.0053	0.356	0.0064	1.18	0.83
550	1.200	45.92	0.23	0.0050	0.330	0.0060	1.63	0.84
600	1.362	42.37	0.21	0.0050	0.298	0.0052	—	0.96

In Eq. (73) all the fluid properties are evaluated at the mean temperature between fluid and tubes except for the expansion coefficient, which is evaluated at the fluid temperature.

Fluid properties at saturation conditions are given in Table 4.26.

EXAMPLE

1-in. pipes are used to maintain boiler water at 100°F in a tank using steam at 212°F, which is condensed inside the tubes. Assume that the pipes are at 200°F, and estimate the free convection heat transfer coefficient between pipes and water.

Solution. From Table 4.26, at a mean temperature of 150°F,

$$k = 0.381, \quad \mu = 1.04, \quad \beta = 0.0002, \quad \rho_f = 61.2$$
$$C_p = 1.0, \quad \Delta T = 100, \quad d_o = 1.32$$
$$h_c = 144 \times \left(0.381^3 \times \frac{61.2^2 \times 1.0 \times 0.0002 \times 100}{1.04 \times 1.32} \right)^{0.25}$$
$$= 188 \text{ Btu/ft}^2 \text{ hr } °F$$

4.33

Q: Estimate the surface area of the heat exchanger required to maintain water in a boiler at 100°F using steam at 212°F as in the example of Q4.32. Assume that the heat loss to the cold ambient from the boiler is 0.5 MM Btu/hr. Steam is condensed inside the tubes. 1-in. schedule 40 pipes are used.

A: The overall heat transfer coefficient can be estimated from

$$\frac{1}{U_o} = \frac{1}{h_o} + \frac{1}{h_i} + R_m + \text{ff}_i + \text{ff}_o$$

where R_m = metal resistance, and ff_i and ff_o are inside and outside fouling factors; see Eq. (3).

h_o, the free convection heat transfer coefficient between the tubes and boiler water, obtained from Q4.32, = 188 Btu/ft^2 hr °F. Assume $h_i = 1500$, $\text{ff}_i = \text{ff}_o = 0.001$, and

$$\text{Metal resistance } R_m = \frac{d_o}{24K} \ln \left(\frac{d_o}{d_i} \right) = 0.0005$$

Then

$$\frac{1}{U_o} = \frac{1}{188} + \frac{1}{1500} + 0.0025 = 0.00849,$$

or $U_o = 117$ Btu/ft^2 hr °F

$\Delta T = $ log-mean temperature difference $= 212 - 100 = 112$ °F.
Then,

$$\text{Surface area } A = \frac{Q}{U_o \Delta T} = \frac{500,000}{117 \times 112} = 38 \text{ ft}^2.$$

4.34

Q: Can we determine gas or steam temperature profiles in a heat recovery steam generator (HRSG) without actually designing it?

A: Yes. One can stimulate the design as well as the off-design performance of an HRSG without designing it in terms of tube size, surface area, etc. The methodology has several applications. Consultants and plant engineers can determine for a given set of gas inlet conditions for an HRSG how much steam can be generated, what the gas/steam temperature profile will look like, and hence write better specifications for the HRSG or select auxiliaries based on this simulation without going to a boiler firm for this information. Thus several options can be ruled out or ruled in depending on the HRSG performance. The methodology has applications in complex, multipressure cogeneration or combined cycle plant evaluation with gas turbines. More information on HRSG simulation can be found in refs. 8, 11, and 12.

EXAMPLE
140,000 lb/hr of turbine exhaust gases at 980°F enter an HRSG generating saturated steam at 200 psig. Determine the steam generation and temperature profiles if feedwater temperature is 230°F and blowdown $= 5\%$. Assume that average gas specific heat is 0.27 at the evaporator and 0.253 at the economizer.

Two important terms that determine the design should be defined here (see Figure 4.12). *Pinch point* is the difference between the gas temperature leaving the evaporator and satura-

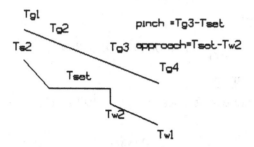

Figure 4.12 Pinch and approach points.

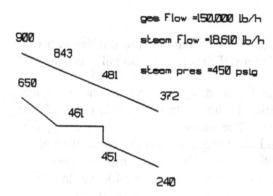

Figure 4.13 Temperature profile in an HRSG.

tion temperature. *Approach point* is the difference between the saturation temperature and the water temperature entering the evaporator. More information on how to select these important values and how they are influenced by gas inlet conditions is available in the references.

For unfired gas turbine HRSGs, pinch and approach points lie in the range of 15 to 30°F. The higher these values, the smaller will be the boiler size and cost, and vice versa.

Let us choose a pinch point of 20°F and an approach point of 15°F. Saturation temperature = 388°F. Figure 4.13 shows the temperature profile. The gas temperature leaving the evaporator = 388 + 20 = 408°F, and water temperature entering it = 388 − 15 = 373°F.

Evaporator duty $= 140,000 \times 0.99 \times 0.27 \times (980$
$$- 408) = 21.4 \text{ MM Btu/hr}$$

(0.99 is the heat loss factor with a 1% loss.)

Enthalpy absorbed by steam in evaporator $= (1199.3$
$$- 345) + 0.05 \times (362.2 - 345)$$
$$= 855.2 \text{ Btu/lb}$$

(1199.3, 345, and 362.2 are the enthalpies of saturated steam, water entering the evaporator, and saturated water, respectively. 0.05 is the blowdown factor for 5% blowdown.)

Hence

$$\text{Steam generated} = \frac{21.4 \times 10^6}{855.2} = 25,000 \text{ lb/hr}$$

Economizer duty $= 25,000 \times 1.05 \times (345 - 198.5)$
$$= 3.84 \text{ MM Btu/hr}$$

$$\text{Gas temperature drop} = \frac{3,840,000}{140,000 \times 0.253 \times 0.99}$$
$$= 109°F$$

Hence gas temperature leaving economizer $= 408 - 109$ $= 299°F$. Thus the thermal design of the HRSG is simulated.

4.35

Q: Simulate the performance of the HRSG designed in Q4.34 when a gas flow of 165,000 lb/hr enters the HRSG at 880°F. The HRSG will operate at 150 psig. Feedwater temperature remains at 230°F.

A: Gas turbine exhaust flow and temperature change with ambient conditions and load. As a result the HRSG has to operate at different gas parameters, and hence simulation is necessary to determine how the HRSG behaves under different gas and steam parameters.

The evaporator performance can be determined using Eq. (34). Based on design conditions, compute K.

$$\ln \left[\frac{980 - 388}{408 - 388} \right] = K \times (140{,}000)^{-0.4} = 3.388;$$

$$K = 387.6$$

Under the new conditions,

$$\ln \left[\frac{880 - 366}{t_{g2} - 366} \right] = 387.6 \times (165{,}000)^{-0.4} = 3.1724$$

Hence $t_{g2} = 388°F$.

Evaporator duty $= 165{,}000 \times 0.99 \times 0.27 \times (880$
$$- 388) = 21.70 \text{ MM Btu/hr}$$

In order to estimate the steam flow, the feedwater temperature leaving the economizer must be known. This is arrived at through a series of iterations. Try $t_{w2} = 360°F$. Then

$$\text{Steam flow} = \frac{21.70 \times 10^6}{(1195.7 - 332) + 0.05 \times (338.5 - 332)}$$
$$= 25{,}110 \text{ lb/hr}$$

Economizer assumed duty $Q_a = 25{,}110 \times 1.05 \times (332$
$$- 198.5) = 3.52 \text{ MM Btu/hr}$$

Compute the term $(US)_{\text{design}} = Q/\Delta T$ for economizer based on design conditions.

$$Q = 3.84 \times 10^6 \quad \text{and}$$

$$\Delta T = \frac{(299 - 230) - (408 - 373)}{\ln (69/35)} = 50°F$$

Hence $(US)_{\text{design}} = 3{,}840{,}000/50 = 76{,}800$. Correct this for off-design conditions.

$$(US)_{\text{perf}} = (US)_{\text{design}} \times \left(\frac{\text{gas flow, perf}}{\text{gas flow, design}} \right)^{0.65}$$

$$= 76{,}800 \times \left(\frac{165{,}000}{140{,}000} \right)^{0.65} = 85{,}200$$

The economizer transferred duty is then $(US)_{\text{perf}} \times \Delta T$. Based on 360°F water leaving the economizer, $Q_a = 3.52$ MM Btu/hr and the exit gas temperature is

$$t_{g2} = \frac{3,520,000}{165,000 \times 0.99 \times 0.253} = 85°F$$

Hence $t_{g2} = 388 - 85 = 303°F$, and

$$\Delta T = \frac{(303 - 230) - (388 - 360)}{\ln (73/28)} = 47°F.$$

Transferred duty $Q_t = 85,200 \times 47 = 4.00$ MM Btu/hr

Since the assumed and transferred duty to not match, another iteration is required. We can show that at duty of 3.55 MM Btu/hr the assumed and transferred duty match. Water temperature leaving eco = 366°F (saturation); exit gas temperature = 301°F. Steam generation = 25,310 lb/hr.

Since the calculations are quite involved, I have developed a software program called HRSGS that can simulate the design and off-design performance of complex, multipressure fired and unfired HRSGs. More information can be had by writing to V. Ganapathy, P.O. Box 673, Abilene, Texas 79604.

4.36

Q: Can we assume that a particular exit gas temperature can be obtained in gas turbine HRSGs without doing a temperature profile analysis?

A: No. It is not good practice to assume the HRSG exit gas temperature and compute the duty or steam generation as some consultants and engineers do. The problem is that, depending on the steam pressure and temperature, the exit gas temperature will vary significantly. Often, consultants and plant engineers assume that any stack gas temperature can be achieved. For example, I have seen catalogs published by reputable gas turbine firms suggesting that 300°F stack gas temperature can be obtained irrespective of the steam pressure or parameters. Now this may be possible at low pressures but not at all steam conditions. In order to arrive at the correct temperature profile, several heat balance calculations have to be performed, as explained below.

It will be shown that one cannot arbitrarily fix the stack gas temperature or the pinch point.

Looking at the superheater and evaporator of Figure 4.12,

$$W_g \times C_{pg} \times (T_{g1} - T_{g3}) = W_s \times (h_{so} - h_{w2}) \qquad (74)$$

Looking at the entire HRSG,

$$W_g \times C_{pg} \times (T_{g1} - T_{g4}) = W_s \times (h_{so} - h_{w1}) \qquad (75)$$

Blowdown was neglected in the above equations for simplicity. Dividing (74) by (75) and neglecting variations in C_{pg}, we have

$$\frac{T_{g1} - T_{g3}}{T_{g1} - T_{g4}} = \frac{h_{so} - h_{w2}}{h_{so} - h_{w1}} = X \qquad (76)$$

Factor X depends only on steam parameters and approach point used. T_{g3} depends on the pinch point selected. Hence if T_{g1} is known, T_{g4} can be calculated.

It can be concluded from the above analysis that one cannot assume that any HRSG exit gas temperature can be obtained. To illustrate, Table 4.27 shows several operating steam conditions and X values and exit gas temperatures. As the steam pressure or steam temperature increases, so does the exit gas temperature, with the result that less energy is transferred to steam. This also tells us why we need to go in for multiple-pressure-level HRSGs when the main steam pressure is high. Note that even with

Table 4.27   HRSG Exit Gas Temperatures[a]

Pressure (psig)	Steam temp. (°F)	Sat. temp. (°F)	X	Exit gas temp. (°F)
100	sat	338	0.904	300
150	sat	366	0.8754	313
250	sat	406	0.8337	332
400	sat	448	0.7895	353
400	600	450	0.8063	367
600	sat	490	0.740	373
600	750	492	0.7728	398

[a]Based on 15°F approach point, 20°F pinch point, 900°F gas inlet temperature, and no blowdown. Feedwater temperature is 230°F. Similar data can be generated for other conditions.

infinite surface areas we cannot achieve low temperatures, as this is a thermodynamic limitation.

EXAMPLE 1

Determine the HRSG exit gas temperature when the gas inlet temperature is 900°F and the steam pressure is 100 psig sat. *Solution.* $X = 0.904$. Saturation temp $= 338°F$. Hence with a 20°F pinch point, $T_{g3} = 358°F$, and $t_{w2} = 323°F$ with a 15°F approach point.

$$\frac{900 - T_{g4}}{900 - 358} = 0.904, \quad \text{or} \quad T_{g4} = 300°F$$

EXAMPLE 2

What is T_{g4} when steam pressure is 600 psig and temperature is 750°F? *Solution.* $X = 0.7728$. Saturation temperature $= 492°F$; $t_{w2} = 477°F$; $T_{g3} = 512°F$.

$$\frac{900 - 512}{900 - T_{g4}} = 0.7728, \quad \text{or} \quad T_{g4} = 398°F$$

So a 300°F stack temperature is not thermodynamically feasible. Let us see what happens if we try to achieve that.

EXAMPLE 3

Can you obtain 300°F stack gas temperature with 900°F inlet gas temperature and at 600 psig, 750°F and 15°F approach temperature? *Solution.* $X = 0.7728$. Let us see, using Eq. (76), what T_{g3} results in a T_{g4} of 300°F, as that is the only unknown.

$(900 - T_{g3})/(900 - 300) = 0.7728$, or $T_{g3} = 436°F$, which is not thermodynamically feasible as the saturation temperature at 615 psig is 492°F! This is the reason one has to be careful in specifying HRSG exit gas temperatures or computing steam generation based on a particular exit gas temperature.

EXAMPLE 4

What should be done to obtain a stack gas temperature of 300°F in the situation described in Example 3? *Solution.* One of the options is to increase the gas inlet temperature to the HRSG by supplementary firing. If T_{g1} is increased, then it is possible to get a lower T_{g4}. Say $T_{g1} = 1600°F$. Then

$$\frac{1600 - T_{g3}}{1600 - 300} = 0.7728, \quad \text{or } T_{g3} = 595°F$$

This is a feasible temperature as the pinch point is now $(595 - 492) = 103°F$. This brings us to another important rule: *Pinch point and exit gas temperature cannot be arbitrarily selected in the fired mode*. It is preferable to analyze the temperature profiles in the unfired mode and evaluate the off-design performance using available simulation methods [8].

EXAMPLE 5

If gas inlet temperature in Example 1 is 800°F instead of 900°F, what happens to the exit gas temperature at 100 psig sat? *Solution.*

$$\frac{800 - 358}{800 - T_{g4}} = 0.904,$$

or $T_{g4} = 312°F$ versus 300°F when the inlet gas temperature was 900°F. We note that the exit gas temperature increases when the gas inlet temperature decreases, and vice versa. This is another important basic fact.

Once the exit gas temperature is arrived at, one can use Eq. (75) to determine how much steam can be generated.

4.37

Q: How can HRSG simulation be used to optimize gas and steam temperature profiles?

A: HRSG simulation is a method of arriving at the design or off-design performance of HRSGs without physically designing them as shown in Q4.34. By using different pinch and approach points and different configurations, particularly in multipressure HRSGs, one can maximize heat recovery. We will illustrate this with an example [8, 12].

EXAMPLE

A gas turbine exhausts 300,000 lb/hr of gas at 900°F. It is desired to generate about 20,500 lb/hr of superheated steam at 600 psig and 650°F and as much as 200 psig saturated steam using feedwater at 230°F. Using the method discussed in Q4.34, we can arrive at the gas/steam temperature profiles and steam flows. Figure 4.14 shows results obtained with HRSGS software. In option 1, we have the high-pressure (HP) section consisting of the superheater, evaporator, and economizer followed by the low-pressure (LP) section consisting of the LP evaporator and economizer. By using a pinch point of 190°F and approach point of 15°F, we generate 20,438 lb/hr of high-pressure steam at

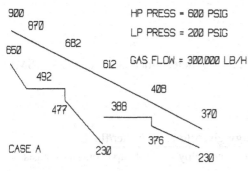

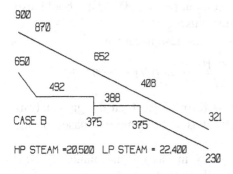

Figure 4.14　Optimizing temperature profiles.

650°F. Then, using a pinch point of 20°F and approach of 12°F, we make 18,670 lb/hr low-pressure steam. The stack gas temperature is 370°F. In option 2, we have the HP section consisting of the superheater and evaporator and the LP section consisting of only the evaporator. A common economizer feeds both the HP and LP sections with feedwater at 375°F. Due to the larger heat sink available beyond the LP evaporator, the stack gas temperature reduces to 321°F. The HP steam generation is adjusted using the pinch point to make 20,488 lb/hr while the LP steam is allowed to float. With a pinch point of 20°F, we see that we can make 22,400 lb/hr in comparison with the 18,673 lb/hr earlier. The ASME system efficiency is much higher now. Thus by manipulating the HRSG configuration, one can maximize the heat recovery.

4.38

Q: How is the HRSG efficiency determined according to ASME Power Test Code 4.4?

A: The efficiency E is given by

$$E = \frac{\text{energy given to steam/water/fluids}}{\text{gas flow} \times \text{inlet enthalpy} + \text{fuel input on LHV basis}}$$

To evaluate the efficiency, the enthalpy of the turbine exhaust gas should be known. The Appendix gives the enthalpy based on a particular gas analysis. Fuel input on an LHV basis should also be known if auxiliary firing is used.

In Q4.37 the efficiency in the design case is

$$E = \frac{(21.4 + 3.84) \times 10^6}{140,000 \times 242} = 0.715, \text{ or } 71.5\%$$

If steam or water injection is resorted to, then the gas analysis will change, and the enthalpy has to be computed based on the actual analysis.

The HRSG system efficiency in gas turbine plants will improve with the addition of auxiliary fuel, which increases the gas temperature to the HRSG and hence increases its steam generation. There are two reasons for this.

1. Addition of auxiliary fuel reduces the effective excess air in the exhaust gases, as no air is added, only fuel. Hence the exhaust gas loss in relation to steam production is reduced.
2. With increased steam generation, usually the HRSG exhaust gas temperature decreases. This is due to the increased flow of water in the economizer, which offers a larger heat sink, which in turn pulls down the gas temperature further. In gas turbine units, the gas flow does not vary much with steam output as in conventional steam generators, which accounts for the larger temperature drop.

More information on HRSG temperature profiles can be found in Ref. 8.

Table 4.28 shows the performance of an HRSG under different operating conditions. Case 1 is the unfired case, while cases 2 and 3 have different firing conditions. It can be seen that the system efficiency is higher when more fuel is fired, for reasons explained above.

Table 4.28 Data for Supplementary Fired Boiler[a]

	Case 1	Case 2	Case 3
Gas flow, pph	150,000	150,000	150,000
Inlet gas temp., °F	900	900	900
Firing temperature, °F	900	1290	1715
Burner duty, LHV, MM Btu/hr	0	17.30	37.60
Steam flow, pph	22,780	40,000	60,000
Steam pressure, psig	200	200	200
Feedwater temp., °F	240	240	240
Exit gas temp., °F	327	315	310
Steam duty, MM Btu/hr	22.67	39.90	59.90
System efficiency %	68.7	79.2	84.90

[a]Gas analysis (in % *vol*) $CO_2 = 3$, $H_2O = 7$, $N_2 = 75$, $O_2 = 15$; blowdown = 3%.

4.39

Q: In some cogeneration plants with gas turbines, a forced draft fan is used and to send atmospheric air to the HRSG into which fuel is fired to generate steam when the gas turbine is not in operation. What should be the criteria for the fan size?

A: The air flow should be large enough to have turbulent flow regimes in the HRSG and at the same time be small enough to minimize the loss due to exiting gases. If the air flow is high, the firing temperature will be low, but the system efficiency will be lower and the fuel input will be higher. This is illustrated for a simple case of two fans generating 130,000 and 150,000 lb/hr of air flow in the HRSG discussed above. The HRSGS program was used in the simulation. See Table 4.29 [8].

It can be seen that though the firing temperature is higher with the smaller fan, the efficiency is higher due to the lower exit gas losses considering the lower mass flow and exit gas temperature. It should be noted that as the firing temperature increases, the exit gas temperature will decrease when an economizer is used. Also, with the smaller fan the initial and operating costs are lower. One should ensure that the firing temperature does not increase to the point of changing the basic design concept of the HRSG. For example, insulated casing design is used up to 1700°F firing temperature, beyond which a water-cooled membrane wall design is required. See Ref. 8.

Table 4.29   Fresh Air Firing Performance

Air flow, pph	130,000	150,000
Inlet temp., °F	60	60
Firing temp., °F	1424	1294
Exit gas temp., °F	308	314
Steam flow, pph	40,000	40,000
Burner duty, MM Btu/hr	48.72	50.32
Efficiency, %	81.8	79.25

4.40

Q: How does one evaluate the operating costs of an HRSG? Would an HRSG that generates more steam but costs more be preferred to a lower cost alternative that generates less steam?

A: Let us illustrate the situation with the design of two HRSGs for gas turbine exhaust application, the two having different pinch points. A larger pinch point naturally means a smaller boiler and hence lower cost. The performance of both options is unfired and fired modes when 30,000 lb/hr of steam is to be generated is presented in Table 4.30. The larger boiler, design B, has a larger gas pressure drop but a smaller fuel input to generate the same amount of steam in the fired mode. Let us assume that each boiler operates for 50% of the time in both the unfired and fired modes.

Table 4.30   Performance of Alternative Designs

	Design A		Design B	
	Unfired	Fired	Unfired	Fired
Gas flow, pph	150,000		150,000	
Gas temp to evap, °F	900	1086	900	1062
Temp to eco, °F	407	419	388	393
Stack temp, °F	332	329	309	302
Gas d_p, in. WC	4.20	4.60	5.40	5.80
Steam (150 psig sat)	22,107	30,000	22,985	30,000
Feedwater temp, °F	230		230	
Temp to evap, °F	351	337	352	340
Burner duty, MM Btu/hr	0	8.10	0	6.90
Gas turbine output, kW (nominal)	4500		4500	
Surface area, evap, ft^2	13,227		16,534	
eco, ft^2	5948		8922	
Pinch point, °F	41	53	22	27
Approach point, °F	15	29	14	26

Let fuel cost = $2.7/MM Btu (LHV), cost of steam = $3/1000 lb, and electricity 5¢/kWh. Assume 8000 hr of operation per year. Assume also that an additional 4 in. WC of gas pressure drop is equivalent to 1.1% decrease in gas turbine output, which is a nominal 4500 kW.

Design B has the following edge over design A in operating costs:

Due to higher steam generation in unfired mode:

$$(22{,}985 - 22{,}107) \times 3 \times \frac{4000}{1000} = \$10{,}536$$

Due to lower fuel consumption in fired mode:

$$(8.1 - 6.9) \times 2.7 \times 4000 = \$12{,}960$$

Due to higher gas pressure drop of 1.2 in. WC:

$$1.1 \times \frac{1.2}{4} \times 4500 \times 0.05 \times \frac{8000}{100} = -\$5940$$

Hence the net benefit of design B over A = (10,536 + 12,960 − 5940) = $17,556 per year. If the additional cost of design B is $35,000, the payback is less than 2 years. Detailed economic evaluation may be done to select the option. If the HRSG operates for only, say, 3000 hr per year, option B will take a much longer time to pay back and hence may not be viable.

4.41

Q: What is steaming, and why is it likely in gas turbine HRSGs and not in conventional fossil fuel–fired boilers?

A: When the economizer in a boiler or HRSG starts generating steam, particularly with downward flow of water, problems can arise in the form of water hammer, vibration, etc. With upward water flow design, a certain amount of steaming, 3 to 5%, can be tolerated as the bubbles have a natural tendency to go up along with the water. However, steaming should generally be avoided. To understand why the economizer is likely to steam, we should first look at the characteristics of a gas turbine as a function of ambient temperature and load (see Table 4.31).

In single-shaft machines, which are widely used, as the ambient temperature or load decreases, the exhaust gas temperature decreases. The variation in mass flow is marginal compared to fossil fuel–fired boilers, while the steam or water flow drops off significantly. (The effect of mass flow increase in most cases does not offset the effect of lower exhaust gas temperature.) The energy-transferring ability of the economizer, which is governed by the gas-side heat transfer coefficient, does not change much with gas turbine load or ambient temperature, and hence nearly the same duty is transferred with a smaller water flow through the economizer, which results in a water exit temperature approaching saturation temperature as seen in Q4.35. Hence we should design the economizer such that it does not steam in the lowest unfired ambient case, which will ensure that steaming does not occur at other ambient conditions. A few other steps may also be taken, such as designing the economizer [8] with a horizontal gas flow with horizontal tubes (Figure 4.15). This ensures that the last few rows of the economizer, which are likely to steam, have a vertical flow of steam/water mixture.

In conventional fossil fuel–fired boilers the gas flow decreases in proportion to the water flow, and the energy-transferring ability of the economizer is also lower at lower loads. Hence steaming is not a concern in these boilers; usually the approach point increases at lower loads in fired boilers, while it is a concern in HRSGs.

4.42

Q: Why are water tube boilers generally preferred to fire tube boilers for gas turbine exhaust applications?

A: Fire tube boilers require a lot of surface area to reduce the temperature of gas leaving the evaporator to within 15 to 25°F of saturation temperature (pinch point). They have lower heat transfer coefficients than those of bare tube water tube boilers (see

Table 4.31 Typical Exhaust Gas Flow, Temperature Characteristics of a Gas Turbine

Ambient Temp.	(F)	20.0	40.0	59.0	80.0	100.0	120.0
Power Output	(KW)	38150	38600	35020	30820	27360	24040
Heat Rate	(BTU/KW.HR)	9384	9442	9649	9960	10257	10598
Water Flow Rate	(Lb/hr)	16520	17230	15590	13240	10540	6990
Turbine Inlet Temp.	(F)	1304	1363	1363	1363	1363	1363
Exhaust Temp.	(F)	734	780	797	820	843	870
Exhaust Flow	(Lb/sec)	312	304	286	264	244	225

Fuel: Natural Gas, Elevation: Sea Level, Relative Humidity: 60%, Inlet Loss: 4 Inch H_2O, Exhaust Loss: 15 Inch H_2O, Speed: 3600 rpm, Output Terminal: Generator

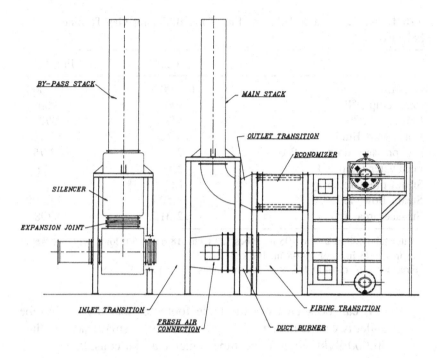

Figure 4.15   Horizontal gas flow economizer.

Q4.10), which do not compare well with finned tube boilers. Water tube boilers can use extended surfaces to reduce the pinch point to 15 to 25°F in the unfired mode and hence be compact. The tubes will be very long if fire tube boilers are used, and hence the gas pressure drop will be higher. (A fire tube boiler can be made into a two-pass boiler to reduce the length; however, this will increase the shell diameter and the labor cost, as twice the number of tubes will have to be welded to the tube sheets.) The fire tube boiler will have to be even larger if the same gas pressure drop is to be maintained. Table 4.32 compares the performance of water tube and fire tube boilers for the same duty and pressure drop.

Table 4.32   Water Tube vs. Fire Tube Boiler for Gas Turbine
Exhaust

	Water tube[a]	Fire tube[b]
Gas flow, lb/hr	100,000	100,000
Inlet temp., °F	900	900
Exit temp., °F	373	373
Duty, MM Btu/hr	13.72	13.72
Gas pressure drop, in. WC	2.75	2.75
Feedwater temp., °F	220	220
Steam pressure, psig	125	125
Steam flow, lb/hr	13,500	13,500
Surface area, ft^2	12,315	9798

[a]Water tube boiler: 2 × 0.105 in. tubes, 20 wide, 18 deep, 6 ft long, with 5 serrated
fins/in., 0.75 in. high, 0.05 in. thick.
[b]Fire tube boiler: 1400 1.5 × 0.105 in. tubes, 21 ft long

It can be seen from the table footnotes that the water tube
boiler is very compact. If the gas flow is very small, say less than
50,000 lb/hr, then a fire tube boiler may be considered.

4.43

Q: Does addition of 10% surface area to a boiler increase its duty
    by 10%?

A: No. The additional surface area increases the duty only slightly.
    The increased temperature drop across the boiler and the tem-
    perature rise of water or steam (if single-phase) due to the higher
    duty results in a lower log-mean temperature difference. This
    results in lower transferred duty, even assuming that the overall
    heat transfer coefficient U remains unchanged. If the larger
    surface area results in lower gas velocities, the increase in duty
    will be marginal as U is further reduced.
        As an example, consider the performance of a fire tube boiler
    with 10% and 20% increase in surface area as shown in Table
    4.33. As can be seen, a 10% increase in surface area increases
    the duty by only 3%, and a 20% increase in surface area in-

Table 4.33 Boiler Performance with Increased Surface Area[a]

Case	No. of tubes	Length (ft)	Surface (ft²)	Duty (MM Btu/hr)	Exit gas temp. (°F)
1	390	16	2839	20.53	567
2	390	17.6	3123	21.16	533
3	390	19.2	3407	21.68	505

[a]Gas flow = 70,000 lb/hr; inlet gas temperature = 1600°F. Gas analysis (% vol): CO_2 = 7, H_2O = 12, N_2 = 75, O_2 = 6; steam pressure = 125 psig saturated. Tubes: 2 × 0.120 carbon steel.

creases the duty by only 6%. Similar trends may be shown for water tube boilers, superheaters, economizers, etc.

4.44

Q: How do we estimate the time required to heat a boiler?

A: A boiler can take a long time to get heated, depending on the initial temperature of the system, mass of steel, and amount of water stored. The following procedure gives a quick estimate of the time required to warm up a boiler. The methodology is applicable to either fire tube or water tube boilers.

Gas at a temperature of T_{g1} enters the unit, which is initially at a temperature of t_1 (both the water and the boiler tubes/metal). The following energy balance equation can then be written neglecting heat losses:

$$M_c \frac{dt}{dz} = W_g C_{pg} \times (T_{g1} - T_{g2}) = UA \, \Delta T \qquad (77)$$

where

$\quad M_c$ = water equivalent of the boiler
$\qquad$ = mass of steel × specific heat of steel + mass of water × specific heat of water
$\qquad$ (Weight of the boiler tubes, drum, casing, etc., is included in the steel weight.)
$\quad dt/dz$ = rate of change of temperature, °F/hr

$$W_g = \text{gas flow, lb/hr}$$

$$C_{pg} = \text{gas specific heat, Btu/lb °F}$$

$$T_{g1}, T_{g2} = \text{entering and exit boiler gas temperature, °F}$$

$$U = \text{overall heat transfer coefficient, Btu/ft}^2 \text{ hr°F}$$

$$A = \text{surface area, ft}^2$$

$$\Delta T = \text{log-mean temperature difference, °F}$$

$$= \frac{(T_{g1} - t) - (T_{g2} - t)}{\ln\,[(T_{g1} - t)/(T_{g2} - t)]}$$

$$t = \text{temperature of the water/steam in boiler, °F}$$

From (77) we have

$$\ln\left[\frac{T_{g1} - t}{T_{g2} - t}\right] = \frac{UA}{W_g C_{pg}} \tag{78}$$

or

$$T_{g2} = t + \frac{T_{g1} - t}{e\,UA/W_g C_{pg}} = t + \frac{T_{g1} - t}{K} \tag{79}$$

Substituting (79) into (77), we get

$$M_c\,\frac{dt}{dz} = W_g C_{pg}\,(T_{g1} - t) \times \frac{K - 1}{K}$$

or

$$\frac{dt}{T_{g1} - t} = \frac{W_g C_{pg}}{M_c} \times \frac{K - 1}{K}\,dz \tag{80}$$

To estimate the time to heat up the boiler from an initial temperature t_1 to t_2, we have to integrate dt between the limits t_1 and t_2.

$$\ln\left[\frac{T_{g1} - t_1}{T_{g1} - t_2}\right] = \frac{W_g C_{pg}}{M_c} \times \frac{(K - 1)z}{K} \tag{81}$$

The above equation can be used to estimate the time required to heat up the boiler from a temperature of t_1 to t_2, using flue gases entering at T_{g1}. However, in order to generate steam, we must first bring the boiler to the boiling point at atmospheric pressure and slowly raise the steam pressure through manipulation of vent valves, drains, etc; the first term of Eq. (77) would involve the term for steam generation and flow in addition to metal heating.

EXAMPLE

A water tube waste heat boiler of weight 50,000 lb and containing 30,000 lb of water is initially at a temperature of 100°F. 130,000 lb of flue gases at 1400°F enter the unit. Assume the following:

Gas specific heat = 0.3 Btu/lb °F
Steel specific heat = 0.12 Btu/lb °F
Surface area of boiler = 21,000 ft²
Overall heat transfer coefficient = 8 Btu/ft² hr °F

Estimate the time required to bring the boiler to 212°F.

Solution

$$\frac{U}{W_g C_{pg}} = \frac{8 \times 21,000}{130,000 \times 0.3} = 4.3$$

$$K = e^{4.3} = 74$$

$$M_c = 50,000 \times 0.12 + 30,000 \times 1 = 36,000$$

$$\ln\left[\frac{1400 - 100}{1400 - 212}\right] = 0.09 = \frac{130,000 \times 0.3}{36,000} \times \frac{73}{74} \times z$$

or $z = 0.084$ hr = 5.1 min.

One could develop a computer program to solve (77) to include steam generation and pressure-raising terms. In real-life boiler operation, the procedure is corrected by factors based on operating data of similar units.

It can also be noted that, in general, fire tube boilers with the same capacity as water tube boilers would have a larger water equivalent and hence the startup time for fire tube boilers would be larger.

4.45

Q: Discuss the parameters influencing the test results of an HRSG during performance testing.

A: The main variables affecting the performance of an HRSG are the gas flow, inlet gas temperature, gas analysis, and steam parameters. Assuming that an HRSG has been designed for a given set of gas conditions, in reality several of the parameters could be different at the time of testing. In the case of a gas turbine HRSG in particular, ambient temperature also influences the exhaust gas conditions. The HRSG could, as a result, be receiving a different gas flow at a different temperature, in which case the steam production would be different from that predicted.

Even if the ambient temperature and the gas turbine load were to remain the same, it is difficult to ensure that the HRSG receives the design gas flow at the design temperature. This is due to instrument errors. Typically, in large ducts, the gas measurement could be off by 3 to 5% and the gas temperatures could differ by 10 to 20°F according to ASME Power Test Code 4.4. As a result it is possible that the HRSG is receiving 5% less flow at 10°F lower gas temperature than design conditions, although the instruments record design conditions. As a result, the HRSG steam generation and steam temperature would be less than predicted through no fault of the HRSG design. Figure 4.16 shows the performance of an HRSG designed for 500,000 lb/hr gas flow at 900°F; steam generation is 57,000 lb/hr at 650 psig and 750°F. The chart shows how the same HRSG behaves when the mass flow changes from 485,000 to 515,000 lb/hr while the exhaust temperature varies from 880°F to 902°F. The steam temperature falls to 741°F with 880°F gas temperature, while it is 758°F at 920°F. The steam flow increases from 52,900 to 60,900 lb/hr as the gas mass flow increases. Thus the figure shows the map of performance of the HRSG for possible instrument error variations only. Hence HRSG designers and plant users should mutually agree upon possible variations in gas parameters and their influence on HRSG performance before conducting such tests.

4.46

Q: Estimate the boiling heat transfer coefficient inside tubes for water and the tube wall temperature rise for a given heat flux and steam pressure.

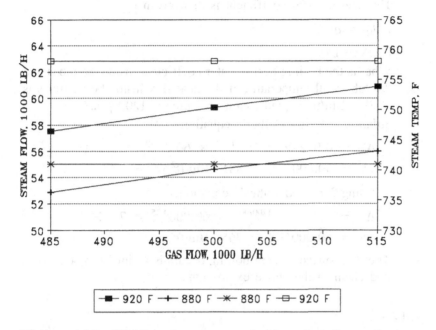

Figure 4.16 HRSG performance as a function of gas flow and exhaust temperature.

A: Subcooled boiling heat transfer coefficient inside tubes for water can be estimated by the following equations:
According to Collier [13],

$$\Delta T = 0.072e^{-P/1260} q^{0.5} \tag{82a}$$

According to Jens and Lottes [13],

$$\Delta T = 1.9e^{-P/900} q^{0.25} \tag{82b}$$

where

ΔT = difference between saturation temperature and tube wall temperature, °F

P = steam pressure, psia

q = heat flux inside tubes, Btu/ft² hr

The heat transfer coefficient is then given by

$$h_i = q/\Delta T$$

EXAMPLE

What is the boiling heat transfer coefficient inside the tubes and the tube wall temperature if the heat flux inside boiler tubes is 60,000 Btu/ft^2 hr and steam pressure $= 1200$ psia?

Solution. Using Collier's equation,

$$\Delta T = 0.072 \times e^{-1200/1260} \times 60,000^{0.5} = 6.8°F$$
$$h_i = 60,000/6.8 = 8817 \text{ Btu/ft}^2 \text{ hr } °F$$

Using Jens and Lottes's equation,

$$\Delta T = 1.9 \times e^{-1200/900} \times 60,000^{0.25} = 7.8°F$$
$$h_i = 60,000/7.8 = 7650 \text{ Btu/ft}^2 \text{ hr } °F$$

The tube surface where boiling occurs is assumed to be smooth and clean in the above expressions.

4.47a

Q: What is the relation among critical heat flux, steam pressure, quality, and flow in water tube boilers?

A: Several variables influence the critical heat flux or the departure from nucleate boiling (DNB) condition. These are

Steam pressure
Mass velocity of mixing inside the tubes
Steam quality
Tube roughness and cleanliness
Tube size and orientation

Correlations such as the Macbeth correlation are available in the literature [13]. The Macbeth correlation is

$$q_c = 0.00633 \times 10^6 \times h_{fg}\, d_i^{-0.1}\, (G_i/10^6)^{0.51} \times (1 - x) \quad (83)$$

where

q_c = critical heat flux, Btu/ft^2 hr
h_{fg} = latent heat of steam, Btu/lb

G_i = mass velocity inside tubes, lb/ft^2 hr
x = steam quality, expressed as a fraction
d_i = tube inner diameter, in.

EXAMPLE

Estimate the critical heat flux under the following conditions: steam pressure = 1000 psia, tube inner diameter = 1.5 in. mass velocity = 600,000 lb/ft^2 hr, and steam quality = 0.20.

$$q_c = 0.00633 \times 10^6 \times 650 \times 1.50^{-0.1} \times 0.6^{0.51}$$
$$\times (1 - 0.2) = 2.43 \times 10^6 \text{ Btu/ft}^2 \text{ hr}$$

In real-life boilers, the allowable heat flux to avoid DNB is much lower, say 20 to 30% lower, than the values obtained by laboratory tests under controlled conditions due to factors such as roughness of tubes, water quality, and safety considerations. Boiler suppliers have their own data and design boilers accordingly.

4.47b

Q: How is the critical heat flux q_c determined in pool boiling situations as in fire tube boilers?

A: Several correlations are available in the literature, but only two will be cited. Motsinki suggests the simple equation [13]

$$q_c = 803P_c \times \left(\frac{P_s}{P_c}\right)^{0.35} \times \left(\frac{1 - P_s}{P_c}\right)^{0.9} \tag{84a}$$

where P_s, P_c are the steam pressure and critical pressure, both in psia.

Zuber's correlation takes the form [13]

$$\frac{q_c}{\rho_g h_{fg}} = 0.13 \times \left(\frac{\sigma(\rho_f - \rho_g)gg_o}{\rho_g^2}\right)^{0.25}$$
$$\times \left(\frac{\rho_f}{\rho_g + \rho_f}\right)^{0.5} \tag{84b}$$

where

σ = surface tension

ρ = density
h_{fg} = latent heat
g, g_o = acceleration due to gravity and conversion factor
g and force units

all in metric units.

EXAMPLE

Determine the critical heat flux for steam at 400 psia under pool boiling conditions.

Solution. The following data can be obtained from steam tables:

Saturation temperature at 400 psia = 445°F
Density of liquid = 51 lb/cu ft (827 kg/m³)
Density of vapor = 0.86 lb/cu ft (13.8 kg/m³)
Latent heat of vaporization = 780 Btu/lb (433 kcal/kg)

From Table 4.26 at saturation temperature of 445°F, surface tension is 0.0021 lb$_f$/ft (0.31 kg$_f$/m)

$$g = 9.8 \times 3600^2 \text{ m/hr}^2, \qquad g_o = 9.8 \times 3600^2$$
$$\text{kg}_m/\text{kg}_f \text{ hr}^2$$

Substituting into (84a):

$$q_c = 803 \times 3208 \times \left(\frac{400}{3208}\right)^{0.35} \times \left(1 - \frac{400}{3208}\right)^{0.9}$$

$$= 1.102 \text{ MM Btu/ft}^2 \text{ hr}$$

Using Eq. (84b),

$$q_c = 13.8 \times 433 \times 0.13 \times \left(0.0031 \times 813 \times 9.8^2\right.$$

$$\times \left.\frac{3600^4}{(13.8)^2}\right)^{0.25} \times \left(\frac{827}{827 + 13.8}\right)^{0.5}$$

$$= 2.95 \times 10^6 \text{ kcal/m}^2 = 1.083 \text{ MM Btu/ft}^2 \text{ hr}$$

Again, as before, factors such as surface roughness, water quality, scale formation, and bundle configuration play a role, and for conservative estimates, boiler designers use a value that is 20 to 30% of these values.

4.48

Q: Discuss the simplified approach to designing fire tube boilers.

A: Engineers often must estimate the size of heat transfer equipment, such as heat exchangers, gas coolers, boilers, and economizers, for preliminary costing and to check space requirements. With the approach presented here, one can quickly determine one or more configurations to accomplish a certain amount of heat transfer. One can also size equipment so as to limit the pressure drop, without performing lengthy calculations. Life-cycle costing can then be applied to select the optimum design.

Several situations will be discussed [8].

The tube-side heat transfer coefficient governs the overall heat transfer. Examples: fire tube boilers; gas coolers; heat exchangers in which a medium such as air or flue gas flows on the tube side and a fluid with a high heat transfer coefficient flows on the outside. Phase changes can also occur on the outside of the tubes.

The shell side governs. Examples: water-tube boilers, steam-air exchangers, and gas–liquid heat transfer equipment. See Q4.49.

TUBE-SIDE TRANSFER GOVERNS

In a fire tube boiler, gas flows inside the tubes; a steam–water mixture flows on the outside. The gas heat transfer coefficient is small, about 10 to 20 Btu/ft^2 hr °F, compared to the outside coefficient of 2000 to 3000 Btu/ft^2 hr °F. The metal resistance is also small, and hence the gas-side coefficient governs the overall coefficient and the size of the equipment.

The energy transferred is given by

$$Q = UA \, \Delta T = W_i C_p \times (T_1 - T_2) \qquad (85)$$

The overall heat transfer coefficient is obtained from Eq. (4),

$$\frac{1}{U} = \frac{d_o}{h_i d_i} + \frac{1}{h_o} + \frac{d}{24K_m} \ln\left(\frac{d_o}{d_i}\right)$$

$$+ \text{ff}_i \times \left(\frac{d_o}{d_i} \right) + \text{ff}_o$$

Since the inside coefficient governs U, we can rewrite Eq. (4) as follows (neglecting lower-order resistances, such as h_o, metal resistance, and fouling factors, which contribute to about 5% of U):

$$U = 0.95 h_i \times \frac{d_i}{d_o} \tag{86}$$

The value of the tube-side coefficient is obtained from the familiar Dittus-Boelter equation, Eq. (8),

$$\text{Nu} = 0.023 \ \text{Re}^{0.8} \ \text{Pr}^{0.4}$$

where

$$\text{Nu} = \frac{h_i d_i}{12k}, \qquad \text{Re} = 15.2 \ \frac{w}{d_i \mu}$$

The fluid transport properties are evaluated at the bulk temperature.

Substituting Eqs. (8) to (11) into Eq. (86) and simplifying, we have the following expression [Eq. (12)]:

$$h_i = 2.44 w^{0.8} F_1 / d_i^{1.8}$$

where

$$F_1 = (C_p/\mu)^{0.4} k^{0.6} \tag{87}$$

Combining Eqs. (86), (85), and (87), we have, after substituting $A = 3.14 d_i LN/12$, and for flow per tube $w = W_i/N$,

$$\frac{Q}{\Delta T \ F_1 W_i^{0.8}} = 0.606 \times \frac{LN^{0.2}}{d_i^{0.8}} \tag{88}$$

This simple equation relates several important variables. Given Q, ΔT, W_i, and F_1, one can try combinations of L, d_i, and N to arrive at a suitable configuration. Also, for given thermal data, $LN^{0.2}/d_i^{0.8}$ is constant in Eq. (88).

F_1 is shown in Table 4.34 for flue gas and air. For other gases, F_1 can be computed from Eq. (87).

Table 4.34 Factors F_1/C_p, F_2/C_p F_2 and F_3 for Air and Flue Gas[a]

Temp. (°F)	F_1/C_p	F_2	F_2/C_p	F_3
Air				
100	0.6660	0.0897	0.3730	0.5920
200	0.6870	0.0952	0.3945	0.6146
300	0.7068	0.1006	0.4140	0.6350
400	0.7225	0.1056	0.4308	0.6528
600	0.7446	0.1150	0.4591	0.6810
1000	0.7680	0.1220	0.4890	0.6930
1200	0.7760	0.1318	0.5030	0.7030
		0.1353		0.7150
Flue gas[a]				
200	0.6590	0.0954	0.3698	0.5851
300	0.6780	0.1015	0.3890	0.6059
400	0.6920	0.1071	0.4041	0.6208
600	0.7140	0.1170	0.4300	0.6457
800	0.7300	0.1264	0.4498	0.6632
1000	0.7390	0.1340	0.4636	0.6735
1200	0.7480	0.1413	0.4773	0.6849

[a]Flue gas is assumed to have 12% water vapor by volume.

When a phase change occurs, as in a boiler, ΔT is written as

$$\Delta T = \frac{(T_1 - t_s) - (T_2 - t_s)}{\ln[(T_1 - t_s) - (T_2 - t_s)]} \tag{89}$$

Combining Eqs. (88) and (89) and simplifying, we arrive at the expression

$$\ln\left[\frac{T_1 - t_s}{T_2 - t_s}\right] = 0.606 \times \frac{F_1}{C_p} \times N^{0.2} \times \frac{L}{W_i^{0.2} d_i^{0.8}} \tag{90}$$

Factor F_1/C_p is also given in Table 4.34.

Equation (90) relates the major geometric parameters to thermal performance. Using this method, one need not evaluate heat transfer coefficients.

GAS PRESSURE DROP

Now, consider gas pressure drop. The equation that relates the geometry to tube-side pressure drop in in. H_2O is

$$\Delta P_i = 9.3 \times 10^{-5} f \times \left(\frac{W_i}{N}\right)^2 (L + 5d_i) \times \frac{v}{d_i^5}$$

$$= 9.3 \times 10^{-5} \times \left(\frac{W_i}{N}\right)^2 K_2 v \qquad (91)$$

where

$$K_2 = f(L + 5d_i)/d_i^5 \qquad (92)$$

Combining Eqs. (90) to (92) and eliminating N,

$$\ln\left(\frac{T_1 - t_s}{T_2 - t_s}\right) = 0.24 \times \frac{F_1}{C_p} \times K_1 \frac{v^{0.1}}{\Delta P_i^{0.1}} \qquad (93)$$

where

$$K_1 = (L + 5d_i)^{0.1} L f^{0.1}/d_i^{1.3} \qquad (94)$$

K_1 and K_2 appear in Tables 4.35 and 4.36, respectively, as a function of tube ID and length. In the turbulent range, the

Table 4.35   Values of K_1 as a Function of Tube Diameter and Length

L (ft)	\multicolumn{9}{c}{d_i (in.)}
	1.00	1.25	1.50	1.75	2.00	2.25	2.50	2.75	3.00
8	7.09	5.33	4.22	3.46	2.92	2.52	2.20	1.95	1.75
10	8.99	6.75	5.34	4.38	3.70	3.17	2.78	2.46	2.21
12	10.92	8.20	6.48	5.31	4.48	3.85	3.36	2.98	2.67
14	12.89	9.66	7.63	6.25	5.27	4.53	3.95	3.50	3.14
16	14.88	11.14	8.80	7.21	6.07	5.21	4.55	4.02	3.61
18	16.89	12.65	9.98	8.17	6.88	5.91	5.15	4.56	4.10
20	18.92	14.16	11.17	9.14	7.70	6.60	5.76	5.10	4.56
22	20.98	15.70	12.38	10.12	8.52	7.31	6.37	5.64	5.05
24	23.05	17.24	13.59	11.11	9.35	8.02	6.99	6.19	5.54
26	25.13	18.80	14.81	12.11	10.19	8.74	7.61	6.74	6.03
28	27.24	20.37	16.05	13.11	11.00	9.46	8.74	7.30	6.52

Table 4.36   Values of K_2 as a Function of Tube Diameter and Length

L (ft)	\multicolumn{9}{c}{d_i (in.)}
	1.00	1.25	1.50	1.75	2.00	2.25	2.50	2.75	3.00
8	0.2990	0.1027	0.0428	0.0424	0.0109	0.0062	0.0037	0.0024	0.0016
10	0.3450	0.1171	0.0484	0.0229	0.0121	0.0069	0.0041	0.0027	0.0018
12	0.3910	0.1315	0.0539	0.0252	0.0134	0.0075	0.0045	0.0029	0.0019
14	0.4370	0.1460	0.0595	0.0277	0.0146	0.0082	0.0049	0.0031	0.0021
16	0.4830	0.1603	0.0650	0.0302	0.0158	0.0088	0.0053	0.0033	0.0022
20	0.5750	0.1892	0.0760	0.0350	0.0183	0.0101	0.0060	0.0038	0.0025
22	0.6210	0.2036	0.0816	0.0375	0.0195	0.0108	0.0064	0.0040	0.0027
24	0.6670	0.2180	0.0870	0.0400	0.0207	0.0114	0.0067	0.0042	0.0028
26	0.7130	0.2320	0.0926	0.0423	0.0219	0.0121	0.0071	0.0045	0.0030
28	0.7590	0.2469	0.0982	0.0447	0.0231	0.0217	0.0075	0.0047	0.0031

friction factor for cold-drawn tubes is a function of inner diameter.

Using Eq. (93), one can quickly figure the tube diameter and length that limit tube pressure drop to a desired value. Any two of the three variables N, L, and d_i, determine thermal performance as well as gas pressure drop. Let us discuss the conventional design procedure:

1. Assume w, calculate N.
2. Calculate U, using Eqs. (4) and (86).
3. Calculate L after obtaining A from (85).
4. Calculate ΔP_i from Eq. (91).

If the geometry or pressure drop obtained is unsuitable, repeat steps 1–4. This procedure is lengthy.

Some examples will illustrate the simplified approach. The preceding equations are valid for single-pass design. However, with minor changes, one can derive the relationships for multipass units (e.g., use length = $L/2$ for two-pass units).

EXAMPLE 1

A fire tube waste heat boiler will cool 66,000 lb/hr of flue gas from 1160°F to 440°F. Saturation temperature is 350°F. Molecular weight is 28.5, and gas pressure is atmospheric. If L is to be limited to 20 ft due to layout, determine N and ΔP_i for two tube sizes: (1) 2 × 1.77 in. (2 in. OD, 1.77 in. ID) and (2) 1.75 × 1.521 in.

Solution. Use Eq. (88) to find N. Use 2-in tubes. (F_1/C_p) from Table 4.34 is 0.73 for flue gas at the average gas temperature of 0.5 × (1160 + 440) = 800°F.

$$\ln\left[\frac{1160 - 350}{440 - 350}\right] = 2.197$$

$$2.197 = 0.606 \times 0.73 \times N^{0.2} \times \frac{20}{(66{,}000)^{0.2} \times (1.77)^{0.8}}$$

$$= 0.6089 N^{0.2}, \qquad N = 611$$

Compute ΔP_i using Eq. (91). From Table 4.36, K_2 is 0.035. Compute the gas specific volume.

$$\text{Density}\,(\rho) = 28.5 \times \frac{492}{359 \times (460 + 800)} = 0.031 \text{ lb/ft}^3,$$

or $v = 32.25 \text{ ft}^3/\text{lb}$

Substituting in Eq. (91), we have

$$\Delta P_i = 9.3 \times 10^{-5} \times \left(\frac{66{,}000}{611}\right)^2 \times 0.035 \times 32.25$$

$$= 1.23 \text{ in. H}_2\text{O}$$

Repeat the exercise with 1.75-in tubes; length remains at 20 ft. From Eq. (88), we note that for the same thermal performance and gas flow, $N^{0.2}L/d_i^{0.8}$ = a constant. The above concept comes in handy when one wants to quickly figure the effect of geometry on performance. Hence,

$$611^{0.2} \times \frac{20}{(1.77)^{0.8}} = N^{0.2} \times \frac{20}{(1.521)^{0.8}}$$

or $N = 333$

With smaller tubes, one needs fewer tubes for the same duty. This is due to a higher heat transfer coefficient; however, the gas pressure drop would be higher. From Table 4.36, $K_2 = 0.076$ for 1.521-in. tubes. From Eq. (91),

$$\Delta P_i = 9.3 \times 10^{-5} \times \left(\frac{66,000}{333}\right)^2 \times 0.076 \times 32.25$$

$$= 8.95 \text{ in. } H_2O$$

EXAMPLE 2

Size the heat exchanger for 2.0-in. tubes, with a pressure drop of 3.0 in. H_2O. For the same thermal performance, determine the geometry.

Solution. The conventional approach would take several trials to arrive at the right combination. However, with Eq. (93), one can determine the geometry rather easily:

$$\ln\left[\frac{1160 - 350}{440 - 350}\right] = 2.197 = 0.24 \times \frac{F_1}{C_p} \times \frac{K_1 v^{0.1}}{\Delta P_i^{0.1}}$$

From Table 4.34, $(F_1/C_p) = 0.73$; $\Delta P_i = 3$, $v = 32.25$. Then

$$\ln\left[\frac{1160 - 350}{440 - 350}\right] = 2.197 = 0.24 K_1 \times (32.25)^{0.1}$$

$$\times \frac{0.73}{3^{0.1}} = 0.222 K_1, \qquad K_1 = 9.89$$

From Table 4.35, we can obtain several combinations of tube diameter and length that have the same K_1 value and would yield the same thermal performance and pressure drop. For the 1.77-in. ID tube, L is 21.75 ft. Use Eq. (88) to calculate the number of tubes,

$$2.197 = 0.606 \times 0.73 \times 21.75 \times \frac{N^{0.2}}{(66,000)^{0.2} \times (1.77)^{0.8}},$$

or $N = 402$

Thus, several alternative tube geometries can be arrived at for the same performance, using the preceding approach. One saves a lot of time by not calculating heat transfer coefficients and gas properties.

LIFE-CYCLE COSTING

Such techniques determine the optimum design, given several alternatives. Here, the major operating cost is from moving the gas through the system, and the installed cost is that of the equipment and auxiliaries such as the fan. The life-cycle cost is the sum of the capitalized cost of operation and the installed cost:

$$LCC = C_{c_o} + I_c$$

The capitalized cost of operation is

$$C_{c_o} = C_a Y \, \frac{1 - Y^T}{1 - Y}$$

where

$$Y = (1 + e)/(1 + i)$$

The annual cost of operating the fan is estimated as

$$C_a = 0.001 \times PHC_e$$

where the fan power consumption in kW is

$$P = 1.9 \times 10^{-6} \times W_i \times \frac{\Delta P_i}{\rho \eta}$$

The above procedure is used to evaluate LCC. The alternative with the lowest LCC is usually chosen if the geometry is acceptable. [C_e is cost of electricity.]

4.49

Q: Discuss the simplified approach to designing water tube boilers.

A: Whenever gas flows outside a tube bundle—as in water tube boilers, economizers, and heat exchangers with high heat transfer coefficients on the tube side—the overall coefficient is governed by the gas-side resistance. Assuming that the other resistances contribute about 5% to the total, and neglecting the effect of nonluminous transfer coefficients, one may write the expression for U as

$$U = 0.95h_o \qquad\qquad\qquad\qquad (95a)$$

where the outside coefficient, h_o, is obtained from

$$Nu = 0.35 \, Re^{0.6} Pr^{0.3} \qquad\qquad\qquad (95b)$$

where, Eqs. (16) to (18), and

$$Nu = \frac{h_o d_o}{12k}, \qquad Re = \frac{Gd}{12\mu}, \qquad Pr = \frac{\mu C_p}{k}$$

$$G = 12 \frac{W_o}{N_w L (S_T - d_o)} \tag{21}$$

Equation (95) is valid for both in-line (square or rectangular pitch) and staggered (triangular pitch) arrangements. For bare tubes, the difference in h_o between in-line and staggered arrangements at Reynolds numbers and pitches found in practice is 3 to 5%. For finned tubes, the variation is significant.

Substituting Eqs. (17) to (21) into Eq. (95) and simplifying,

$$h_o = 0.945 G^{0.6} F_2/d_o^{0.4} \tag{96}$$

$$U = 0.9 G^{0.6} F_2/d_o^{0.4} \tag{97}$$

where

$$F_2 = k^{0.7}(C_p/\mu)^{0.3} \tag{98}$$

F_2 is given in Table 4.34. Gas Transport properties are computed at the film temperature.

$$A = \pi d_o N_w N_d L/12$$

Combining the above with Eq. (85) and simplifying gives

$$Q/\Delta T = UA = \pi 0.9 G^{0.6} F_2 d_o N_w N_d L/12 d_o^{0.4}$$
$$= 0.235 F_2 G^{0.6} N_w N_d L d_o^{0.6}$$

Substituting for G from Eq. (21)

$$Q/\Delta T = 1.036 F_2 W_o^{0.6} \frac{N_w^{0.4} L^{0.4} N_d}{(S_T/d_o - 1)^{0.6}} \tag{99}$$

The above equation relates thermal performance to geometry. When there is a phase change, as in a boiler, further simplification leads to

$$\ln\left[\frac{T_1 - t_s}{T_2 - T_s}\right] = 2.82 \times \frac{F_2}{C_p} \times \frac{N_d}{G^{0.4}(S_T/d_o - 1)d_o^{0.4}} \tag{100}$$

If the tube diameter and pitch are known, one can estimate N_d or G for a desired thermal performance.

Let us now account for gas pressure drop. The equation that relates the gas pressure drop to G is Eq. (28) of Chapter 3:

$$\Delta P_o = 9.3 \times 10^{-10} \times G^2 f \, \frac{N_d}{\rho}$$

For in-line arrangements, the friction factor is obtained from Eq. (29) of Chapter 3:

$$f = Re^{-0.15} X$$

where

$$X = 0.044 + \frac{0.08 S_L/d_o}{(S_T/d_o - 1)^{0.43 + 1.13 d_o/S_L}}$$

Another form of Eq. (3.28) is

$$\Delta P_o = 1.34 \times 10^{-7} \times \frac{W_o^{1.85} v N_d^{2.85} \mu^{015} X}{N_w^{1.85} L^{1.85} d_o^{0.15} (S_T - d_o)^{1.85}} \tag{101}$$

Substituting for f in Eq. (3.28) and combining with Eq. (100), we can relate ΔP_o to performance in a single equation:

$$\Delta P_o = 4.78 \times 10^{-10} \times G^{2.25} (S_T - d_o) \ln \left[\frac{T_1 - t_s}{T_2 - t_s} \right]$$

$$\times \frac{X}{d_o^{0.75} F_3 \rho} \tag{102}$$

where

$$F_3 = (F_2/C_p) \mu^{-0.15} \tag{103}$$

F_3 is given in Table 4.34. With the above equation, one can easily calculate the geometry for a given tube bank so as to limit the pressure drop to a desired value. An example will illustrate the versatility of the technique.

EXAMPLE

In a water tube boiler, 66,000 lb/hr of flue gas is cooled from 1160 to 440°F. Saturation temperature is 350°F. Tube outside diameter is 2 in., and an in-line arrangement is used with $S_T = S_T = 4$ in. Determine a suitable configuration to limit the gas pressure to 3 in. H_2O.

Let us use Eq. (102). Film temperature is $0.5 \times (800 + 350)$ = 575°F. Interpolating from Table 4.34 at 475°F, $F_3 = 0.643$. Gas density at 800°F is 0.031 lb/ft³ from Example 1.

$$\Delta P_o = 4.78 \times 10^{-10} \times G^{2.25} \times (4 - 2)$$

$$\times \ln \left[\frac{1160 - 350}{440 - 350} \right]$$

$$\times \frac{(0.044 + 0.08 \times 2)}{2^{0.75} \times 0.643 \times 0.031}$$

$$= 128 \times 10^{-10} \times G^{2.25} = 3$$

Hence, $G = 5200$ lb/ft² hr. From Eq. (21) one can choose different combinations of N_w and L:

$$N_w L = 66{,}000 \times 12/(2 \times 5200) = 76$$

If $N_w = 8$, then $L = 9.5$ ft.

Calculate N_d from Eq. (100):

$$\ln \left[\frac{1160 - 350}{440 - 350} \right] = 2.197 = 2.82 \, \frac{F_2}{C_p}$$

$$\times \frac{N_d}{G^{0.4}(S_T/d_o - 1)d_o^{0.4}}$$

$$2.197 = 2.82 \times 0.426 N_d/(5200^{0.4} \times 1 \times 2^{0.4}) \quad \text{or} \quad N_d = 74$$

Thus, the entire geometry has been arrived at.

4.50

Q: How is the bundle diameter of heat exchangers or fire tube boilers determined?

A: Tubes of heat exchangers and fire tube boilers are typically arranged in triangular or square pitch (Figure 4.17). The ratio of tube pitch to diameter could range from 1.25 to 2 depending on the tube size and the manufacturer's past practice.

Looking at the triangular pitch arrangement, we see that *half* of a tube area is located within the triangle, whose area is given by

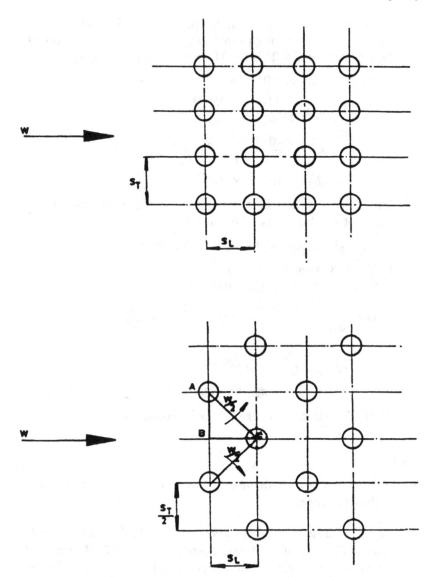

Figure 4.17   Triangular and square pitch for boiler/exchanger tubes.

Area of triangle $= 0.5 \times 0.866p^2 = 0.433p^2$

If there are N tubes in the bundle,

Total area occupied $= 0.866Np^2$

If the bundle diameter is D, then $3.14 \times D^2/4 =$ area of bundle $= 0.866Np^2$ or

$$D = 1.05pN^{0.5} \tag{104}$$

Similarly, for the square pitch, the area occupied by one tube $= p^2$. Hence bundle area $= 3.14 \times D^2/4 = Np^2$, or

$$D = 1.128pN^{0.5} \tag{105}$$

In practice, a small clearance is added to the above number for manufacturing purposes.

EXAMPLE

If 500 tubes of 2-in. diameter are located in a fire tube boiler shell at a triangular pitch of 3 in., the bundle diameter would be

$$D = 1.05 \times 3 \times 500^{0.5} = 70.5 \text{ in.}$$

If the pitch were square, the bundle diameter would be

$$D = 1.128 \times 3 \times 500^{0.5} = 75.7 \text{ in.}$$

Sometimes tubes have to be located within a given sector of a circle. In such cases, it is helpful to know the area of a sector of a circle given its height and diameter. Table 4.37 gives the factor C, which when multiplied by D_2 gives the sector area.

EXAMPLE

Find the area of a sector of height 10 in. and diameter 24 in. *Solution.* For $h/D = 10/24 = 0.4167$, C from Table 4.37 $= 0.309$. Hence

$$\text{Area} = C \times D^2 = 0.309 \times 24 \times 24 = 178 \text{ in.}^2$$

4.51

Q: How is the thickness of insulation for a flat or curved surface determined? Determine the thickness of insulation to limit the

Table 4.37 C Factors for Finding Area of Sector (Area $= CD^2$) of a Circle

h/D	C	h/D	C	h/D	C	h/D	C	h/D	C
0.001	0.00004	0.050	0.01468	0.100	0.04087	0.150	0.07387	0.200	0.11182
0.002	0.00012	0.051	0.01512	0.101	0.04148	0.151	0.07459	0.201	0.11262
0.003	0.00022	0.052	0.01556	0.102	0.04208	0.152	0.07531	0.202	0.11343
0.004	0.00034	0.053	0.01601	0.103	0.04269	0.153	0.07603	0.203	0.11423
0.004	0.00034	0.054	0.01646	0.104	0.04330	0.154	0.07675	0.204	0.11504
0.005	0.00047	0.055	0.01691	0.105	0.04391	0.155	0.07747	0.205	0.11584
0.006	0.00062	0.056	0.01737	0.106	0.04452	0.156	0.07819	0.206	0.11665
0.007	0.00078	0.057	0.01783	0.107	0.04514	0.157	0.07892	0.207	0.11746
0.008	0.00095	0.058	0.01830	0.108	0.04578	0.158	0.07965	0.208	0.11827
0.009	0.00113	0.059	0.01877	0.109	0.04638	0.159	0.08038	0.209	0.11908
0.010	0.00133	0.060	0.01924	0.110	0.04701	0.160	0.08111	0.210	0.11990
0.011	0.00153	0.061	0.01972	0.111	0.04763	0.161	0.08185	0.211	0.12071
0.012	0.00175	0.062	0.02020	0.112	0.04826	0.162	0.08258	0.212	0.12153
0.013	0.00197	0.063	0.02068	0.113	0.04889	0.163	0.08332	0.213	0.12235
0.014	0.00220	0.064	0.02117	0.114	0.04953	0.164	0.08406	0.214	0.12317
0.015	0.00244	0.065	0.02166	0.115	0.05016	0.165	0.08480	0.215	0.02399
0.016	0.00268	0.066	0.02215	0.116	0.05080	0.166	0.08554	0.216	0.12481
0.017	0.00294	0.067	0.02265	0.117	0.05145	0.167	0.08629	0.217	0.12563
0.018	0.00320	0.068	0.02315	0.118	0.05209	0.168	0.08704	0.218	0.12646
0.019	0.00347	0.069	0.02366	0.119	0.05274	0.169	0.08779	0.219	0.12729
0.020	0.00375	0.070	0.02417	0.120	0.05338	0.170	0.08854	0.220	0.12811
0.021	0.00403	0.071	0.02468	0.121	0.05404	0.171	0.08929	0.221	0.12894
0.022	0.00432	0.072	0.02520	0.122	0.05469	0.172	0.09004	0.222	0.12977
0.023	0.00462	0.073	0.02571	0.123	0.05535	0.173	0.09080	0.223	0.13060

0.024	0.00492	0.074	0.02624	0.124	0.05600	0.174	0.09155	0.224	0.13144
0.025	0.00523	0.075	0.02676	0.125	0.05666	0.175	0.09231	0.225	0.13277
0.026	0.00555	0.076	0.02729	0.126	0.05733	0.176	0.09307	0.226	0.13311
0.027	0.00587	0.077	0.02782	0.127	0.05799	0.177	0.09384	0.227	0.13395
0.028	0.00619	0.078	0.02836	0.128	0.05866	0.178	0.09460	0.228	0.13478
0.029	0.00653	0.079	0.02889	0.129	0.05933	0.179	0.09537	0.229	0.13562
0.030	0.00687	0.080	0.02943	0.130	0.06000	0.180	0.09613	0.230	0.13646
0.031	0.00721	0.081	0.02998	0.131	0.06067	0.181	0.09690	0.231	0.13731
0.032	0.00756	0.082	0.03053	0.132	0.06135	0.182	0.09767	0.232	0.13815
0.033	0.00791	0.083	0.03108	0.133	0.06203	0.183	0.09845	0.233	0.13900
0.034	0.00827	0.084	0.03163	0.134	0.06271	0.184	0.09922	0.234	0.13984
0.035	0.00864	0.085	0.03219	0.135	0.06339	0.185	0.10000	0.235	0.14069
0.036	0.00901	0.086	0.03275	0.136	0.06407	0.186	0.10077	0.236	0.14154
0.037	0.00938	0.087	0.03331	0.137	0.06476	0.187	0.10155	0.237	0.14239
0.038	0.00976	0.088	0.03387	0.138	0.06545	0.188	0.10233	0.238	0.14324
0.039	0.01015	0.089	0.03444	0.139	0.06614	0.189	0.10312	0.239	0.14409
0.040	0.01054	0.090	0.03501	0.140	0.06683	0.190	0.10390	0.240	0.14494
0.041	0.01093	0.091	0.03559	0.141	0.06753	0.191	0.10469	0.241	0.14580
0.042	0.01133	0.092	0.03616	0.142	0.06822	0.192	0.10547	0.292	0.14666
0.043	0.01173	0.093	0.03674	0.143	0.06892	0.193	0.10626	0.243	0.14751
0.044	0.01214	0.094	0.03732	0.144	0.06963	0.194	0.10705	0.244	0.14837
0.045	0.01255	0.095	0.03791	0.145	0.07033	0.195	0.10784	0.245	0.14923
0.046	0.01297	0.096	0.03850	0.146	0.07103	0.196	0.10864	0.246	0.15009
0.047	0.01339	0.097	0.03909	0.147	0.07174	0.197	0.10943	0.247	0.15095
0.048	0.01382	0.098	0.03968	0.148	0.07245	0.198	0.11023	0.248	0.15182
0.049	0.01425	0.099	0.04028	0.149	0.07316	0.199	0.11102	0.249	0.15268

Table 4.37 Continued

h/D	C	h/D	C	h/D	C	h/D	C	h/D	C
0.250	0.15355	0.300	0.19817	0.350	0.24498	0.400	0.29337	0.450	0.34278
0.251	0.15441	0.301	0.19908	0.351	0.24593	0.401	0.29435	0.451	0.34378
0.252	0.15528	0.302	0.20000	0.352	0.24689	0.402	0.29533	0.452	0.34477
0.253	0.15615	0.303	0.20092	0.353	0.24784	0.403	0.29631	0.453	0.34577
0.254	0.15702	0.304	0.20184	0.354	0.24880	0.404	0.29729	0.454	0.34676
0.255	0.15789	0.305	0.20276	0.355	0.24976	0.405	0.29827	0.455	0.34776
0.256	0.15876	0.306	0.20368	0.356	0.25071	0.406	0.29926	0.456	0.34876
0.257	0.15964	0.307	0.20460	0.357	0.25167	0.407	0.30024	0.457	0.34975
0.258	0.16501	0.308	0.20553	0.358	0.25263	0.408	0.30122	0.458	0.35075
0.259	0.16139	0.309	0.20645	0.359	0.25359	0.409	0.30220	0.459	0.35175
0.260	0.16226	0.310	0.20738	0.360	0.25455	0.410	0.30319	0.460	0.35274
0.261	0.16314	0.311	0.20830	0.361	0.25551	0.411	0.30417	0.461	0.35374
0.262	0.16402	0.312	0.20923	0.362	0.25647	0.412	0.30516	0.462	0.35474
0.263	0.16490	0.313	0.21015	0.363	0.25743	0.413	0.30614	0.463	0.35573
0.264	0.16578	0.314	0.21108	0.364	0.25839	0.414	0.30712	0.464	0.35673
0.265	0.16666	0.315	0.21201	0.365	0.25936	0.415	0.30811	0.465	0.35773
0.266	0.16755	0.316	0.21294	0.366	0.26032	0.416	0.30910	0.466	0.35873
0.267	0.16843	0.317	0.21387	0.367	0.26128	0.417	0.31008	0.467	0.35972
0.268	0.16932	0.318	0.21480	0.368	0.26225	0.418	0.31107	0.468	0.36072
0.269	0.17020	0.319	0.21573	0.369	0.26321	0.419	0.31205	0.469	0.36172
0.270	0.17109	0.320	0.21667	0.370	0.26418	0.420	0.31304	0.470	0.36272
0.271	0.17198	0.321	0.21760	0.371	0.26514	0.421	0.31403	0.471	0.36372
0.272	0.17287	0.322	0.21853	0.372	0.26611	0.422	0.31502	0.472	0.36471
0.273	0.17376	0.323	0.21947	0.373	0.26708	0.423	0.31600	0.473	0.36571

x		x		x		x		x	
0.274	0.17465	0.324	0.22040	0.374	0.26805	0.424	0.31699	0.474	0.36671
0.275	0.17554	0.325	0.22134	0.375	0.26901	0.425	0.31798	0.475	0.36771
0.276	0.17644	0.326	0.22228	0.376	0.26998	0.426	0.31897	0.476	0.36871
0.277	0.17733	0.327	0.22322	0.377	0.27095	0.427	0.31996	0.477	0.36971
0.278	0.17823	0.328	0.22415	0.378	0.27192	0.428	0.32095	0.478	0.37071
0.279	0.17912	0.329	0.22509	0.379	0.27289	0.429	0.32194	0.479	0.37171
0.280	0.18002	0.330	0.22603	0.380	0.27386	0.430	0.32293	0.480	0.37270
0.281	0.18092	0.331	0.22697	0.381	0.27483	0.431	0.32392	0.481	0.37370
0.282	0.18182	0.332	0.22792	0.382	0.27580	0.432	0.32491	0.482	0.37470
0.283	0.18272	0.333	0.22886	0.383	0.27678	0.433	0.32590	0.483	0.37570
0.284	0.18362	0.334	0.22980	0.384	0.27775	0.434	0.32689	0.484	0.37670
0.285	0.18452	0.335	0.23074	0.385	0.27872	0.435	0.32788	0.485	0.37770
0.286	0.18542	0.336	0.23169	0.386	0.27969	0.436	0.32887	0.486	0.37870
0.287	0.18633	0.337	0.23263	0.387	0.28067	0.437	0.32987	0.487	0.37970
0.288	0.18723	0.338	0.23358	0.388	0.28164	0.438	0.33086	0.488	0.38070
0.289	0.18814	0.339	0.23453	0.389	0.28262	0.439	0.33185	0.489	0.38170
0.290	0.18905	0.340	0.23547	0.390	0.28359	0.440	0.33284	0.490	0.38270
0.291	0.18996	0.341	0.23642	0.391	0.28457	0.441	0.33384	0.491	0.38370
0.292	0.19086	0.342	0.23737	0.392	0.28554	0.442	0.33483	0.492	0.38470
0.293	0.19177	0.343	0.23832	0.393	0.28652	0.443	0.33582	0.493	0.38570
0.294	0.19268	0.344	0.23927	0.394	0.28750	0.444	0.33682	0.494	0.38670
0.295	0.19360	0.345	0.24022	0.395	0.28848	0.445	0.33781	0.495	0.38770
0.296	0.19451	0.346	0.24117	0.396	0.28945	0.446	0.33880	0.496	0.38870
0.297	0.19542	0.347	0.24212	0.397	0.29043	0.447	0.33980	0.497	0.38970
0.298	0.19634	0.348	0.24307	0.398	0.29141	0.448	0.34079	0.498	0.39070
0.299	0.19725	0.349	0.24403	0.399	0.29239	0.449	0.34179	0.499	0.39070
								0.500	0.39270

casing surface temperature of a pipe operating at 800°F to 200°F, when

Ambient temperature $t_a = 80°F$
Thermal conductivity of insulation K_m at average temperature
   of 500°F = 0.35 Btu in./ft² hr. °F
Pipe outer diameter $d = 12$ in.
Wind velocity $V = 264$ ft/min (3 mph)
Emissivity of casing = 0.15 (oxidized)

A: The heat loss q from the surface is given by [7]

$$q = 0.174\epsilon \left[\left(\frac{t_s + 459.6}{100} \right)^4 - \left(\frac{t_a + 459.6}{100} \right)^4 \right]$$
$$+ 0.296 (t_s - t_a)^{1.25} \times \left(\frac{V + 68.9}{68.9} \right)^{1/2} \qquad (106)$$

ϵ may be taken as 0.9 for oxidized steel, 0.05 for polished aluminum, and 0.15 for oxidized aluminum. Also,

$$q = \frac{K_m(t - t_s)}{[(d + 2L)/2] \times \ln [(d + 2L)/d]} = \frac{K_m(t - t_s)}{L_e} \qquad (107)$$

where t is the hot face temperature, °F, and L_e is the equivalent thickness of insulation for a curved surface such as a pipe or tube.

$$L_e = \frac{d + 2L}{2} \ln \frac{d + 2L}{d} \qquad (108)$$

Substituting $t_s = 200$, $t_a = 80$, $V = 264$, and $\epsilon = 0.15$ into Eq. (106), we have

$$q = 0.173 \times 0.15 \times (6.6^4 - 5.4^4) + 0.296$$
$$\times (660 - 540)^{1.25} \times \left(\frac{264 + 69}{69} \right)^{0.5}$$

$$= 285 \text{ Btu/ft}^2 \text{ hr}$$

From Eq. (107),

$$L_e = 0.35 \times \frac{800 - 200}{285} = 0.74 \text{ in.}$$

Table 4.38 Equivalent Thickness of Insulation, L_e (in.)[a]

Tube diam. d (in.)	Thickness of insulation L (in.)							
	0.5	1	1.5	2.0	3.0	4.0	5.0	6.0
1	0.69	1.65	2.77	4.0	6.80	9.90	13.2	16.7
2	0.61	1.39	2.29	3.30	5.50	8.05	10.75	13.62
3	0.57	1.28	2.08	2.97	4.94	7.15	9.53	12.07
4	0.56	1.22	1.96	2.77	4.55	6.60	8.76	11.10
5	0.55	1.18	1.88	2.65	4.34	6.21	8.24	10.40
6	0.54	1.15	1.82	2.55	4.16	5.93	7.85	9.80
8	0.53	1.12	1.75	2.43	3.92	5.55	7.50	9.15
10	0.52	1.09	1.70	2.35	3.76	5.29	6.93	8.57
12	0.52	1.08	1.67	2.30	3.65	5.11	6.65	8.31
16	0.52	1.06	1.63	2.23	3.50	4.86	6.31	7.83
20	0.51	1.05	1.61	2.19	3.41	4.70	6.10	7.52

[a]$L_e = \dfrac{d + 2L}{2} \ln \dfrac{d + 2L}{d}$. For example, for $d = 3$ and $L = 1.5$, $L_e = 2.08$.

We can solve for L given L_e and d using Eq. (108) by trial and error or use Table 4.38. It can be shown that $L = 0.75$ in. The next standard thickness available will be chosen. A trial-and-error method as discussed next will be needed to solve for the surface temperature t_s. (Note that L is the actual thickness of insulation.)

4.52

Q: Determine the surface temperature of insulation in Q4.51 when 1.0-in-thick insulation is used on the pipe. Other data are as given earlier.

A: Calculate the equivalent thickness L_e. From Eq. (108),

$$L_e = \frac{12 + 2}{2} \ln \frac{14}{12} = 1.08 \text{ in.}$$

Assume that for the first trial t_s = 150°F. Let K_m at a mean temperature of (800 + 150)/2 = 475 be 0.34 Btu in./ft^2 hr °F. From Eq. (106),

$$q = 0.173 \times 0.15 \times (6.1^4 - 5.4^4) + 0.296$$
$$\times (610 - 540)^{1.25} \times \left(\frac{264 + 69}{69}\right)^{0.5}$$
$$= 145 \text{ Btu/ft}^2 \text{ hr}$$

From Eq. (107),

$$q = 0.34 \times \frac{800 - 150}{1.08} = 205 \text{ Btu/ft}^2 \text{ hr}$$

Since these two values of q do not agree, we must go for another trial. Try t_s = 170°F. Then, from Eq. (106),

$$q = 200 \text{ Btu/ft}^2 \text{ hr}$$

and from Eq. (107),

$$q = 198 \text{ Btu/ft}^2 \text{ hr}$$

These two are quite close. Hence the final surface temperature is 170°F, and the heat loss is about 200 Btu/ft^2 hr.

4.53

Q: A horizontal flat surface is 10°F. The ambient dry bulb temperature is 80°F and the relative humidity is 80%. Determine the thickness of fibrous insulation that will prevent condensation of water vapor on the surface. Use K_m = 0.28 Btu/ft hr °F. The wind velocity is zero. Use a surface emissivity of 0.9 for the casing.

A: The surface temperature must be above the dew point of water to prevent condensation of water vapor. Q1.10 shows how the dew point can be calculated. The saturated vapor pressure is 80°F, from the steam tables in the Appendix, is 0.51 psia. At 80% relative humidity, the vapor pressure will be 0.8 × 0.51 = 0.408 psia. From the steam tables, this corresponds to a saturation temperature of 73°F, which is also the dew point. Hence we must design the insulation so the casing temperature is above 73°F.

From Eq. (106),

$$q = 0.173 \times 0.9 \times (5.4^4 - 5.33^4) + 0.296$$
$$\times (80 - 73)^{1.25} = 10.1 \text{ Btu/ft}^2 \text{ hr}$$

Also, from Eq. (107),

$$q = (t_d - t_s) \times \frac{K_m}{L} = (73 - 10) \times \frac{0.28}{L}$$

(In this case of a flat surface, $L_e = L$.)

Note that the heat flow is from the atmosphere to the surface. t_d and t_s are the dew point and surface temperature, °F. Solving for L, we get $L = 1.75$ in.

Hence, by using the next standard insulation thickness available, we can ensure that the casing is above the dew point. To obtain the exact casing temperature with the standard thickness of insulation, a trial-and-error procedure as discussed in Q4.52 may be used. But this is not really necessary, as we have provided a safe design thickness.

4.54a

Q: A 1½-in. schedule 40 pipe 1000 ft long carries hot water at 300°F. What is the heat loss from its surface if it is not insulated (case 1) or if it has 1-in.-, 2-in.-, and 3-in.-thick insulation (case 2)?

The thermal conductivity of insulation may be assumed as 0.25 Btu in./ft² hr °F. The ambient temperature is 80°F, and the wind velocity is zero.

A: *Case 1* Equation (106) can be used to determine the heat loss. For the bare pipe surface, assume that ϵ is 0.90. Then

$$q = 0.173 \times 0.9 \times (7.6^4 - 5.4^4) + 0.296$$
$$\times (300 - 80)^{1.25} = 638 \text{ Btu/ft}^2 \text{ hr}.$$

Case 2   Determination of the surface temperature given the insulation thickness involves a trial-and-error procedure as discussed in Q4.52 and will be done in detail for the 1-in. case.

Various surface temperatures are assumed and q is computed from Eqs. (106) and (107). Let us use a ϵ value of 0.15. The following table gives the results of the calculations.

t_s	q from Eq. (106)	q from Eq. (107)
110	26	34
120	37	32
140	61	28

We can draw a graph of t_s versus υ with these values and obtain the correct t_s. However, we see from the table, by interpolation, that at $t_s = 115°F$, q, from both equations, is about 33 Btu/ft²hr.

$$\text{Total heat loss} = 3.14 \times \frac{3.9}{12} \times 1000 \times 33$$

$$= 33,\ 675\ \text{Btu/hr}$$

Similarly, we may solve for q when the thicknesses are 2 and 3 in. It can be shown that at $L = 2$ in., $q = 15$ Btu/ft² hr, and at $L = 3$ in., $q = 9$ Btu/ft² hr. Also, when $L = 2$ in., $t_s = 98°F$ and total heat loss = 23,157 Btu/hr. When $L = 3$ in., $t_s = 92°F$ and total loss = 18,604 Btu/hr.

4.54b

Q: Estimate the drop in water temperature of 1-in.-thick insulation were used in Q4.54a. The water flow is 7500 lb/hr.

A: The total heat loss has been shown to be 33,675 Btu/hr. This is lost by the water and can be written as 7500 ΔT, where ΔT is the drop in temperature, assuming that the specific heat is 1. Hence

$$\Delta T = \frac{33{,}675}{7500} = 4.5°F$$

By equating the heat loss from insulation to the heat lost by the fluid, be it air, oil, steam, or water, one can compute the drop in temperature in the pipe or duct. This calculation is particularly important when oil lines are involved, as viscosity is affected, leading to pumping and atomization problems.

4.55

Q: In Q4.54 determine the optimum thickness of insulation with the following data.

Cost of energy = $3/MM Btu
Cost of operation = $8000/year
Interest and escalation rates = 12% and 7%
Life of the plant = 15 years
Total cost of 1-in.-thick insulation, including labor and material = $5200; for 2-in. insulation, $7100; and for 3-in. insulation, $10,500.

A: Let us calculate the capitalization factor F from Q1.22.

$$F = \frac{1.07}{1.12} \times \frac{1 - (1.07/1.12)^{15}}{1 - (1.07/1.12)} = 10.5$$

Let us calculate the annual heat loss.
For $L = 1$ in.,

$$C_a = 33{,}675 \times 3 \times \frac{8000}{10^6} = \$808$$

For $L = 2$ in.,

$$C_a = 23{,}157 \times 3 \times \frac{8000}{10^6} = \$555$$

For $L = 3$ in.,

$$C_a = 18{,}604 \times 3 \times \frac{8000}{10^6} = \$446$$

Calculate capitalized cost $= C_aF$.
For $L = 1$ in.,

$C_aF = 808 \times 10.5 = \$8484$

For $L = 2$ in.,

$C_aF = 555 \times 10.5 = \$5827$

For $L = 3$ in.,

$C_aF = 446 \times 10.5 = \$4683$

Calculate total capitalized cost or life-cycle cost (LCC):
For $L = 1$ in.,

$LCC = 8484 + 5200 = \$13,684$

For $L = 2$ in., $LCC = \$12,927$; and for $L = 3$ in.,
$LCC = \$15,183$.

Hence we see that the optimum thickness is about 2 in. With higher thicknesses, the capital cost becomes more than the benefits from savings in heat loss. A trade-off would be to go for 2-in.-thick insulation.

Several factors enter into calculations of this type. If the period of operation were less, probably a lesser thickness would be adequate. If the cost of energy were more, we might have to go for a higher thickness. Thus each case must be evaluated before we decide on the optimum thickness. This examples gives only a methodology, and the evaluation can be as detailed as desired by the plant engineering personnel.

If there were no insulation, the annual heat loss would be

$$3.14 \times \frac{1.9}{12} \times 1000 \times 638 \times 3 \times \frac{8000}{10^6} = \$7600$$

Hence simple payback with even 1-in.-thick insulation is $5200/(7600 - 808) = 0.76$ year, or 9 months.

4.56

Q: What is a hot casing? What are its uses?

A: Whenever hot gases are contained in an internally refractory-
     lined (or insulated) duct, the casing temperature can fall below
     the dew point of acid gases, which can seep through the refrac-
     tory cracks and cause acid condensation, which is a potential
     problem. To avoid this, some engineers prefer a "hot casing"
     design, which ensures that the casing or the vessel or duct
     containing the gases is maintained at a high enough temperature
     to minimize or prevent acid condensation. At the same time, the
     casing is also externally insulated to minimize the heat losses to
     the ambient (see Figure 4.18). A "hot casing" is a combination
     of internal plus external insulation used to maintain the casing at
     a high enough temperature to avoid acid condensation while
     ensuring that the heat losses to the atmosphere are low.
         Consider the use of a combination of two refractories inside
     the boiler casing: 4 in. of KS4 and 2 in. of CBM. The hot gases
     are at 1000°F. Ambient temperature = 60°F, and wind velocity
     is 100 ft/min. Casing emissivity is 0.9. To keep the boiler casing

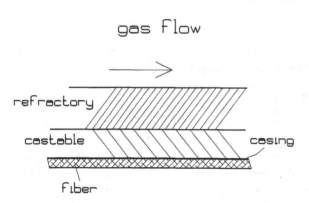

Figure 4.18   Arrangement of hot casing.

hot, an external 0.5 in. of mineral fiber is added. Determine the boiler casing temperature, the outer casing temperature, and the heat loss.

One can perform the calculations discussed earlier to arrive at the temperatures and heat loss. For the sake of illustrating the point, a computer printout of the result is shown in Figure 4.19. It can be seen that the boiler casing is at 392°F, while the outermost casing is at 142°F. The heat loss is 180 Btu/ft^2 hr. The boiler casing is hot enough to avoid acid condensation, while the heat losses are kept low.

4.57

Q: What happens if ducts or stacks handling flue gases are not insulated? What would the gas or stack wall temperature be?

A: This question faces engineers involved in engineering of boiler plants. If ducts and stacks are not insulated, the heat loss from the casing can be substantial. Also, the stack wall temperature can drop low enough to cause acid dew point corrosion.

Let the flue gas flow = W lb/hr at a temperature of t_{g1} at the inlet to the duct or stack (Figure 4.20). The heat loss from the casing wall is given by Eq. (106),

```
RESULTS-INSULATION PERFORMANCE-          flat surface

_____Project:  HOT CASING

NAME            THICK-IN    TEMP-F    TEMP1     COND1   TEMP2     COND2

Casing          0.00        142.27    0.00      0.00    0.00      0.00
deltbd/mf/fib   0.50        392.02    200.00    0.32    400.00    0.45
cbm             2.00        880.12    200.00    0.57    600.00    0.72
ks4             4.00        1001.42   800.00    6.02    1600.00   6.20

HEAT LOSS -BTU/ft2h= 179.5997 Number of layers of insulation= 3

AMB TEMP= 70 WIND VEL-fpm= 100 EMISS= .9 MAX LOSS-BTU/FT2H= 9330.736
```

Figure 4.19   Results of printout on casing temperature.

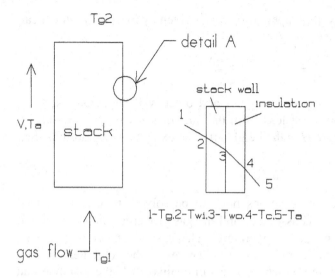

Figure 4.20   Stack wall temperature.

$$q = 0.174 \, \epsilon \times \left[\left(\frac{t_c + 460}{100} \right)^4 - \left(\frac{t_a + 460}{100} \right)^4 \right]$$
$$+ \; 0.296 \, (t_c - t_a)^{1.25} \times \left[\frac{V + 69}{69} \right]^{0.5}$$

The temperature drop across the gas film is given by

$$t_g - t_{w1} = q \; \frac{d_o/d_i}{h_c}$$

where

h_c = convective heat transfer coefficient Btu/ft² hr °F

d_o, d_i = outer and inner diameter of the stack, in.

$$h_c = 2.44 \times \frac{W^{0.8} C}{d_i^{1.8}}$$

where, from Eq. (12),

$$C = \left(\frac{C_p}{\mu} \right)^{0.4} k^{0.6}$$

The duct wall temperature drop is given by Eq. (107), which can be rearranged to give

$$t_{w1} - t_{wo} = qd_o \frac{\ln(d_o/d_i)}{24K_m}$$

where t_{w1}, t_{wo} are the inner and outer wall temperatures, °F.

The total heat loss from the duct or stack is $Q = 3.14d_o \times H/12$, where H is the height, ft. The exit gas temperature is then

$$t_{g2} = t_{g1} - \frac{Q}{W_g \times C_p} \tag{109}$$

The above equations have to be solved iteratively. A trial value for t_{g2} is assumed, and the gas properties are computed at the average gas temperature. The casing temperature is also obtained through an iterative process. The total heat loss is computed and which t_{g2} is again evaluated. If the assumed and calculated t_{g2} values agree, then iteration stops. A computer program can be developed to obtain accurate results, particularly if the stack is tall and calculations are better done in several segments.

EXAMPLE

110,000 lb/hr of flue gases at 410°F enter a 48-in. ID stack that is 50 ft long and 1 in. thick. If the ambient temperature is 70°F and wind velocity is 125 ft/min, determine the casing temperature, total heat loss, and exit gas temperature.

Flue gas properties can be assumed to be as follows at 400°F (or computed from methods discussed in Q4.12 if analysis is known): $C_p = 0.265$, $\mu = 0.058$ lb/ft hr, $k = 0.0211$ Btu/ft hr °F. Let the gas temperature drop in the stack $= 20$°F; hence the exit gas temperature $= 390$°F.

The gas-side heat transfer coefficient is

$$2.44 \times (110,000)^{0.8} \times \left(\frac{0.265}{0.058}\right)^{0.4} \times (0.0211)^{0.6}$$

$$= 4.5 \text{ Btu/ft}^2 \text{ hr °F}$$

Let the casing temperature t_c ($= t_{wo}$ without insulation) $= 250$°F.

$$q = 0.174 \times 0.9 \times [(7.1)^4 - (5.3)^4] + 0.296$$
$$\times (710 - 530)^{1.25} \times \left(\frac{125 + 69}{69} \right)^{0.5}$$
$$= 601 \text{ Btu/ft}^2 \text{ hr}$$

Gas temperature drop across gas film = 601/4.5 = 134°F

Temperature drop across the stack wall = 601 × 50

$$\times \frac{\ln (50/48)}{(24 \times 25)} = 2°F$$

Hence stack wall outer temperature = 400 − 134 − 2 = 264°F.

It can be shown that at a casing or wall temperature of 256 °F, the heat loss through gas film matches the loss through the stack wall. The heat loss = 629 Btu/ft² hr, and total heat loss = 411,400 Btu/hr.

$$\text{Gas temperature drop} = \frac{411,400}{110,000/0.265} = 14 \text{ °F}.$$

The average gas temperature = 410 − 14 = 396°F, which is close to the 400°F assumed. With a computer program, one can fine tune the calculations to include fouling factors.

4.58

Q: What are the effects of wind velocity and casing emissivity on heat loss and casing temperature?

A: Using the method described earlier, the casing temperature and heat loss were determined for the case of an insulated surface at 600°F using 3 in. of mineral fiber insulation. (Aluminum casing has an emissivity of about 0.15, and oxidized steel 0.9.) The results are shown in Table 4.39.

It can be seen that the wind velocity does not result in reduction of heat losses though the casing temperature is significantly reduced. Also the use of lower emissivity casing does not affect the heat loss, though the casing temperature is increased, particularly at low wind velocity.

Table 4.39   Results of Insulation Performance

Casing	Emissivity	Wind vel. (fpm)	Heat loss	Casing temp. (°F)
Aluminum	0.15	0	67	135
Aluminum	0.15	1760	71	91
Steel	0.90	0	70	109
Steel	0.90	1760	70	88

4.59a

Q: How does one check heat transfer equipment for possible noise and vibration problems?

A: A detailed procedure is outlined in Refs. 1 and 8. Here only a brief reference to the methodology will be made.

Whenever a fluid flows across a tube bundle such as boiler tubes in an economizer, air heater, or superheater (see Figure 4.21), vortices are formed and shed in the wake beyond the tubes. This shedding on alternate sides of the tubes causes a harmonically varying force on the tube perpendicular to the normal flow of the fluid. It is a self-excited vibration. If the frequency of the von Karman vortices as they are called coincides with the natural frequency of vibration of the tubes, resonance occurs and tubes vibrate, leading to leakage and damage at supports. Vortex shedding is more prevalent in the range of Reynolds numbers from 300 to 2×10^5. This is the range in which many boilers, economizers, and superheaters operate. Another mechanism associated with vortex shedding is acoustic oscillation, which is normal to both fluid flow and tube length. This is observed with gases and vapors only. These oscillations coupled with vortex shedding lead to resonance and excessive noise. Standing waves are formed inside the duct.

Hence in order to analyze tube bundle vibration and noise, three frequencies must be computed: natural frequency of vibration of tubes; vortex shedding frequency; and acoustic frequency. When these are apart by at least 20%, vibration and noise may be absent. Q4.59b to Q4.59e show how these values are computed and evaluated.

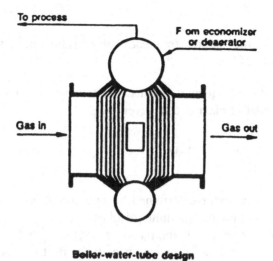

Boiler-water-tube design

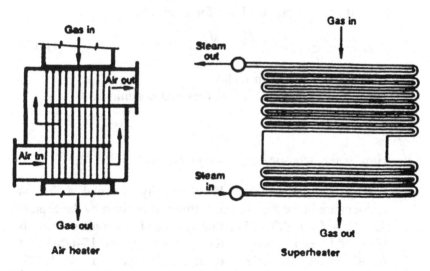

Air heater   Superheater

Figure 4.21   Crossflow of gas over tube bundles.

4.59b

Q: How is the natural frequency of vibration of a tube bundle determined?

A: The natural frequency of transverse vibrations of a uniform beam supported at each end is given by

$$f_n = \frac{C}{2\pi} \times \left(\frac{EIg_o}{M_e L^4}\right)^{0.5}$$ (110a)

where

$\quad C$ = a factor determined by end conditions
$\quad E$ = Young's modulus of elasticity
$\quad I$ = moment of inertia = $\pi (d_o^4 - d_i^4)/64$
$\quad M_e$ = mass per unit length of tube, lb/ft (including ash deposits, if any, on the tube.)
$\quad L$ = tube length, ft

Simplifying (110a), we have for steel tubes

$$f_n = \frac{90C}{L^2} \times \left(\frac{d_o^4 - d_i^4}{M_e}\right)^{0.5}$$ (110b)

where d_o and d_i are in inches.
Table 4.40 gives C for various end conditions.

4.59c

Q: How is the acoustic frequency computed?

A: f_a is given by V_s/λ, where V_s = velocity of sound at the gas temperature in the duct or shell, ft/sec. It is given by the expression $V_s = (g_o \upsilon RT)^{0.5}$. For flue gases and air, sonic velocity is obtained by substituting 32 for g_o, 1.4 for υ, and 1546/MW for R, where the molecular weight for flue gases is nearly 29. Hence,

$$V_s = 49 \times T^{0.5}$$ (111)

Wavelength $\lambda = 2W/n$, where W is the duct width, ft, and n is the mode of vibration.

Table 4.40   Values of C for Eq. (110b)

End support conditions	Mode of vibration		
	1	2	3
Both ends clamped	22.37	61.67	120.9
One clamped, one hinged	15.42	49.97	104.2
Both hinged	9.87	39.48	88.8

4.59d

Q: How is the vortex shedding frequency f_e determined?

A: f_e is obtained from the Strouhal number S:

$$S = f_e d_o / 12V \qquad (112)$$

where

d_o = tube outer diameter, in.
V = gas velocity, ft/sec

S is available in the form of charts for various tube pitches; it typically ranges from 0.2 to 0.3 (see Figure 4.22) [1, 8].

Q4.59e shows how a tube bundle is analyzed for noise and vibration.

4.59e

Q: A tubular air heater 11.7 ft wide, 12.5 ft deep, and 13.5 ft high is used in a boiler. Carbon steel tubes of 2 in. O.D. and 0.08 in. thickness are used in in-line fashion with a transverse pitch of 3.5 in. and longitudinal pitch of 3.0 in. There are 40 tubes wide (3.5 in. pitch) and 60 tubes deep (2.5 in. pitch). Air flow across the tubes is 300,000 lb/hr at an average temperature of 219°F. The tubes are fixed at both ends in tube sheets. Check if bundle vibrations are likely. Tube mass per unit length = 1.67 lb/ft.

A: First compute f_a, f_e, and f_n. L = 13.5 ft, d_o = 2 in., d_i = 1.84 in., M_e = 1.67 lb/ft, and, from Table 4.40, C = 22.37.

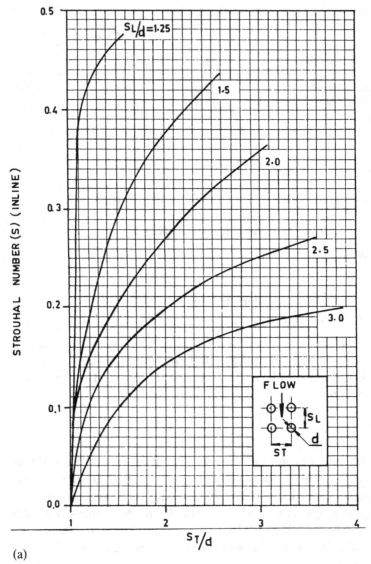

(a)

Figure 4.22 Strouhal number (a) for in-line bank of tubes; (b) for staggered bank of tubes; (c) for staggered bank of tubes; (d) for in-line bank of tubes.

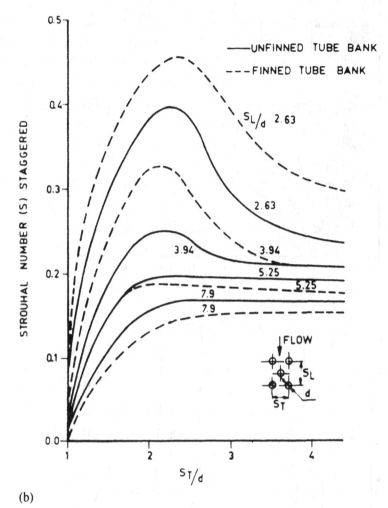

(b)

Using Eq. (110b), we have

$$f_n = \frac{90 \times 22.37}{(13.5)^2} \times \frac{(2^4 - 1.84^4)^{0.5}}{(1.67)^{0.5}} = 18.2 \text{ Hz}$$

This is in mode 1. In mode 2, $C = 61.67$; hence f_{n2} is 50.2 Hz. (The first two modes are important.)

Let us compute f_e. S from Figure 4.22 for $S_T/d_o = 3.5/2 = 1.75$ and a longitudinal pitch of $3.0/2 = 1.5$ is 0.33.

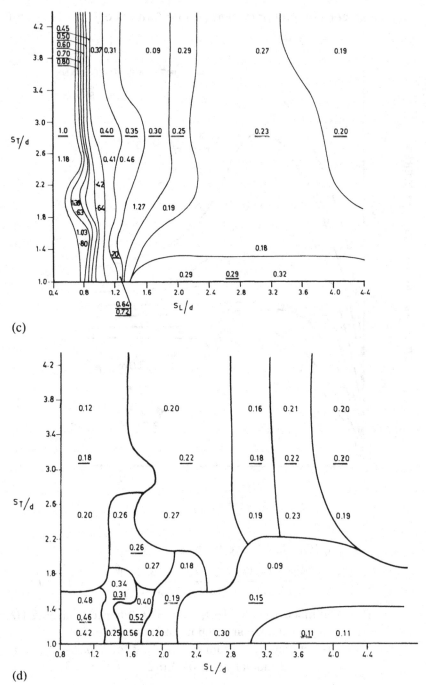

(c)

(d)

Figure 4.22 Continued

From Eq. (1) of Chapter 1, $\rho = 40/(219 + 460) = 0.059$ lb/ cu ft.

Free gas area $= 40 \times (3.5 - 2) \times 13.5/12 = 67.5$ lb/ft^2 hr. (13.5 is the tube length, and 40 tubes wide is used with a pitch of 3.5 in.) Hence air velocity across tubes is

$$V = \frac{300,000}{67.5 \times 3600 \times 0.059} = 21 \text{ ft/sec}$$

Hence

$$f_e = \frac{12SV}{d_o} = 12 \times \frac{0.33 \times 21}{2} = 41.6 \text{ Hz}$$

Let us compute f_a. $T = (219 + 460) = 679°R$. Hence $V_s = 49 \times 679^{0.5} = 1277$ ft/sec. Width $W = 11.7$ ft, and $\lambda = 2 \times 11.7 = 23.4$ ft. For mode 1 or $n = 1$,

$$f_{a1} = 1277/23.4 = 54.5 \text{ Hz}$$

For $n = 2$,

$$f_{a2} = 54.5 \times 2 = 109 \text{ Hz}$$

The results for modes 1 and 2 are summarized in Table 4.41. It can be seen that without baffles the frequencies f_a and f_e are within 20% of each other. Hence noise problems are likely to arise. If a baffle or plate is used to divide the duct width into two regions, the acoustic frequency is doubled as the wavelength or width is halved. This is a practical solution to acoustic vibration problems.

Table 4.41   Summary of Frequencies
for Modes 1 and 2

Mode of vibration n	1	2
f_n, (cps or Hz)	18.2	50.2
f_e, (cps or Hz)	41.6	41.6
f_a (without baffles)	54.5	109
f_a (with one baffle)	109	218

4.60

Q: How are the gas properties C_p, μ, and k estimated for a gaseous mixture? Determine C_p, μ, and k for a gas mixture having the following analysis at 1650°F and 14.7 psia.

Gas	Vol%	C_p	μ	k	MW
N_2	80	0.286	0.108	0.030	28
O_2	12	0.270	0.125	0.043	32
SO_2	8	0.210	0.105	0.040	64

Mixture properties are needed to evaluate heat transfer coefficients. For flue gas obtained from the combustion of fossil fuels, in the absence of flue gas analysis, one can use the data on air.

A: For a gaseous mixture at atmospheric pressure, the following relations apply. For high gas pressures, readers are referred to Ref. 1.

$$\mu_m = \frac{\Sigma y_i \mu_i \sqrt{MW_i}}{\Sigma y_i \sqrt{MW_i}} \tag{113a}$$

$$k_m = \frac{\Sigma y_i k_i \sqrt[3]{MW_i}}{y_i \sqrt[3]{MW_i}} \tag{113b}$$

$$C_{pm} = \frac{\Sigma C_{pi} \, MW \times y_i}{\Sigma \, MW \times y_i} \tag{113c}$$

where

$\qquad$ MW = molecular weight

$\qquad\quad$ y = volume fraction of any constituent

Subscript m stands for mixture.

$\qquad$ Substituting in Eqs. (113), we have

$$C_{pm} = \frac{0.286 \times 0.8 \times 28 + 0.27 \times 0.12 \times 32 + 0.21 \times 0.08 \times 64}{0.8 \times 28 + 0.12 \times 32 + 0.08 \times 64}$$

$$= 0.272 \text{ Btu/lb °F}$$

$$k_m = \frac{0.03 \times 28^{1/3} \times 0.80 + 0.043 \times 32^{1/3} \times 0.12 + 0.04 \times 64^{1/3} \times 0.08}{28^{1/3} \times 0.80 + 32^{1/3} \times 0.12 + 64^{1/3} \times 0.08}$$

$$= 0.032 \text{ Btu/ft hr } °F$$

$$\mu_m = \frac{0.108 \times \sqrt{28} \times 0.8 + 0.125 \times \sqrt{32} \times 0.12 + 0.105 \times \sqrt{64} \times 0.08}{\sqrt{28} \times 0.8 + \sqrt{32} \times 0.12 + \sqrt{64} \times 0.105}$$

$$= 0.109 \text{ lb/ft hr}$$

4.61

Q: How do gas analysis and pressure affect heat transfer performance?

A: The presence of gases such as hydrogen and water vapor increases the heat transfer coefficient significantly, which can affect the heat flux and the boiler size. Also, if the gas is at high pressure, say 100 psi or more, the mass velocity inside the tubes (fire tube boilers) or outside the boiler tubes (water tube boilers) can be much higher due to the higher density, which also contributes to the higher heat transfer coefficients. Table 4.42 compares two gas streams, reformed gases from a hydrogen plant and flue gases from combustion of natural gas.

Factors C and F used in the estimation of heat transfer coefficients inside and outside the tubes are also given in Table 4.42. It can be seen that the effect of gas analysis is very significant. Even at low gas pressures of reformed gases (50 to 100 psig), the factors C and F would be very close to the values shown, within 2 to 5%.

4.62

Q: How does gas pressure affect the heat transfer coefficient?

A: The effect of gas pressure on factors C and F for some common gases is shown in Figures 4.23 and 4.24. It can be seen that the

Table 4.42　Effect of Gas Analysis on Heat Transfer

	Reformed gas		Flue gas	
CO_2, % vol	5.0		17.45	
H_2O, % vol	38.0		18.76	
N_2, % vol	—		62.27	
O_2, % vol	—		1.52	
CO, % vol	9.0		—	
H_2, % vol	45.0		—	
CH_4, % vol	3.0		—	
Gas pressure, psia	400		15	
Temp, °F	1550	675	1540	700
C_p, Btu/lb °F	0.686	0.615	0.320	0.286
μ, lb/ft hr	0.087	0.056	0.109	0.070
k, Btu/ft hr °F	0.109	0.069	0.046	0.028
Factor C[a]	0.571		0.225	
Factor F[a]	0.352		0.142	

[a]$C = (C_p/\mu)^{0.4} k^{0.6};$　$F = C_p^{0.33} k^{0.67}/\mu^{0.27}.$

pressure effect becomes smaller at high gas temperatures, while at low temperatures there is a significant difference. Also, the pressure effect is small and can be ignored up to a gas pressure of 200 psia.

4.63

Q: How do we convert gas analysis in % by weight to % by volume?

A: One of the frequent calculations performed by heat transfer engineers is the conversion from weight to volume basis and vice versa. The following example shows how this is done.

EXAMPLE

A gas contains 3% CO_2, 6% H_2O, 74% N_2, and 17% O_2 by weight. Determine the gas analysis in % by volume.

Solution.

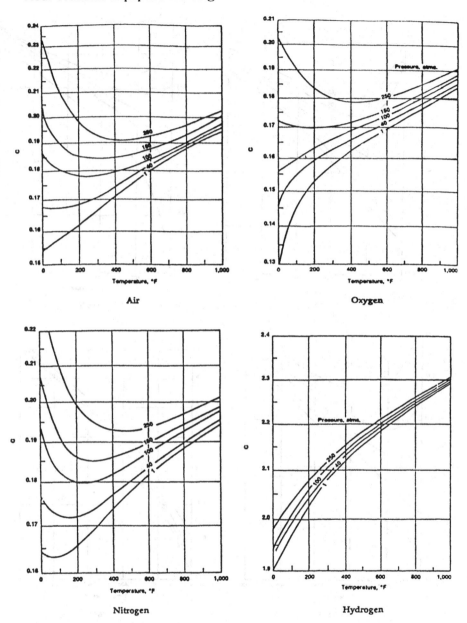

Figure 4.23 Effect of pressure on heat transfer—flow inside tubes. (From Ref. 1.)

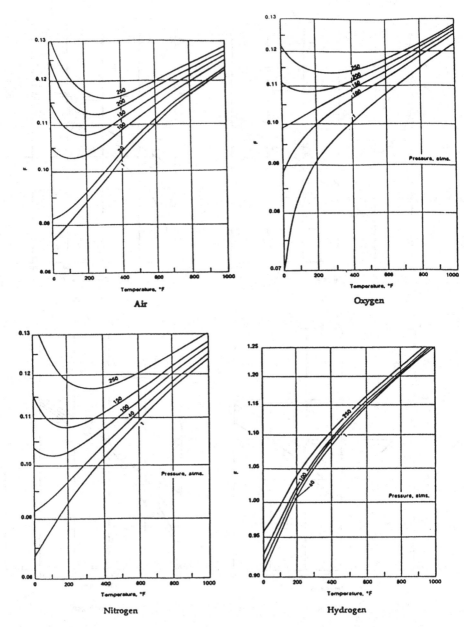

Figure 4.24 Effect of pressure on heat transfer—flow outside tubes. (From Ref. 1.)

Gas	% Weight	MW	Moles	% Volume
CO_2	3	44	0.06818	1.91
H_2O	6	18	0.3333	9.32
N_2	74	28	2.6429	73.91
O_2	17	32	0.5312	14.86
		Total	3.57563	100

Moles of a gas are obtained by dividing the weight by the molecular weight; moles of CO_2 = 3/44 = 0.06818.

The volume of each gas, then, is the mole fraction × 100. Percent volume of O_2 = (0.5312/3.57563) × 100 = 14.86, and so on. One can work in reverse and convert from volume (or mole) basis to weight basis.

NOMENCLATURE

A	Surface area, ft^2
A_f, A_t, A_i, A_o	fin, total, inside, and obstruction surface areas, ft^2/ft
A_w	Area of tube wall, ft^2/ft
B	Factor used in Grimson's correlation
b	Fin thickness, in.
C	Factor used to estimate heat transfer coefficient
C_p	Specific heat, Btu/lb °F; subscripts g, w, m stand for gas, water, and mixture
C1–C6	Factors used in heat transfer and pressure drop calculations for finned tubes
D	Exchanger diameter, in.
d, d_i	Tube outer and inner diameter, in.
e	Escalation factor used in life-cycling costing calculations; base of natural logarithm
E	Efficiency of HRSG or fins
f	Frequency, Hz or cps; subscripts a, e, n stand for acoustic, vortex-shedding, and natural
ff	Fouling factor, ft^2 hr °F/Btu; subscripts i and o stand for inside and outside

F	Factor used in the estimation of outside heat transfer coefficient and in the estimation of capitalized costs
G	Gas mass velocity, lb/ft^2 hr
h	Fin height, in.
h_c	Convective heat transfer coefficient, Btu/ft^2 hr °F
h_i, h_o	Heat transfer coefficients inside and outside tubes, Btu/ft^2 hr °F
h_{lf}	Heat loss factor, fraction
h_N	Nonluminous heat transfer coefficient, Btu/ft^2 hr °F
Δh	Change in enthalpy, Btu/lb
i	Interest rate
k	Thermal conductivity, Btu/ft hr °F or Btu in./ft^2 hr °F; subscript m stands for mixture
K_m	Metal thermal conductivity, Btu/ft hr °F
K_1, K_2	Constants
L	Length, ft; thickness of insulation, in.; or beam length
L_e	Equivalent thickness of insulation, in.
m	Factor used in Eq. (25b)
M_c	Water equivalent, Btu/°F
M_e	Weight of tube, lb/ft
MW	Molecular weight
n	Number of fins per inch
N	Constant used in Grimson's correlation; also number of tubes
Nu	Nusselt number
NTU	Number of transfer units
P	Term used in temperature cross-correction
p_w, p_c	Partial pressure of water vapor and carbon dioxide
Pr	Prandtl number
Q	Energy transferred, Btu/hr; heat flux, Btu/ft^2 hr
q	Heat flux, heat loss, Btu/ft^2 hr
q_c	Critical heat flux, Btu/ft^2 hr
R	Thermal resistance, ft^2 hr °F/Btu; subscripts i, o, and t stand for inside, outside, and total
Re	Reynolds number
R_m	Metal thermal resistance, ft^2 hr °F/Btu
S	Fin clearance, in.; Strouhal number

S_T, S_L	Transverse and longitudinal pitch, in.
t	Fluid temperature, °F; subscripts a, s, b, stand for ambient, surface, fin base
t_f	Fin tip temperature, °F
t_m	Metal temperature, °F
t_{sat}	Saturation temperature, °F
T	Absolute temperature, K or °R; subscripts g and w stand for gas and wall
ΔT	Log-mean temperature difference, °F
U	Overall heat transfer coefficient, Btu/ft^2 hr °F
V	Fluid velocity, ft/sec or ft/min
V_s	Sonic velocity, ft/sec
W	Fluid flow, lb/hr; subscripts g, s, w stand for gas, steam, and water
w	Flow per tube, lb/hr
x	Steam quality, fraction
y	Volume fraction of gas
ϵ	Effectiveness factor
ϵ_c, ϵ_w	Emissivity of CO_2 and H_2O
ϵ_g	Gas emissivity
$\Delta\epsilon$	Emissivity correction term
η	Fin effectiveness
μ	Viscosity, lb/ft hr; subscript m stands for mixture
ϕ	Fin efficiency
ρ	gas density, lb/cu ft
λ	wavelength, ft
ν	ratio of specific heats

REFERENCES

1. V. Ganapathy, *Applied Heat Transfer*, PennWell Books, Tulsa, Okla., 1982.
2. D. Q. Kern, *Process Heat Transfer*, McGraw-Hill, New York, 1950.
3. V. Ganapathy, Nomogram determines heat transfer coefficient for water flowing pipes or tubes, *Power Engineering*, July 1977, p. 69.
4. V. Ganapathy, Charts simplify spiral finned tube calculations, *Chemical Engineering*, Apr. 25, 1977, p. 117.

5.   V. Ganapathy, Estimate nonluminous radiation heat transfer coefficients, *Hydrocarbon Processing*, April 1981, p. 235.

6.   V. Ganapathy, Evaluate the performance of waste heat boilers, *Chemical Engineering*, Nov. 16, 1981, p. 291.

7.   W. C. Turner and J. F. Malloy, *Thermal Insulation Handbook*, McGraw-Hill, New York, 1981, pp. 40–45.

8.   V. Ganapathy, *Waste Heat Boiler Deskbook*, Fairmont Press, Atlanta, 1991.

9.   ESCOA Corp., *ESCOA Fintube Manual*, Tulsa, Okla., 1979.

10. V. Ganapathy, Evaluate extended surfaces carefully, *Hydrocarbon Processing*, October 1990, p. 65.

11. V. Ganapathy, Fouling—the silent heat transfer thief, *Hydrocarbon Processing*, October 1992, p. 49.

12. V. Ganapathy, HRSG temperature profiles guide energy recovery, *Power*, September 1988.

13. W. Roshenow and J. P. Hartnett, *Handbook of Heat Transfer*, McGraw-Hill, New York, 1972.

Fans, Pumps, and Steam Turbines

ing motor current consumption to pump flow and head; analyzing for pump problems

5.01

Q: How is the steam rate for steam turbines determined?

A: The actual steam rate (ASR) for a turbine is given by the equation

$$ASR = \frac{3413}{\eta_t \times (h_1 - h_{2s})} \tag{1}$$

where ASR is the actual steam rate in lb/kWh. This is the steam
flow in lb/hr required to generate 1 kW of electricity. h_1 is the
steam enthalpy at inlet to turbine, Btu/lb, and h_{2s} is the steam
enthalpy at turbine exhaust pressure if the expansion is assumed
to be isentropic, Btu/lb. That is, the entropy is the same at inlet
condition and at exit. Given h_1, h_{2s} can be obtained either from
the Mollier chart or by calculation using steam table data (see the
Appendix). η_t is the efficiency of the turbine, expressed as a
fraction. Typically, η_t ranges from 0.65 to 0.80.

Another way to estimate ASR is to use published data on
turbine theoretical steam rates (TSRs) (see Table 5.1).

$$TSR = \frac{3413}{h_1 - h_{2s}} \tag{2}$$

TSR divided by η_t gives ASR. The following example shows
how the steam rate can be used to find required steam flow.

Table 5.1 Theoretical Steam Rates for Steam Turbines at Some Common Conditions (lb/kWh)

Exhaust pressure	150 psig, 366°F, saturated	200 psig, 388°F, saturated	200 psig, 500°F, 94°F, superheat	400 psig, 750°F, 302°F, superheat	600 psig, 750°F, 261°F, superheat	600 psig, 825°F, 336°F, superheat	850 psig, 825°F, 298°F, superheat
				Inlet			
2 in. Hg	10.52	10.01	9.07	7.37	7.09	6.77	6.58
4 in. Hg	11.76	11.12	10.00	7.99	7.65	7.28	7.06
0 psig	19.37	17.51	15.16	11.20	10.40	9.82	9.31
10 psig	23.96	21.09	17.90	12.72	11.64	10.96	10.29
30 psig	33.6	28.05	22.94	15.23	13.62	12.75	11.80
50 psig	46.0	36.0	28.20	17.57	15.36	14.31	13.07
60 psig	53.9	40.4	31.10	18.75	16.19	15.05	13.66
70 psig	63.5	45.6	34.1	19.96	17.00	15.79	14.22
75 psig	69.3	48.5	35.8	20.59	17.40	16.17	14.50

Source: R. H. Perry and C. H. Chilton, *Chemical Engineers' Handbook*, 5th ed., McGraw-Hill, New York, 1974, pp. 18–20.

EXAMPLE

How many lb/hr of superheated steam at 1000 psia, 900°F, is required to generate 7500 kW in a steam turbine if the back pressure is 200 psia and the overall efficiency of the turbine generator system is 70%?

Solution. From the steam tables, at 1000 psia, 900°F, $h_1 = 1448.2$ Btu/lb and entropy $s_1 = 1.6121$ Btu/lb °F. At 200 psia, corresponding to the same entropy, we must calculate h_{2s} by interpolation. We can note that steam is in superheated condition. $h_{2s} = 1257.7$ Btu/lb. Then

$$\text{ASR} = \frac{3413}{0.70 \times (1448 - 1257.7)} = 25.6 \text{ lb/kWh}$$

Hence, to generate 7500 kW, the steam flow required is

$$W_s = 25.6 \times 7500 = 192,000 \text{ lb/hr}$$

5.02a

Q: What is cogeneration? How does it improve the efficiency of the plant?

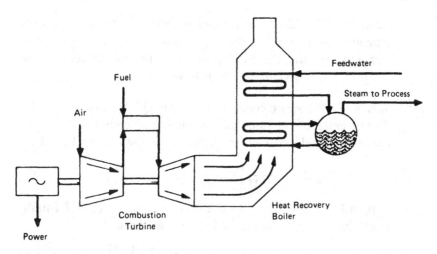

Figure 5.1   Cogeneration produces power and steam from the same fuel source by converting the turbine exhaust heat in a boiler, which produces steam for process.

A: Cogeneration is the term used for simultaneous generation of power and process steam from a single full source, as in a system of gas turbine and process waste heat boiler, wherein the gas turbine generates electricity and the boiler generates steam for process (see Figure 5.1).

   In a typical power plant that operates at 35 to 43% overall efficiency, the steam pressure in the condenser is about 2 to 4 in. Hg. A lot of energy is wasted in the cooling water, which condenses the steam in the condenser.

   If, instead, the steam is generated at a high pressure and expanded in a steam turbine to the process steam pressure, we can use the steam for process, and electricity is also generated. A full credit for the process steam can be given if the steam is used— hence the improvement in overall energy utilization. Q5.02b explains this in detail.

5.02b

Q: 50,000 lb/hr of superheated steam at 1000 psia and 900°F is available in a process plant. One alternative is to expand this in a steam turbine to 200 psia and use the 200-psia steam for process (cogeneration). Another alternative is to expand the superheated steam in a steam turbine to 1 psia, generating electricity alone in a power plant. Evaluate each scheme.

A: *Scheme 1*. The steam conditions are as in Q5.01, so let us use the data on enthalpy. Assume that the turbine efficiency is 70%. The electricity produced can be written as follows using Eq. (1):

$$P = W_s \eta_t \times \frac{h_1 - h_{2s}}{3413} \tag{3}$$

P is in kilowatts. $h_1 = 1448$ Btu/lb and $h_{2s} = 1257.7$ Btu/lb, from Q5.01. Substituting in Eq. (3), we have

$$P = 50,000 \times (1448 - 1257.7) \times \frac{0.70}{3413} = 1954 \text{ kW}$$

Now let us calculate the final enthalpy at condition 2, h_2. Using the equation

$$\eta_t(h_1 - h_{2s}) = h_1 - h_2 \tag{4}$$

we obtain

$$0.70 \times (1448 - 1257.7) = 1448 - h_2$$

or

$$h_2 = 1315 \text{ Btu/lb}$$

This enthalpy is available for process in the cogeneration mode. The energy Q available is the cogeneration mode is the sum of the electricity produced and the energy to process, all in Btu/hr. Hence the total energy is

$$Q = 1954 \times 3413 + 50,000 \times 1315 = 72.4 \times 10^6 \text{ Btu/hr}$$

Scheme 2. Let us take the case when electricity alone is generated. Let us calculate the final steam conditions at a pressure of 1 psia. $s_1 = 1.6121 = s_{2s}$. At 1 psia, from the steam tables, at saturated conditions, $s_f = 0.1326$ and $s_g = 1.9782$. s_f and s_g are entropies of saturated liquid and vapor. Since the entropy s_{2s} is in between s_f and s_g, the steam at isentropic conditions is wet. Let us estimate the quality x. From basics,

$$0.1326(1 - x) + 1.9782x = 1.6121$$

Hence

$$x = 0.80$$

The enthalpy corresponding to this condition is

$$h = (1 - x)h_f + xh_g$$

or

$$h_{2s} = 0.80 \times 1106 + 0.2 \times 70 = 900 \text{ Btu/lb}$$

(h_f and h_g are 70 and 1106 at 1 psia.) Using a turbine efficiency of 75%, from Eq. (3) we have

$$P = 50,000 \times (1448 - 900) \times \frac{0.75}{3413} = 6023 \text{ kW}$$

$$= 20.55 \times 10^6 \text{ Btu/hr}$$

Hence we note that there is a lot of difference in the energy pattern between the two cases, with the cogeneration scheme using much more energy than that used in Scheme 2.

Even if the steam in Scheme 1 were used for oil heating, the latent heat of 834 Btu/lb at 200 psia could be used.

$$\text{Total output} = 1954 \times 3413 + 50,000 \times 834$$

$$= 48.3 \times 10^6 \text{ Btu/hr}$$

This is still more than the output in the case of power generation alone.

Note, however, that if the plant electricity requirement were more than 2000 kW, Scheme 1 should have more steam availability, which means that a bigger boiler should be available. Evaluation of capital investment is necessary before a particular scheme is chosen. However, it is clear that in cogeneration the utilization of energy is better.

5.03

Q: Which is a better location for tapping steam for deaeration in a cogeneration plant with an extraction turbine, the HRSG or the steam turbine?

A: When steam is taken for deaeration from the HRSG and not from an extraction point in a steam turbine, there is a net loss to the system power output as the steam is throttled and not expanded to the lower deaerator pressure. Throttling is a mere waste of energy, whereas steam generates power while it expands to a lower pressure. To illustrate, consider the following example.

EXAMPLE

An HRSG generates 80,000 lb/hr of steam at 620 psig and 650°F from 550,000 lb/hr of turbine exhaust gases at 975°F. The steam is expanded in an extraction-condensing steam turbine. Figure 5.2 shows the two schemes. The condenser operates at 2.5 in. Hg abs. The deaerator is at 10 psig. Blowdown losses = 2%. Neglecting flash steam and vent flow, we can show that when steam is taken for deaeration from the HRSG,

$$81,700 \times 208 = 1700 \times 28 + (80,000 - X) \times 76 + 1319X$$

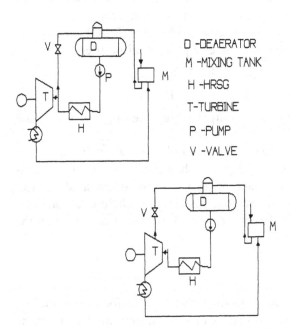

D –DEAERATOR

M –MIXING TANK

H –HRSG

T–TURBINE

P –PUMP

V –VALVE

Figure 5.2   Options for taking steam for deaeration.

where 208, 28, 76, and 1319 are enthalpies of feedwater at 240°F, makeup water at 60°F, condensate at 108°F, and steam at 620 psig, 650°F.

The deaeration steam X = 8741 lb/hr; use 8785 to account for losses. Now compute the actual steam rate (ASR) in the steam turbine (see Q5.01). It can be shown that ASR = 11.14 lb/kWh at 70% expansion efficiency; hence power output of the turbine generator = 0.96 × (80,000 − 8785)/11.14 = 6137 kW, assuming 4% loss in the generator.

Similarly, when steam is taken at 30 psia from the extraction point in the steam turbine, the enthalpy of steam for deaeration is 1140.6 Btu/lb. An energy balance around the deaerator shows

$$81,700 \times 208 = 1140.6X + (80,000 - X) \times 76$$
$$+ 1700 \times 28$$

Hence X = 10,250 lb/hr. Then ASR for expansion from 620 psig to 30 psia = 19 lb/kWh and 11.14 for the remaining flow. The power output is

$$P = 0.96 \times \left(\frac{10,250}{19} + \frac{80,000 - 10,250}{11.14} \right) = 6528 \text{ kW}$$

Thus a significant difference in power output can be seen. However, one has to review the cost of extraction machine versus the straight condensing type and associated piping, valves, etc.

5.04

Q: A fan develops an 18-in. WC static head when the flow is 18,000 acfm and static efficiency of the fan is 75%. Determine the brake horsepower required, the horsepower consumed when the motor has an efficiency of 90%, and the annual cost of operation if electricity costs 5 cents/kWh and the annual period of operation is 7500 hr.

A: The power required when the flow is q acfm and the head is H_w in. WC is

$$\text{BHP} = q \times \frac{H_w}{6356\eta_f} \tag{5}$$

where η_f is the efficiency of the fan, fraction; in this case, $\eta_f = 0.75$.

The horsepower consumed is

$$HP = \frac{BHP}{\eta_m} \tag{6}$$

where η_m is the motor efficiency, fraction. Substituting the data, we have

$$BHP = 18,000 \times \frac{18}{0.75 \times 6356} = 68 \text{ hp}$$

and

$$HP = \frac{68}{0.9} = 76 \text{ hp}$$

The annual cost of operation will be

$$76 \times 0.74 \times 0.05 \times 7500 = \$21,261$$

(0.74 is the conversion factor from hp to kW.)

5.05

Q: A fan develops 18,000 acfm at 18 in. WC when the ambient conditions are 80°F and the elevation is 1000 ft (case 1). What are the flow and the head developed by the fan when the temperature is 60°F and the elevation is 5000 ft (case 2)?

A: The head developed by a fan would vary with density as follows:

$$\frac{H_{w1}}{\rho_1} = \frac{H_{w2}}{\rho_2} \tag{7}$$

where ρ is the density, lb/cu ft, and the subscripts 1 and 2 refer to any two ambient conditions.

The flow q in acfm developed by a fan would remain the same for different ambient conditions; however, the flow in lb/hr would vary as the density changes.

Let us use Table 5.2 for quick estimation of density as a function of elevation and temperature. $\rho = 0.075/\text{factor}$ from Table 5.2. At 80°F and 1000 ft. elevation,

$$\rho_1 = \frac{0.075}{1.06} = 0.0707 \text{ lb/cu ft}$$

At 60°F and 5000 ft,

$$\rho_2 = \frac{0.075}{1.18} = 0.0636 \text{ lb/cu ft}$$

Substitution into Eq. (7) yields

$$\frac{18}{0.0707} = \frac{H_{w2}}{0.0636}$$

$H_{w2} = 16.1$ in. WC

In case 1 flow will be

$18,000 \times 0.0707 \times 60 = 76,356$ lb/hr

and in case 2 the flow will be

$18,000 \times 0.0636 \times 60 = 68,638$ lb/hr

The exact operating point of the fan can be obtained after plotting the new H_w versus q characteristic and noting the point of intersection of the new curve with the system resistance curve.

5.06a

Q: Why should the capacity of forced draft fans for boilers be reviewed at the lowest density condition?

A: For the same heat input to boilers, the air quantity required in mass flow units (lb/hr) remains the same irrespective of the ambient conditions.

$$W = 60 \rho q$$

where

W = mass flow, lb/hr
ρ = density, lb/cu ft
q = volumetric flow, acfm

Fans discharge constant volumetric flow at any density. Hence if the fan is sized to give a particular volumetric flow at the high-

Table 5.2 Temperature and Elevation Factors

Temp. (°F)	Altitude (ft) and barometric pressure (in. Hg)												
	0	500	1000	1500	2000	2500	3000	3500	4000	4500	5000	5500	6000
	29.92	29.38	28.86	28.33	27.82	27.31	26.82	26.32	25.84	25.36	24.90	24.43	23.96
−40	.79	.81	.82	.84	.85	.87	.88	.90	.92	.93	.95	.97	.99
0	.87	.88	.90	.92	.93	.95	.97	.99	1.00	1.02	1.04	1.06	1.08
40	.94	.96	.98	1.00	1.01	1.03	1.05	1.07	1.09	1.11	1.13	1.16	1.18
70	1.00	1.02	1.04	1.06	1.08	1.10	1.12	1.14	1.16	1.18	1.20	1.22	1.25
80	1.02	1.04	1.06	1.08	1.10	1.12	1.14	1.16	1.18	1.20	1.22	1.25	1.27
100	1.06	1.08	1.10	1.12	1.14	1.16	1.18	1.20	1.22	1.25	1.27	1.29	1.32
120	1.09	1.11	1.13	1.16	1.18	1.20	1.22	1.24	1.27	1.29	1.31	1.34	1.37
140	1.13	1.15	1.17	1.20	1.22	1.24	1.26	1.29	1.31	1.34	1.36	1.39	1.41
160	1.17	1.19	1.21	1.24	1.26	1.28	1.31	1.33	1.35	1.38	1.41	1.43	1.46
180	1.21	1.23	1.25	1.28	1.30	1.32	1.35	1.37	1.40	1.42	1.45	1.48	1.51
200	1.25	1.27	1.29	1.32	1.34	1.36	1.39	1.42	1.44	1.47	1.50	1.53	1.55

	1.34	1.36	1.39	1.41	1.44	1.47	1.49	1.52	1.55	1.58	1.61	1.64	1.67
250	1.34	1.36	1.39	1.41	1.44	1.47	1.49	1.52	1.55	1.58	1.61	1.64	1.67
300	1.43	1.46	1.49	1.51	1.54	1.57	1.60	1.63	1.66	1.69	1.72	1.76	1.79
350	1.53	1.56	1.58	1.61	1.64	1.67	1.70	1.74	1.77	1.80	1.84	1.87	1.91
400	1.62	1.65	1.68	1.71	1.75	1.78	1.81	1.84	1.88	1.91	1.95	1.99	2.02
450	1.72	1.75	1.78	1.81	1.85	1.88	1.92	1.95	1.99	2.03	2.06	2.10	2.14
500	1.81	1.84	1.88	1.91	1.95	1.98	2.02	2.06	2.10	2.14	2.18	2.22	2.26
550	1.91	1.94	1.98	2.01	2.05	2.09	2.13	2.17	2.21	2.25	2.29	2.33	2.38
600	2.00	2.04	2.07	2.11	2.15	2.19	2.23	2.27	2.32	2.36	2.40	2.45	2.50
650	2.09	2.13	2.17	2.21	2.25	2.29	2.34	2.38	2.43	2.47	2.52	2.56	2.61
700	2.19	2.23	2.27	2.31	2.35	2.40	2.44	2.49	2.53	2.58	2.63	2.68	2.73
750	2.28	2.32	2.37	2.41	2.46	2.50	2.55	2.60	2.64	2.69	2.74	2.80	2.85
800	2.38	2.42	2.46	2.51	2.56	2.60	2.65	2.70	2.75	2.80	2.86	2.91	2.97
850	2.47	2.52	2.56	2.61	2.66	2.71	2.76	2.81	2.86	2.92	2.97	3.03	3.08
900	2.57	2.61	2.66	2.71	2.76	2.81	2.86	2.92	2.97	3.03	3.08	3.14	3.20
950	2.66	2.71	2.76	2.81	2.86	2.91	2.97	3.02	3.08	3.14	3.20	3.26	3.32
1000	2.76	2.81	2.86	2.91	2.96	3.02	3.07	3.13	3.19	3.25	3.31	3.37	3.44

density condition, the mass flow would decrease when density decreases as can be seen in the equation above. Hence the fan must be sized to deliver the volumetric flow at the lowest density condition, in which case the output in lb/hr will be higher at the higher density condition, which can be then controlled.

Also, the gas pressure drop ΔP in in. WC across the wind box is proportional to W^2/ρ. If the air density decreases as at high-temperature conditions, the pressure drop increases, because W remains unchanged for a given heat input. Considering the fact that H/ρ is a constant for a given fan, where H is the static head in in. WC, using the lowest ρ ensures that the head available at higher density will be larger.

5.06b

Q: How does the horsepower of a forced draft fan for boilers or heaters change with density?

A: Equation (5) gives the fan horsepower:

$$\text{BHP} = \frac{q \, H_w}{6356 \, \eta_f}$$

Using the relation $W = 60 \, q \, \rho$, we can rewrite the above as

$$\text{BHP} = \frac{W \, H_w}{381{,}360 \, \rho \, \eta_f}$$

For a boiler at a given duty, the air flow in lb/hr and the head in in. WC, H_w, remain unchanged, and hence as the density decreases, the horsepower increases. This is yet another reason to check the fan power at the lowest density condition. However, if the application involves an uncontrolled fan that delivers a given volume of air at all densities, then the horsepower should be evaluated at the highest density case as the mass flow would be higher as well as the gas pressure drop.

5.07

Q: A triplex reciprocating pump is used for pumping 40 gpm (gallons per minute) of water at 100°F. The suction pressure is 4 psig and the discharge pressure is 1000 psig. Determine the BHP required.

A: Use the expression

$$\mathrm{BHP} = q \times \frac{\Delta P}{1715\eta_p} \tag{8}$$

where

q = flow, gpm
ΔP = differential pressure, psi
η_p = pump efficiency, fraction

In the absence of data on pumps, use 0.9 for triplex and 0.92 for quintuplex pumps.

$$\mathrm{BHP} = 40 \times \frac{1000 - 4}{1715 \times 0.90} = 25.8 \text{ hp}$$

A 30-hp motor can be used.

The same expression can be used for centrifugal pumps. The efficiency can be obtained from the pump characteristic curve at the desired operating point.

5.08

Q: A pump is required to develop 230 gpm of water at 60°F at a head of 970 ft. Its efficiency is 70%. There are two options for the drive: an electric motor with an efficiency of 90% or a steam turbine drive with a mechanical efficiency of 95%. Assume that the exhaust is used for process and not wasted.

If electricity costs 50 mills/kWh, steam for the turbine is generated in a boiler with an efficiency of 85% (HHV basis), and fuel costs $3/MM Btu (HHV basis), determine the annual cost of operation of each drive if the plant operates for 6000 hr/year.

A: Another form of Eq. (8) is

$$\mathrm{BHP} = W \times \frac{H}{1,980,000\eta_p} \tag{9}$$

where

W = flow, lb/hr
H = head developed by the pump, ft of liquid

For relating head in ft with differential pressure in psi or flow in lb/hr with gpm, refer to Q1.01. Substituting into Eq. (9) and assuming that $s = 1$, $W = 230 \times 500$ lb/hr,

$$BHP = 230 \times 500 \times \frac{970}{0.70 \times 1,980,000} = 81 \text{ hp}$$

The annual cost of operation with an electric motor drive will be

$$81 \times 0.746 \times 0.05 \times \frac{6000}{0.90} = \$20,142$$

(0.746 is the conversion factor from hp to kW.)
   If steam is used, the annual cost of operation will be

$$81 \times 2545 \times 6000 \times \frac{3}{0.85 \times 0.90 \times 10^6} = \$4595$$

(2545 Btu/hr = 1 hp; 0.85 is the boiler efficiency; 0.95 is the mechanical efficiency.) Hence the savings in cost of operation is $(20,142 - 4545) = \$15,547$/year.

   Depending on the difference in investment between the two drives, payback can be worked out. In the calculation above it was assumed that the back-pressure steam was used for process. If it was wasted, the economics may not work out the same way.

5.09a

Q: How does the specific gravity or density of liquid pumped affect the BHP, flow, and head developed?

A: A pump always delivers the same flow in gpm (assuming that viscosity effects can be neglected) and head in feet of liquid at any temperature. However, due to changes in density, the flow in lb/hr, pressure in psi, and BHP would change. A variation of Eq. (9) is

$$BHP = \frac{q \times \Delta P}{1715\eta_p} = \frac{W \times \Delta P}{857,000\eta_p s} \qquad (10)$$

where

q = liquid flow, gpm
W = liquid flow, lb/hr
s = specific gravity
ΔP = pressure developed, psi
H = head developed, ft of liquid

Also,

$$H = 2.31 \times \frac{\Delta P}{s} \tag{11}$$

EXAMPLE

If a pump can develop 1000 gpm of water at 40°F through 1000 ft, what flow and head can it develop when the water is at 120°F? Assume that pump efficiency is 75% in both cases.

Solution. s_1 at 40°F (from the steam tables; see the Appendix) is 1. s_2 at 120°F is 0.988.

$$\Delta P_1 = 1000 \times \frac{1}{2.31} = 433 \text{ psi}$$

From Eq. (11),

$$BHP_1 = 1000 \times \frac{433}{0.75 \times 1715} = 337 \text{ hp}$$

$$W_1 = 500 q_1 s_1 = 500 \times 1000 \times 1 = 500,000 \text{ lb/hr}$$

At 120°F,

$$\Delta P_2 = 1000 \times \frac{0.988}{2.31} = 427 \text{ psi}$$

$$BHP_2 = 1000 \times \frac{427}{0.75 \times 1715} = 332 \text{ hp}$$

$$W_2 = 500 \times 0.988 \times 1000 = 494,000 \text{ lb/hr}$$

If the same W is to be maintained, BHP must increase.

5.09b

Q: How does the temperature of water affect pump power consumption?

A: The answer can be obtained by analyzing the following equations for pump power consumption. One is based on flow in gpm and the other in lb/hr.

$$BHP = \frac{Q\,H\,s}{3960\eta_p} \tag{12}$$

where

Q = flow, gpm
H = head, ft of water
s = specific gravity
η_p = efficiency

In boilers, one would like to maintain a constant flow in lb/hr, *not* in gpm, and at a particular pressure in psi. The relationships are

$$Q = \frac{W}{500s} \quad \text{and} \quad H = 2.31 \times \frac{\Delta P}{s}$$

where

W = flow, lb/hr
ΔP = pump differential, psi

Substituting these terms into (1), we have

$$BHP = W \times \frac{\Delta P}{857,000\eta_p s} \tag{13}$$

As s decreases with temperature, BHP will increase if we want to maintain the flow in lb/hr and head or pressure in psi. However, if the flow in gpm and head in ft should be maintained, then the BHP will decrease with a decrease in s, which in turn is lower at lower temperatures.

A similar analogy can be drawn with fans in boiler plants, which require a certain amount of air in lb/hr for combustion and a particular head in in. WC.

5.10

Q: A centrifugal pump delivers 100 gpm at 155 ft of water with a 60-Hz supply. If the electric supply is changed to 50 Hz, how will the pump perform?

A: For variations in speed or impeller size, the following equations
    apply:

$$\frac{q_1}{q_2} = \frac{N_1}{N_2} = \frac{\sqrt{H_1}}{\sqrt{H_2}} \tag{14}$$

where

   q = pump flow, gpm
   H = head developed, ft
   N = speed, rpm

Use of Eq. (14) gives us the head and the flow characteristics of a
pump at different speeds. However, to get the actual operating
point, one must plot the new head versus flow curve and note the
point of intersection of this curve with the system resistance
curve. In the case above,

$$q_2 = 100 \times \frac{50}{60} = 83 \text{ gpm}$$

$$H_2 = 155 \times \left(\frac{50}{60}\right)^2 = 107 \text{ ft}$$

In this fashion, the new H versus q curve can be obtained. The
new operating point can then be found.

5.11

Q: How does the performance of a pump change with the viscosity
    of the fluids pumped?

A: The Hydraulic Institute has published charts that give correction
    factors for head, flow, and efficiency for viscous fluids when the
    performance with water is known (see Figures 5.3a and 5.3b).

    EXAMPLE
    A pump delivers 750 gpm at 100 ft head when water is pumped.
    What is the performance when it pumps oil with viscosity 1000
    SSU? Assume that efficiency with water is 82%.
    Solution.   In Figure 5.3b, go up from capacity 750 gpm to cut
    the head line at 100 ft and move horizontally to cut viscosity at
    1000 SSU; move up to cut the various correction factors.

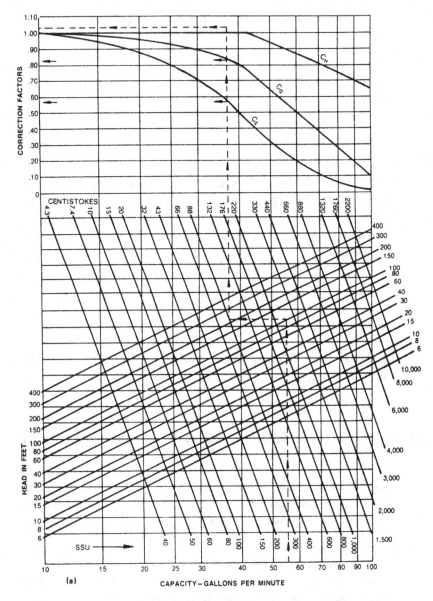

Figure 5.3   (a) Viscosity corrections. (b) Determination of pump performance when handling viscous liquids. [Courtesy of Hydraulic Institute/Gould Pump Manual.]

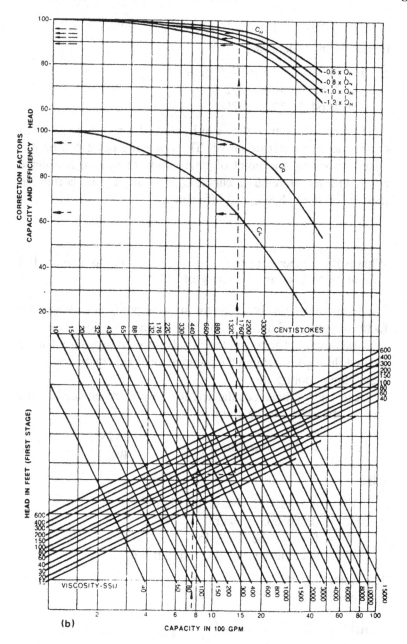

(b)

$$C_q = 0.94, \quad C_H = 0.92, \quad C_E = 0.64$$

Hence the new data are

$$q = 0.94 \times 750 = 705 \text{ gpm}$$
$$H = 0.92 \times 100 = 92 \text{ ft}$$
$$\eta_p = 0.64 \times 82 = 52\%$$

The new H versus q data can be plotted for various flows to obtain the characteristic curve. The operating point can be obtained by noting the point of intersection of system resistance curve with the H versus q curve. C_q, C_H, and C_E are correction factors for flow, head, and efficiency.

5.12

Q: What is the temperature rise of water when a pump delivers 100 gpm at 1000 ft at an efficiency of 60%?

A: The temperature rise of fluids through the pump is an important factor in pump maintenance and performance considerations and must be limited. The recirculation valve is used to ensure that the desired flow goes through the pump at low load conditions of the plant, thus cooling it.

From energy balance, the friction losses are equated to the energy absorbed by the fluid.

$$\Delta T = (\text{BHP} - \text{theoretical power}) \times \frac{2545}{WC_p} \qquad (15a)$$

where

ΔT = temperature rise of the fluid, °F
BHP = brake horsepower
W = flow of the fluid, lb/hr
C_p = specific heat of the fluid, Btu/lb °F

For water, $C_p = 1$.
From Eq. (9),

$$\text{BHP} = W \times \frac{H}{\eta_p \times 3600 \times 550}$$

where η_p is the pump efficiency, fraction. Substituting into Eqs. (15a) and (9) and simplifying, we have

$$\Delta T = H \times \frac{1/\eta_p - 1}{778} \tag{15b}$$

If $H = 100$ ft of water and $\eta_p = 0.6$, then

$$\Delta T = 1000 \times \frac{1.66 - 1}{778} \approx 1°F$$

5.13

Q: How is the minimum recirculation flow through a centrifugal pump determined?

A: Let us illustrate this with the case of a pump whose characteristics are as shown in Figure 5.4. We need to plot the ΔT versus Q characteristics first. Notice that at low flows when the efficiency

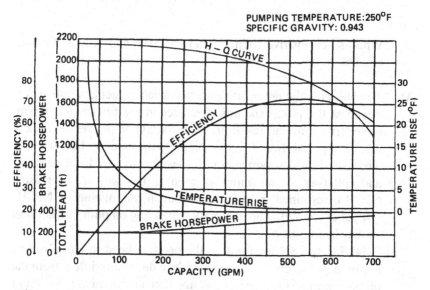

Figure 5.4 Typical characteristic curve of a multistage pump also showing temperature rise versus capacity.

is low, we can expect a large temperature rise. At 100 gpm, for example,

$$\eta_p = 0.23 \quad \text{and} \quad H = 2150 \text{ ft}$$

Then

$$\Delta T = 2150 \times \frac{1/0.23 - 1}{778} = 9°F$$

In a similar fashion, ΔT is estimated at various flows. Note that ΔT is higher at low flows due to the low efficiency and also because of the lesser cooling capacity.

The maximum temperature rise is generally limited to about 20°F, depending on the recommendations of the pump manufacturer. This means that at least 40 gpm must be circulated through the pump in this case. If the load is only 30 gpm, then depending on the recirculation control logic, 70 to 10 gpm could be recirculated through the pump.

5.14

Q: What is net positive suction head (NPSH), and how is it calculated?

A: The NPSH is the net positive suction head in feet absolute determined at the pump suction after accounting for suction piping losses (friction) and vapor pressure. NPSH helps one to check if there is a possibility of cavitation at pump suction. This is likely when the liquid vaporizes or flashes due to low local pressure and collapses at the pump as soon as the pressure increases. NPSH determined from pump layout in this manner is $NPSH_a$ (NPSH available). This will vary depending on pump location as shown in Figure 5.5.

NPSH$_r$ (NPSH required) is the positive head in feet absolute required to overcome the pressure drop due to fluid flow from the pump suction to the eye of the impeller and maintain the liquid above its vapor pressure. NPSH$_r$ varies with pump speed and capacity. Pump suppliers generally provide this information.

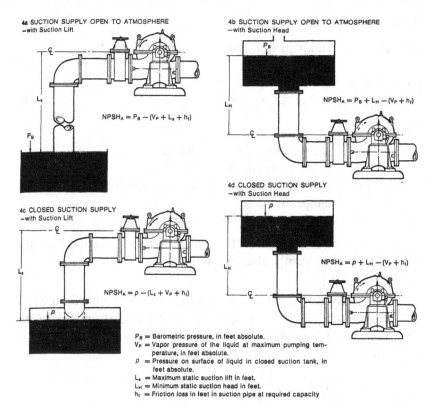

4a SUCTION SUPPLY OPEN TO ATMOSPHERE
—with Suction Lift

$NPSH_A = P_B - (V_P + L_s + h_f)$

4b SUCTION SUPPLY OPEN TO ATMOSPHERE
—with Suction Head

$NPSH_A = P_B + L_H - (V_P + h_f)$

4c CLOSED SUCTION SUPPLY
—with Suction Lift

$NPSH_A = p - (L_s + V_P + h_f)$

4d CLOSED SUCTION SUPPLY
—with Suction Head

$NPSH_A = p + L_H - (V_P + h_f)$

P_B = Barometric pressure, in feet absolute.
V_P = Vapor pressure of the liquid at maximum pumping temperature, in feet absolute.
p = Pressure on surface of liquid in closed suction tank, in feet absolute.
L_s = Maximum static suction lift in feet.
L_H = Minimum static suction head in feet.
h_f = Friction loss in feet in suction pipe at required capacity

Figure 5.5 Calculation of system NPSH available for typical suction conditions.

$NPSH_a$ can be determined by a gauge reading at pump suction:

$$NPSH_a = P_B - VP \pm PG + VH \qquad (16)$$

where

VH = velocity head at the gauge connection, ft
PG = pressure gauge reading, converted to feet
VP = vapor pressure, ft absolute
P_B = barometric pressure, ft (if suction is atmospheric)

To avoid cavitation, $NPSH_a$ must be greater than $NPSH_r$.

5.15

Q: Does the pump suction pressure change $NPSH_a$?

A: $NPSH_a$ is given by

$$NPSH_a = P_s + H - VP - H_f \qquad (17)$$

where

P_s = suction pressure, ft of liquid
H = head of liquid, ft
VP = vapor pressure of the liquid at operating temperature, ft
H_f = friction loss in the suction line, ft

For saturated liquids, $VP \approx P_s$, so changes in suction pressure do not significantly change $NPSH_a$.

EXAMPLE
Determine the $NPSH_a$ for the system shown in Figure 5.5b when $H = 10$ ft, $H_f = 3$ ft, and $VP = 0.4$ psia (from the steam tables). Assume that the water has a density of 62 lb/cu ft.
Solution.

$$VP = 0.4 \times \frac{144}{62} = 0.93 \text{ ft}$$

$$\text{Suction pressure} = 14.6 \text{ psia} = 14.6 \times \frac{166}{62} = 33.9 \text{ ft}$$

$$NPSH_a = 33.9 - 3 - 0.93 + 10 = 40 \text{ ft}$$

5.16

Q: In the absence of information from the pump suppler, can we estimate $NPSH_r$?

A: A good estimate of $NPSH_r$ can be made from the expression for specific speed S.

$$S = N \times \frac{\sqrt{q}}{NPSH_r^{0.75}} \qquad (18)$$

S ranges from 7000 to 12,000 for water.

For example, when $q = 100$ gpm, $N = 1770$, and assuming that $S = 10,000$ for water,

$$\text{NPSH}_r = 1770 \times \left(\frac{\sqrt{100}}{10,000} \right)^{1.33} = 2.2 \text{ ft}$$

Even if we took a conservative value of 7000 for S, we would get

$$\text{NPSH}_r = 3.43 \text{ ft}$$

This information can be used in making preliminary layouts for systems involving pumps.

5.17

Q: How is NPSH_a for a reciprocating pump arrived at?

A: NPSH_a for a reciprocating pump is calculated in the same way as for a centrifugal pump except that the acceleration head H_a is included with the friction losses. This is the head required to accelerate the liquid column on each suction stroke so that there will be no separation of this column in the pump suction line or in the pump [1]:

$$H_a = \frac{LNVC}{K_g} \tag{19}$$

where

    L = length of the suction line, ft (actual length, not developed)
    V = velocity in the suction line, ft/sec
    N = pump speed, rpm

C is a constant: 0.066 for triplex pump, 0.04 for quintuplex, and 0.2 for duplex pumps. K is a factor: 2.5 for hot oil, 2.0 for most hydrocarbons, 1.5 for water, and 1.4 for deaerated water. $g = 32$ ft/sec^2. Pulsation dampeners are used to reduce L significantly. By proper selection, L can be reduced to nearly zero.

EXAMPLE

A triplex pump running at 360 rpm and displacing 36 gpm has a 3-in. suction line 8 ft long and a 2-in. line 18 ft long. Estimate the acceleration head required.

Solution. First obtain the velocity of water in each part of the line. In the 3-in. line, which has an inner diameter of 3.068 in.,

$$V = 0.41 \ \frac{q}{d_i^2} = 0.41 \times \frac{36}{(3.068)^2} = 1.57 \text{ ft/sec}$$

In the 2-in. line having an inner diameter of 2.067 in.,

$$V = 0.41 \times \frac{36}{(2.067)^2} = 3.45 \text{ ft/sec}$$

The acceleration head in the 3-in. line is

$$H_a = 8 \times 360 \times 1.57 \times \frac{0.066}{1.4 \times 32} = 6.7 \text{ ft}$$

In the 2-in. line,

$$H_a = 18 \times 3.45 \times 360 \times \frac{0.066}{1.4 \times 32} = 32.9 \text{ ft}$$

The total acceleration head is $32.9 + 6.7 = 39.6$ ft.

5.18

Q: How can we check the performance of a pump from the motor data?

A: A good estimate of the efficiency of a pump or a fan can be obtained from the current reading if we make a few reasonable assumptions. The efficiency of a motor is more predictable than that of a pump due to its small variations with duty. The pump differential pressure and flow can be obtained rather easily and accurately. By relating the power consumed by the pump with that delivered by the motor, the following can be derived. The pump power consumption, P, in kW from Eq. (8) is

$$P = 0.00043q \times \frac{\Delta P}{\eta_p} \tag{20}$$

Motor power output $= 0.001732EI \cos \phi \ \eta_m$ (21)

Equating Eqs. (20) and (21) and simplifying, we have

$$q \ \Delta P = 4EI \cos \phi \ \eta_p\eta_m \tag{22}$$

where

$$q = \text{flow, gpm}$$
$$\Delta P = \text{differential pressure, psi}$$
$$E = \text{voltage, V}$$
$$I = \text{current, A}$$
$$\eta_p, \eta_m = \text{efficiency of pump and motor, fraction}$$
$$\cos \phi = \text{power factor}$$

From Eq. (22) we can solve for pump efficiency given the other variables. Alternatively, we can solve for the flow by making a reasonable estimate of η_p and check whether the flow reading is good. The power factor $\cos \phi$ typically varies between 0.8 and 0.9, and the motor efficiency between 0.90 and 0.95.

EXAMPLE

A plant engineer observes that at a 90-gpm flow of water and 1000 psi differential, the motor current is 100 A. Assuming that the voltage is 460 V, the power factor is 0.85, and the motor efficiency is 0.90, estimate the pump efficiency.

Solution. Substituting the data in Eq. (22), we obtain

$$90 \times 1000 = 4 \times 460 \times 100 \times 0.85 \times 0.90 \times \eta_p$$

Solving for η_p, we have $\eta_p = 0.65$.

We can use this figure to check if something is wrong with the system. For instance, if the pump has been operating at this flow for some time but the current drawn is more, one can infer that the machine needs attention. One can also check the pump efficiency from its characteristic curve and compare the calculated and predicted efficiencies.

5.19

Q: Derive an expression similar to (22) relating fan and motor.

A: Equating the power consumption of a fan with that delivered by its motor,

$$P = 1.17 \times 10^{-4} \times qH_w = 0.001732EI \cos \phi \, \eta_m \quad (23)$$

where

q = flow, acfm
H_w = static head of fan, in. WC

Other terms are as in Q5.18.

If the efficiency of a fan is assumed to be 65% when its differential head is 4 in. WC, the motor voltage is 460, the current is 7 A, then the power factor is 0.8 and the motor efficiency is 85%. Solving for q, we have

$$1.17 \times 10^{-4} \times q \times 4 = 0.001732 \times 460 \times 7$$
$$\times\ 0.80 \times 0.85 \times 0.65$$

or

$$q = 5267\ \text{acfm}$$

One can check from the fan curve whether the flow is reasonable. Alternatively, if the flow is known, one can check the head from Eq. (23) and compare it with the measured value. If the measured head is lower, for example, we can infer that something is wrong with the fan or its drive or that the flow measured is not correct.

5.20

Q: How is the performance of pumps in series and in parallel evaluated?

A: For parallel operation of two or more pumps, the combined performance curve (H versus q) is obtained by adding horizontally the capacities of the same heads. For series operation, the combined performance curve is obtained by adding vertically the heads at the same capacities.

The operating point is the intersection of the combined performance curve with the system resistance curve. Figure 5.6 explains this. Head and flow are shown as percentages [2]. ABC is the H versus q curve for a single pump, DEF is the H versus q curve for two such pumps in series, and AGH is the H versus q curve for two such pumps in parallel. To obtain the curve DEF,

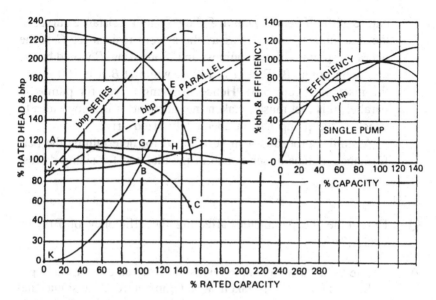

Figure 5.6 Series and parallel operations of pumps with flat head capacity curves. [Courtesy of Karassik Centrifugal Pump Clinic, Marcel Dekker Inc., New York.]

we add the heads at a given flow. For example, at $q = 100\%$, H with one pump is 100%, and with two pumps H will be 200%. Similarly, AGH is obtained by adding flows at a given head. At $H = 100$, q for two pumps will be 200%.

Let the system resistance curve be $KBGE$. When one pump alone operates, the operating point is B. With two pumps in series, E is the operating point. With two pumps in parallel, G is the operating point.

BHP curves also have been plotted and reveal that with series operation, BHP = 250% and with pumps in parallel BHP = 164%, indicating that BHP/q is larger in series operation than in parallel. This varies with pump and system resistance characteristics. NPSH$_r$ also increases with pump capacity.

Note that if the full capacity of the plant were handled by two pumps in parallel, and one tripped, the operating BHP would not be 50% of that with two pumps, but more, depending on the nature of the H versus q curve and the system resistance curve. In the case above, with *KBGE* as the system resistance, G is the operating point with two pumps, and if one trips, B would be the operating point. BHP at G is 142%, while at B it is 100% (see the inset of Figure 5.6). Hence in sizing drives for pumps in parallel, this fact must be taken into account. It is a good idea to check if the pump has an adequately sized drive.

A similar procedure can be adopted for determining the performance of fans in series and in parallel and for sizing drives.

5.21

Q: Determine the parameters affecting the efficiency of Brayton cycle [3].

A: Figure 5.7a shows a simple reversible Brayton cycle used in gas turbine plants. Air is taken at a temperature T_1 absolute and compressed, and the temperature after compression is T_2. Heat is added in the combustor, raising the gas temperature to T_3; the hot

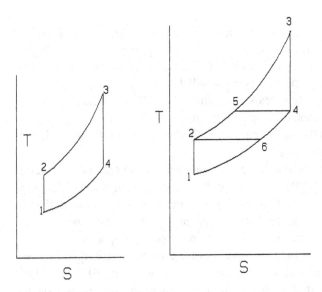

Figure 5.7   Simple and regenerative Brayton cycle.

gases expand to T_4 in the turbine performing work. Following are
some of the terms used to describe the performance.

$$\text{Thermal efficiency TE} = \frac{Q_a - Q_r}{Q_a} \tag{24}$$

where

Q_a = heat added to cycle, Btu/lb
Q_r = heat rejected, Btu/lb

$$Q_a = C_p (T_3 - T_2) \tag{25}$$
$$Q_r = C_p (T_4 - T_1) \tag{26}$$
$$P_2 = P_3 \quad \text{and} \quad P_1 = P_4 \tag{27}$$

Also,

$$\frac{T_2}{T_1} = \frac{T_3}{T_4} = r^{(k-1)/k} \tag{28}$$

where

$$r = \text{pressure ratio} = \frac{P_2}{P_1} = \frac{P_3}{P_4} \tag{29}$$

k = ratio of gas specific heats
C_p = gas specific heat, Btu/lb
T_1 to T_4 = temperatures, °R
P_1 to P_4 = pressures, psia

Using the above, we can write

$$\text{TE} = 1 - \frac{Q_r}{Q_a} = 1 - \frac{T_4 - T_1}{T_3 - T_2}$$

$$= 1 - \frac{T_1}{T_2} \times \frac{T_4/T_1 - 1}{T_3/T_2 - 1} \tag{30a}$$

Since $T_4/T_1 = T_3/T_2$ from (5), we have

$$\text{TE} = 1 - \frac{T_1}{T_2} = 1 - 1/r^{(k-1)/k} \tag{30b}$$

EXAMPLE

A simple cycle takes in air at 80°F and 14.7 psia and compresses
it at constant entropy through a pressure ratio of 4. The combus-

tor raises the gas temperature to 1500°F. The heated air expands to 14.7 psia at constant entropy in the turbine. Assume $k = 1.3$ and $C_p = 0.28$. Find (1) compression work, W_c; (2) heat input to cycle, Q_a; (3) expansion work, Q_e; (4) thermal efficiency, TE.
Solution. From (5),

$$T_2 = (80 + 460) \times 4^{(1.3-1)/1.3} = 742°R$$

Note that $4^{(1.3-1)/1.3} = 1.375$. Hence

$$W_c = C_p \times (T_2 - T_1) = 0.28 \times (742 - 540)$$
$$= 56.6 \text{ Btu/lb}$$

$$\text{Heat input to cycle} = Q_a = C_p \times (T_3 - T_2)$$
$$= 0.28 \times (1500 + 460$$
$$- 742) = 341 \text{ Btu/lb}$$

$$T_4 = \frac{T_3}{1.375} = \frac{1960}{1.375} = 1425°R$$

Expansion work $Q_e = 0.28 \times (1960 - 1425) = 150 \text{ Btu/lb}$

$$TE = \frac{150 - 56.6}{341} = 0.273, \text{ or } 27.3\%$$

Using Eq. (30), TE $= 1 - 1/1.375 = 0.273$.

It can be seen that as the pressure ratio increases, TE increases. Also as inlet air temperature decreases, the efficiency increases. That is why some gas turbine suppliers install chillers or air coolers at the compressor inlet so that during summer months the turbine output does not fall off compared to the winter months.

5.22

Q: How can the efficiency of a simple Brayton cycle be improved?

A: One of the ways of improving the cycle efficiency is to utilize the energy in the exhaust gases (Figure 5.7b) and use it to preheat the air entering the combustor. This is called regeneration.

Assuming 100% regeneration, the exhaust gas at temperature T_4 preheats air from T_2 to T_5 while cooling to T_6. The actual heat rejected corresponds to a temperature drop of $T_6 - T_1$, while the heat added corresponds to $T_3 - T_5$, and hence the cycle is more efficient. Assuming constant C_p,

$$\text{TE} = 1 - \frac{Q_r}{Q_a} = 1 - \frac{T_6 - T_1}{T_3 - T_5},$$

Now $P_2 = P_5 = P_3$ and $P_4 = P_6 = P_1$. Also,

$$\frac{T_2}{T_1} = \frac{T_6}{T_1} = r^{(k-1)/k} = \frac{T_3}{T_4} = \frac{T_3}{T_5}$$

$$\text{TE} = 1 - T_1 \frac{T_6/T_1 - 1}{T_5(T_3/T_5 - 1)}$$

$$= 1 - \frac{T_1}{T_5} \quad \text{(as } T_6/T_1 = T_3/T_5, \text{ from above)}$$

$$\frac{T_1}{T_5} = \frac{T_1}{T_3} \times \frac{T_3}{T_5} = \frac{T_1}{T_3} \times r^{(k-1)/k}$$

Hence

$$\text{TE} = 1 - \frac{T_1}{T_3} \times r^{(k-1)/k} \tag{31}$$

EXAMPLE

Using the same data as above, compute the following for the ideal renerative cycle: (1) Work of compression, W_c; (2) heat added to cycle; (3) heat added to regenerator; (4) expansion work in turbine; (5) cycle efficiency.

Solution. For the same inlet temperature and pressure ratio, $W_c = 56.6$ and $T_2 = 742°R$. Exhaust temperature from above $= 1425°R = T_5$.

Heat added in regenerator $= C_p \times (T_5 - T_2) = 0.28 \times (1425 - 742) = 191.3$ Btu/lb

Heat added in combustor $= Q_a = C_p \times (T_3 - T_5) = 0.28 \times (1960 - 1425) = 150$ Btu/lb

Heat rejected $= Q_r = C_p \times (T_6 - T_1) = 0.28 \times (742 - 540) = 56.6$ Btu/lb

$$TE = 1 - \frac{Q_r}{Q_a} = 1 - \frac{56.6}{150} = 0.622, \text{ or } 62.2\%$$

Using (8),

$$TE = 1 - \frac{540}{1960} \times 4^{(1.3-1)/1.3} = 0.621, \text{ or } 62.1\%$$

It is interesting to note that as the pressure ratio increases, the efficiency decreases. As the combustor temperature increases, the efficiency increases. However, it can be shown that the power output increases with increases in the pressure ratio; hence industrial gas turbines operate at a pressure ratio between 9 to 18 and an inlet gas temperature of 1800 to 2200°F.

NOMENCLATURE

ASR	Actual steam rate, lb/kWh
BHP	Brake horsepower, hp
C_a, C_H, C_E	Factors correcting viscosity effects for flow, head, and efficiency
C_p	Specific heat, Btu/lb °F
d	Tube or pipe diameter, in.; subscript i stands for inner diameter
E	Voltage
h	Enthalpy, Btu/lb; subscripts f and g stand for saturated liquid and vapor
H	Head developed by pump, ft; subscript a stands for acceleration
HP	Horsepower
H_w	Head developed by fan, in. WC
I	Current, A
k	Ratio of gas specific heats, C_p/C_v
L	Length, ft
N	Speed of pump or fan, rpm
NPSH	Net positive suction head, ft; subscripts a and r stand for available and required
P	Power, kW
ΔP	Differential pressure, psi

q	Flow, gpm or acfm
Q_a, Q_r	Heat added, rejected, Btu/lb
r	Pressure ratio
s	Specific gravity
s_f, s_g	Entropy of saturated liquid and vapor, Btu/lb °R
S	Specific speed
ΔT	Temperature rise, °F
TE	Thermal efficiency
T	Temperature, °R
TSR	Theoretical steam rate, lb/kWh
V	Velocity, ft/sec
W	Flow, lb/hr
W_c	Work of compression, Btu/lb
η	Efficiency, fraction; subscripts f, m, p, and t stand for fan, motor, pump, and turbine
ρ	Density, lb/cu ft

REFERENCES

1. Ingersoll Rand, *Cameron Hydraulic Data*, 16th ed., Woodcliff Lake, N.J., 1981, p. 5.1.
2. I. Karassik, *Centrifugal Pump Clinic*, Marcel Dekker, New York, 1981, p. 102.
3. *Power Magazine*, *Power Handbook*, McGraw-Hill, New York, 1983.

Appendix Tables

Table A1   Thermodynamic Properties of Dry Saturated Steam—Pressure Table

Abs press., psi p	Temp., °F t	Specific volume Sat. liquid v_f	Specific volume Sat. vapor v_g	Enthalpy Sat. liquid h_f	Enthalpy Evap. h_{fg}	Enthalpy Sat. vapor h_g	Entropy Sat. liquid s_f	Entropy Evap. s_{fg}	Entropy Sat. vapor s_g	Internal energy Sat. liquid u_f	Internal energy Sat. vapor u_g	Abs press., psi p
1.0	101.74	0.01614	333.6	69.70	1036.3	1106.0	0.1326	1.8456	1.9782	69.70	1044.3	1.0
2.0	126.08	0.01623	173.73	93.99	1022.2	1116.2	0.1749	1.7451	1.9200	93.98	1051.9	2.0
3.0	141.48	0.01630	118.71	109.37	1013.2	1122.6	0.2008	1.6855	1.8863	109.36	1056.7	3.0
4.0	152.97	0.01636	90.63	120.86	1006.4	1127.3	0.2198	1.6427	1.8625	120.85	1060.2	4.0
5.0	162.24	0.01640	73.52	130.13	1001.0	1131.1	0.2347	1.6094	1.8441	130.12	1063.1	5.0
6.0	170.06	0.01645	61.98	137.96	996.2	1134.2	0.2472	1.5820	1.8292	137.94	1065.4	6.0
7.0	176.85	0.01649	53.64	144.76	992.1	1136.9	0.2581	1.5586	1.8167	144.74	1067.4	7.0
8.0	182.86	0.01653	47.34	150.79	988.5	1139.3	0.2674	1.5383	1.8057	150.77	1069.2	8.0
9.0	188.28	0.01656	42.40	156.22	985.2	1141.4	0.2759	1.5203	1.7962	156.16	1070.8	9.0
10	193.21	0.01659	38.42	161.17	982.1	1143.3	0.2835	1.5041	1.7876	161.14	1072.2	10
14.696	212.00	0.01672	26.80	180.07	970.3	1150.4	0.3120	1.4446	1.7566	180.02	1077.5	14.696
15	213.03	0.01672	26.29	181.11	969.7	1150.8	0.3135	1.4415	1.7549	181.06	1077.8	15
20	227.96	0.01683	20.089	196.16	960.1	1156.3	0.3356	1.3962	1.7319	196.10	1081.9	20
25	240.07	0.01692	16.303	208.42	952.1	1160.6	0.3533	1.3606	1.7139	208.34	1085.1	25
30	250.33	0.01701	13.746	218.52	945.3	1164.1	0.3680	1.3313	1.6993	218.73	1087.8	30
35	259.28	0.01708	11.898	227.91	939.2	1167.1	0.3807	1.3063	1.6870	227.80	1090.1	35
40	267.25	0.01715	10.498	236.03	933.7	1169.7	0.3919	1.2844	1.6763	235.90	1092.0	40
45	274.44	0.01721	9.401	243.36	928.6	1172.0	0.4019	1.2650	1.6669	243.22	1093.7	45
50	281.01	0.01727	8.515	250.09	924.0	1174.1	0.4110	1.2474	1.6585	249.93	1095.3	50
55	287.07	0.01732	7.787	256.30	919.6	1175.9	0.4193	1.2316	1.6509	256.12	1096.7	55
60	292.71	0.01738	7.175	262.09	915.5	1177.6	0.4270	1.2168	1.6438	261.90	1097.9	60
65	297.97	0.01743	6.655	267.50	911.6	1179.1	0.4342	1.2032	1.6374	267.29	1099.1	65
70	302.92	0.01748	6.206	272.61	907.9	1180.6	0.4409	1.1906	1.6315	272.38	1100.2	70
75	307.60	0.01753	5.816	277.43	904.5	1181.9	0.4472	1.1787	1.6259	277.19	1101.2	75
80	312.03	0.01757	5.472	282.02	901.1	1183.1	0.4531	1.1676	1.6207	281.76	1102.1	80
85	316.25	0.01761	5.168	286.39	897.8	1184.2	0.4587	1.1571	1.6158	286.11	1102.9	85
90	320.27	0.01766	4.896	290.56	894.7	1185.3	0.4641	1.1471	1.6112	290.27	1103.7	90
95	324.12	0.01770	4.652	294.56	891.7	1186.2	0.4692	1.1376	1.6068	294.25	1104.5	95
100	327.81	0.01774	4.432	298.40	888.8	1187.2	0.4740	1.1286	1.6026	298.08	1105.2	100
110	334.77	0.01782	4.049	303.66	883.2	1188.9	0.4832	1.1117	1.5948	305.80	1106.5	110

Abs Press (psia)	Temp (°F)	v_f	v_g	h_f	h_{fg}	h_g	s_f	s_{fg}	s_g	u_f	u_g	Abs Press (psia)
120	341.25	0.01789	3.728	312.44	877.9	1190.4	0.4916	1.0962	1.5878	312.05	1107.6	120
130	347.32	0.01796	3.455	318.81	872.9	1191.7	0.4995	1.0817	1.5812	318.38	1108.6	130
140	353.02	0.01802	3.220	324.82	868.2	1193.0	0.5069	1.0682	1.5751	324.35	1109.6	140
150	358.42	0.01809	3.015	330.51	863.6	1194.1	0.5138	1.0556	1.5694	330.01	1110.5	150
160	363.53	0.01815	2.834	335.93	859.2	1195.1	0.5204	1.0436	1.5640	335.39	1111.2	160
170	368.41	0.01822	2.675	341.09	854.9	1196.0	0.5266	1.0324	1.5590	340.52	1111.9	170
180	373.06	0.01827	2.532	346.03	850.8	1196.9	0.5325	1.0217	1.5542	345.42	1112.5	180
190	377.51	0.01833	2.404	350.79	846.8	1197.6	0.5381	1.0116	1.5497	350.15	1113.1	190
200	381.79	0.01839	2.288	355.36	843.0	1198.4	0.5435	1.0018	1.5453	354.68	1113.7	200
250	400.95	0.01865	1.8438	376.00	825.1	1201.1	0.5675	0.9588	1.5263	375.14	1115.8	250
300	417.33	0.01890	1.5433	393.84	809.0	1202.8	0.5879	0.9225	1.5104	392.79	1117.1	300
350	431.72	0.01913	1.3260	409.69	794.2	1203.9	0.6056	0.8910	1.4966	408.45	1118.0	350
400	444.59	0.0193	1.1613	424.0	780.5	1204.5	0.6214	0.8630	1.4844	422.6	1118.5	400
450	456.28	0.0195	1.0320	437.2	767.4	1204.6	0.6356	0.8378	1.4734	435.5	1118.7	450
500	467.01	0.0197	0.9278	449.4	755.0	1204.4	0.6487	0.8147	1.4634	447.6	1118.6	500
550	476.94	0.0199	0.8424	460.8	743.1	1203.9	0.6608	0.7934	1.4542	458.8	1118.2	550
600	486.21	0.0201	0.7698	471.6	731.6	1203.2	0.6720	0.7734	1.4454	469.4	1117.7	600
650	494.90	0.0203	0.7083	481.8	720.5	1202.3	0.6826	0.7548	1.4374	479.4	1117.1	650
700	503.10	0.0205	0.6554	491.5	709.7	1201.2	0.6925	0.7371	1.4296	488.8	1116.3	700
750	510.86	0.0207	0.6092	500.8	699.2	1200.0	0.7019	0.7204	1.4223	498.0	1115.4	750
800	518.23	0.0209	0.5687	509.7	688.9	1198.6	0.7108	0.7045	1.4153	506.6	1114.4	800
850	525.26	0.0210	0.5327	518.3	678.8	1197.1	0.7194	0.6891	1.4085	515.0	1113.3	850
900	531.98	0.0212	0.5006	526.6	668.8	1195.4	0.7275	0.6744	1.4020	523.1	1112.1	900
950	538.43	0.0214	0.4717	534.6	659.1	1193.7	0.7355	0.6602	1.3957	530.9	1110.8	950
1000	544.61	0.0216	0.4456	542.4	649.4	1191.8	0.7430	0.6467	1.3897	538.4	1109.4	1000
1100	556.31	0.0220	0.4001	557.4	630.4	1187.8	0.7575	0.6205	1.3780	552.9	1106.4	1100
1200	567.22	0.0223	0.3619	571.7	611.7	1183.4	0.7711	0.5956	1.3667	566.7	1103.0	1200
1300	577.46	0.0227	0.3293	585.4	593.2	1178.6	0.7840	0.5719	1.3559	580.0	1099.4	1300
1400	587.10	0.0231	0.3012	598.7	574.7	1173.4	0.7963	0.5491	1.3454	592.7	1095.4	1400
1500	596.23	0.0235	0.2765	611.6	556.3	1167.9	0.8082	0.5269	1.3351	605.1	1091.2	1500
2000	635.82	0.0257	0.1878	671.7	463.4	1135.1	0.8619	0.4230	1.2849	662.2	1065.6	2000
2500	668.13	0.0287	0.1307	730.6	360.5	1091.1	0.9126	0.3197	1.2322	717.3	1030.6	2500
3000	695.36	0.0346	0.0858	802.5	217.8	1020.3	0.9731	0.1885	1.1615	783.4	972.7	3000
3206.2	705.40	0.0503	0.0503	902.7	0	902.7	1.0580	0	1.0580	872.9	872.9	3206.2

Source: Abridged from Joseph H. Keenan and Frederick G. Keyes, *Thermodynamic Properties of Steam*, John Wiley & Sons, Inc., New York, 1937.

Table A2　Thermodynamic Properties of Dry Saturated Steam—Temperature Table

Temp., °F	Abs press., psi	Specific volume			Enthalpy			Entropy			Temp., °F
t	p	Sat. liquid vf	Evap. vfe	Sat. vapor ve	Sat. liquid hf	Evap. hfe	Sat. vapor he	Sat. liquid sf	Evap. sfe	Sat. vapor se	t
32	0.08854	0.01602	3306	3306	0.00	1075.8	1075.8	0.0000	2.1877	2.1877	32
35	0.09995	0.01602	2947	2947	3.02	1074.1	1077.1	0.0061	2.1709	2.1770	35
40	0.12170	0.01602	2444	2444	8.05	1071.3	1079.3	0.0162	2.1435	2.1597	40
45	0.14752	0.01602	2036.4	2036.4	13.06	1068.4	1081.5	0.0262	2.1167	2.1429	45
50	0.17811	0.01603	1703.2	1703.2	18.07	1065.6	1083.7	0.0361	2.0903	2.1264	50
60	0.2563	0.01604	1206.6	1206.7	28.06	1059.9	1088.0	0.0555	2.0393	2.0948	60
70	0.3631	0.01606	867.8	867.9	38.04	1054.3	1092.3	0.0745	1.9902	2.0647	70
80	0.5069	0.01608	633.1	633.1	48.02	1048.6	1096.6	0.0932	1.9428	2.0360	80
90	0.6982	0.01610	468.0	468.0	57.99	1042.9	1100.0	0.1115	1.8972	2.0087	90
100	0.9492	0.01613	350.3	350.4	67.97	1037.2	1105.2	0.1295	1.8531	1.9826	100
110	1.2748	0.01617	265.3	265.4	77.94	1031.6	1109.5	0.1471	1.8106	1.9577	110
120	1.6924	0.01620	203.25	203.27	87.92	1025.8	1113.7	0.1645	1.7694	1.9339	120
130	2.2225	0.01625	157.32	157.34	97.90	1020.0	1117.9	0.1816	1.7296	1.9112	130
140	2.8886	0.01629	122.99	123.01	107.89	1014.1	1122.0	0.1984	1.6910	1.8894	140
150	3.718	0.01634	98.06	97.07	117.89	1008.2	1126.1	0.2149	1.6537	1.8685	150
160	4.741	0.01639	77.27	77.29	127.89	1002.3	1130.2	0.2311	1.6174	1.8485	160
170	5.992	0.01645	62.04	62.06	137.90	996.3	1134.2	0.2472	1.5822	1.8293	170
180	7.510	0.01651	50.21	50.23	147.92	990.2	1138.1	0.2630	1.5480	1.8109	180
190	9.339	0.01657	40.94	40.96	157.95	984.1	1142.0	0.2785	1.5147	1.7932	190
200	11.526	0.01663	33.62	33.64	167.99	977.9	1145.9	0.2938	1.4824	1.7762	200
210	14.123	0.01670	27.80	27.82	178.05	971.6	1149.7	0.3090	1.4508	1.7598	210
212	14.696	0.01672	26.78	26.80	180.07	970.3	1150.4	0.3120	1.4446	1.7566	212
220	17.186	0.01677	23.13	23.15	188.13	965.2	1153.4	0.3239	1.4201	1.7440	220
230	20.780	0.01684	19.365	19.382	198.23	958.8	1157.0	0.3387	1.3901	1.7288	230
240	24.969	0.01692	16.306	16.323	208.34	952.2	1160.5	0.3531	1.3609	1.7140	240
250	29.825	0.01700	13.804	13.821	216.48	945.5	1164.0	0.3675	1.3323	1.6998	250
260	35.429	0.01709	11.746	11.763	228.64	938.7	1167.3	0.3817	1.3043	1.6860	260
270	41.858	0.01717	10.044	10.061	238.84	931.8	1170.6	0.3958	1.2769	1.6727	270
280	49.203	0.01726	8.628	8.645	249.06	924.7	1173.8	0.4096	1.2501	1.6597	280

290	1.6472	1.2238	0.4234	1176.8	917.5	259.31	7.461	7.444	0.01735	57.556	290
300	1.6350	1.1980	0.4369	1179.7	910.1	269.59	6.466	6.449	0.01745	67.013	300
310	1.6231	1.1727	0.4504	1182.5	902.6	279.92	5.626	5.609	0.01755	77.68	310
320	1.6115	1.1478	0.4637	1185.2	894.9	290.28	4.914	4.896	0.01765	89.66	320
330	1.6002	1.1233	0.4769	1187.7	887.0	300.68	4.307	4.289	0.01776	103.06	330
340	1.5891	1.0992	0.4900	1190.1	879.0	311.13	3.788	3.770	0.01787	118.01	340
350	1.5783	1.0754	0.5029	1192.3	870.7	321.63	3.342	3.324	0.01799	134.63	350
360	1.5677	1.0519	0.5158	1194.0	862.2	332.18	2.957	2.939	0.01811	153.04	360
370	1.5573	1.0287	0.5286	1196.3	853.5	342.79	2.625	2.606	0.01823	173.37	370
380	1.5471	1.0059	0.5413	1198.1	844.6	353.45	2.335	2.317	0.01836	195.77	380
390	1.5371	0.9832	0.5539	1199.6	835.4	364.17	2.0836	2.0651	0.01850	220.37	390
400	1.5272	0.9608	0.5664	1201.0	826.0	374.97	1.8633	1.8447	0.01864	247.31	400
410	1.5174	0.9386	0.5788	1202.1	816.2	385.83	1.6700	1.6312	0.01878	276.75	410
420	1.5078	0.9166	0.5912	1203.1	806.3	396.77	1.5000	1.4811	0.01894	308.83	420
430	1.4982	0.8947	0.6035	1203.8	796.0	407.79	1.3499	1.3308	0.01910	343.72	430
440	1.4887	0.8730	0.6158	1204.3	785.4	418.90	1.2171	1.1979	0.01926	381.59	440
450	1.4793	0.8513	0.6280	1204.6	774.5	430.1	1.0993	1.0799	0.0194	422.6	450
460	1.4700	0.8298	0.6402	1204.6	763.2	441.4	0.9944	0.9748	0.0196	466.9	460
470	1.4606	0.8083	0.6523	1204.3	751.5	452.8	0.9009	0.8811	0.0198	514.7	470
480	1.4513	0.7868	0.6645	1203.7	739.4	464.4	0.8172	0.7972	0.0200	566.1	480
490	1.4419	0.7653	0.6766	1202.8	726.8	476.0	0.7423	0.7221	0.0202	621.4	490
500	1.4325	0.7438	0.6887	1201.7	713.9	487.8	0.6749	0.6545	0.0204	680.8	500
520	1.4136	0.7006	0.7130	1198.2	686.4	511.9	0.5594	0.5385	0.0209	812.4	520
540	1.3942	0.6568	0.7374	1193.2	636.6	536.6	0.4649	0.4434	0.0215	962.5	540
560	1.3742	0.6121	0.7621	1186.4	624.2	562.2	0.3868	0.3647	0.0221	1133.1	560
580	1.3532	0.5659	0.7872	1177.3	588.4	588.9	0.3217	0.2989	0.0228	1325.8	580
600	1.3307	0.5176	0.8131	1165.5	548.5	617.0	0.2668	0.2432	0.0236	1542.9	600
620	1.3062	0.4664	0.8398	1130.3	503.6	646.7	0.2201	0.1955	0.0247	1786.6	620
640	1.2789	0.4110	0.8679	1130.5	452.0	678.6	0.1798	0.1538	0.0260	2059.7	640
660	1.2472	0.3485	0.8987	1104.4	390.2	714.2	0.1442	0.1165	0.0278	2365.4	660
680	1.2071	0.2719	0.9351	1067.2	309.9	757.3	0.1115	0.0810	0.0305	2708.1	680
700	1.1359	0.1484	0.9905	993.4	172.1	823.3	0.0761	0.0392	0.0369	3093.7	700
706.1	1.0580	0	1.0580	902.7	0	902.7	0.0503	0	0.0503	3206.2	705.4

Source: Abridged from Joseph H. Keenan and Frederick G. Keyes, *Thermodynamic Properties of Steam*, John Wiley & Sons, New York, 1937.

Table A3 Thermodynamic Properties of Superheated Steam

Abs press., psi (sat temp)		200	300	400	500	600	700	800	900	1000	1100	1200	1400	1600
1 (101.74)	v	392.6	452.3	512.0	571.6	631.2	690.8	750.4	809.9	869.5	929.1	988.7	1107.8	1227.0
	h	1150.4	1195.8	1241.7	1288.3	1335.5	1383.8	1432.8	1482.7	1533.5	1585.2	1637.7	1745.7	1857.5
	s	2.0612	2.1153	2.1720	2.2233	2.2702	2.3137	2.3542	2.3923	2.4283	2.4625	2.4952	2.5566	2.6137
5 (163.24)	v	78.16	90.25	102.26	114.22	126.16	138.10	150.03	161.95	173.87	185.79	197.71	221.6	245.4
	h	1148.8	1195.0	1241.2	1288.0	1335.4	1383.6	1432.7	1482.6	1533.4	1585.1	1637.7	1745.7	1857.5
	s	1.8718	1.9370	1.9942	2.0456	2.0927	2.1361	2.1767	2.2148	2.2509	2.2851	2.3178	2.3792	2.4363
10 (193.21)	v	38.85	45.00	51.04	57.05	63.03	69.01	74.98	80.95	86.92	92.88	98.84	110.77	122.69
	h	1146.6	1193.9	1240.6	1287.5	1335.1	1383.4	1432.5	1482.4	1533.2	1555.0	1637.6	1745.6	1857.3
	s	1.7927	1.8595	1.9172	1.9689	2.0160	2.0596	2.1002	2.1383	2.1744	2.2086	2.2413	2.3028	2.3598
14.096 (212.00)	v		30.53	34.68	38.78	42.56	46.94	51.00	55.07	59.13	63.19	67.25	75.37	83.48
	h		1192.8	1239.9	1287.1	1334.8	1383.2	1432.3	1482.3	1533.1	1584.8	1637.5	1745.5	1857.3
	s		1.8160	1.8743	1.9261	1.9734	2.0170	2.0676	2.0958	2.1319	2.1662	2.1989	2.2603	2.3174
20 (237.96)	v		22.36	25.43	28.46	31.47	34.47	37.46	40.45	43.44	46.42	49.41	55.37	61.34
	h		1191.6	1239.2	1286.6	1334.4	1382.9	1432.1	1482.1	1533.0	1684.7	1637.4	1745.4	1857.2
	s		1.7808	1.8396	1.8918	1.9392	1.9829	2.0235	2.0618	2.0978	2.1321	2.1648	2.2243	2.2834
40 (267.25)	v		11.040	12.628	14.168	15.688	17.198	18.702	20.20	21.70	23.20	24.69	27.68	30.86
	h		1186.8	1236.5	1234.8	1333.1	1381.9	1431.3	1481.4	1532.4	1584.3	1637.0	1745.1	1857.0
	s		1.6994	1.7006	1.8140	1.8619	1.9058	1.9467	1.9650	2.0212	2.0555	2.0883	2.1498	2.2069
60 (292.71)	v		7.259	8.357	9.403	10.427	11.441	12.449	13.452	14.454	15.453	16.451	18.446	20.44
	h		1181.8	1233.6	1283.0	1331.8	1380.9	1430.5	1480.8	1531.9	1583.3	1636.6	1744.8	1856.7
	s		1.6492	1.7135	1.7678	1.8162	1.8605	1.9015	1.9400	1.9762	2.0106	2.0434	2.1049	2.1621
80 (312.03)	v			6.220	7.020	7.797	8.562	9.322	10.077	10.830	11.582	12.232	13.830	15.325
	h			1230.7	1281.1	1330.5	1379.9	1429.7	1480.1	1531.3	1583.4	1636.2	1744.5	1856.5
	s			1.6791	1.7346	1.7836	1.8281	1.8694	1.9079	1.9442	1.9787	2.0115	2.0731	2.1303
100 (337.81)	v			4.937	5.589	6.218	6.835	7.446	8.052	8.656	9.259	9.860	11.060	12.268
	h			1227.6	1279.1	1329.1	1378.2	1428.9	1479.5	1530.8	1582.9	1635.7	1744.2	1866.2
	s			1.6518	1.7085	1.7581	1.8029	1.8443	1.8829	1.9193	1.9538	1.9867	2.0484	2.1056
120 (341.25)	v			4.081	4.636	5.165	5.683	6.195	6.702	7.207	7.710	8.212	9.214	10.213
	h			1224.4	1277.2	1327.7	1377.8	1428.1	1478.8	1530.2	1582.4	1635.3	1743.9	1856.0
	s			1.6287	1.6869	1.7370	1.7822	1.8237	1.8625	1.8990	1.9335	1.9663	2.0281	2.0664
	v			3.468	3.954	4.413	4.861	5.301	5.738	6.172	6.604	7.035	7.895	8.752

Temp. °F

Press. (Sat. Temp.)												
140 (353.02)	v											
	h	1221.1	1275.2	1326.4	1376.8	1427.3	1476.2	1529.7	1581.9	1634.9	1743.5	1855.7
	s	1.6087	1.6683	1.7190	1.7645	1.8063	1.8151	1.8817	1.9163	1.9493	2.0110	2.0683
160 (363.53)	v	3.008	3.443	3.849	4.244	4.631	5.015	5.396	5.775	6.152	6.906	7.656
	h	1217.6	1273.1	1325.0	1375.7	1426.4	1477.5	1529.1	1581.4	1634.5	1743.2	1855.5
	s	1.5908	1.6519	1.7033	1.7491	1.7911	1.8301	1.8667	1.9014	1.9344	1.9962	2.0535
180 (373.06)	v	2.649	3.044	3.411	3.764	4.110	4.452	4.792	5.129	5.466	6.136	6.804
	h	1214.0	1271.0	1323.5	1374.7	1425.6	1476.8	1528.6	1581.0	1634.1	1742.9	1855.2
	s	1.5745	1.6373	1.6894	1.7355	1.7776	1.8167	1.8534	1.8882	1.9212	1.9831	2.0404
200 (381.79)	v	2.361	2.726	3.040	3.380	3.693	4.002	4.309	4.613	4.917	5.521	6.123
	h	1210.3	1268.9	1322.1	1373.6	1424.8	1476.2	1528.0	1580.5	1633.7	1742.6	1855.0
	s	1.5594	1.6240	1.6767	1.7232	1.7655	1.8048	1.8415	1.8763	1.9094	1.9713	2.0287
220 (389.86)	v	2.125	2.465	2.772	3.066	3.352	3.634	3.913	4.191	4.467	5.017	5.565
	h	1206.5	1266.7	1320.7	1372.0	1424.0	1475.5	1527.5	1580.0	1633.3	1742.3	1854.7
	s	1.5453	1.6117	1.6652	1.7120	1.7545	1.7939	1.8308	1.8656	1.8987	1.9607	2.0181
240 (397.37)	v	1.9276	2.247	2.533	2.804	3.068	3.327	3.584	3.839	4.093	4.597	5.100
	h	1202.5	1264.5	1319.2	1371.5	1423.2	1474.8	1526.9	1579.6	1632.9	1742.0	1854.5
	s	1.5219	1.6003	1.6546	1.7017	1.7444	1.7839	1.8209	1.8553	1.8889	1.9510	2.0064
260 (404.42)	v		2.062	2.330	2.552	2.827	3.067	3.305	3.541	3.776	4.242	4.707
	h		1262.3	1317.7	1370.4	1422.3	1474.2	1526.3	1579.1	1632.5	1741.7	1854.2
	s		1.5897	1.6447	1.6923	1.7362	1.7748	1.8118	1.8467	1.8799	1.9420	1.9905
280 (411.08)	v		1.9047	2.156	2.392	2.621	2.845	3.066	3.286	3.504	3.938	4.370
	h		1260.0	1316.2	1369.4	1421.5	1473.5	1525.8	1578.6	1632.1	1741.4	1854.0
	s		1.5796	1.6354	1.6834	1.7265	1.7662	1.8033	1.8383	1.8716	1.9337	1.9912
300 (417.33)	v		1.7675	2.005	2.227	2.442	2.652	2.859	3.065	3.269	3.674	4.078
	h		1257.6	1314.7	1368.3	1420.6	1472.8	1525.2	1578.1	1631.7	1741.0	1853.7
	s		1.5701	1.6268	1.6751	1.7184	1.7582	1.7954	1.8305	1.8638	1.9260	1.9835
350 (431.72)	v		1.4923	1.7036	1.8980	2.084	2.366	2.445	2.623	2.798	3.147	3.493
	h		1251.5	1310.9	1365.5	1418.5	1471.1	1523.8	1577.0	1630.7	1740.3	1853.1
	s		1.5481	1.6070	1.6563	1.7002	1.7403	1.7777	1.8130	1.8463	1.9065	1.9662
400 (444.59)	v		1.2851	1.4770	1.6508	1.8161	1.9767	2.134	2.290	2.445	2.751	3.055
	h		1245.1	1304.9	1362.7	1416.4	1469.4	1522.4	1575.8	1629.6	1739.5	1852.5
	s		1.5281	1.5894	1.6398	1.6842	1.7247	1.7623	1.7977	1.8311	1.8936	1.9513

Source: Abridged from Thermodynamic Properties of Steam, by Joseph H. Keenan and Frederick G. Keyes, John Wiley & Sons, New York, 1937.

Table A3 Continued

Temp. °F

Abs press., psi (sat temp)		500	550	600	620	640	660	680	700	800	900	1000	1200	1400	1600
450	v	1.1231	1.2155	1.3005	1.3332	1.3652	1.3967	1.4276	1.4384	1.6074	1.7616	1.8928	2.170	2.443	2.714
(456.28)	h	1238.4	1272.0	1302.8	1314.6	1326.2	1337.5	1348.8	1359.9	1414.3	1467.7	1521.0	1628.6	1738.7	1851.9
	s	1.5095	1.5437	1.5735	1.5845	1.5951	1.6054	1.6153	1.6250	1.6599	1.7106	1.7486	1.8177	1.8803	1.9381
500	v	0.9927	1.0600	1.1591	1.1883	1.2186	1.2478	1.2763	1.3044	1.4405	1.5715	1.7096	1.9504	2.197	2.442
(467.01)	h	1231.3	1266.8	1298.6	1310.7	1322.6	1334.2	1345.7	1357.0	1412.1	1466.0	1519.6	1627.6	1737.9	1851.3
	s	1.4919	1.5280	1.5588	1.5701	1.5810	1.5915	1.6016	1.6115	1.6571	1.6962	1.7363	1.8056	1.8683	1.9262
550	v	0.8852	0.9686	1.0431	1.0714	1.0969	1.1259	1.1533	1.1783	1.3068	1.4241	1.5414	1.7704	1.9957	2.219
(476.94)	h	1223.7	1261.2	1294.3	1306.8	1318.9	1330.8	1342.5	1354.0	1409.9	1464.3	1518.2	1626.6	1737.1	1850.6
	s	1.4751	1.5131	1.5451	1.5568	1.5680	1.5787	1.5890	1.5991	1.6452	1.6868	1.7250	1.7946	1.8575	1.9156
600	v	0.7947	0.8753	0.9463	0.9729	0.9988	1.0241	1.0489	1.0732	1.1899	1.3013	1.4096	1.6208	1.8279	2.033
(486.21)	h	1215.7	1255.5	1289.9	1302.7	1315.2	1327.4	1339.3	1351.1	1407.7	1462.5	1516.7	1625.5	1736.3	1850.0
	s	1.4596	1.4990	1.5323	1.5443	1.5558	1.5667	1.5773	1.5875	1.6342	1.6762	1.7147	1.7846	1.8476	1.9068
700	v		0.7277	0.7934	0.8177	0.8411	0.8639	0.8860	0.9077	1.0108	1.1082	1.2024	1.3853	1.5641	1.7405
(503.10)	h		1243.2	1280.6	1294.3	1307.5	1320.3	1332.8	1345.0	1403.2	1459.0	1515.9	1623.5	1734.8	1848.8
	s		1.4722	1.5064	1.5212	1.5323	1.5449	1.5559	1.5665	1.6147	1.6572	1.6962	1.7666	1.8299	1.8881
800	v		0.6154	0.6779	0.7006	0.7223	0.7433	0.7635	0.7833	0.8763	0.9623	1.0470	1.2066	1.3662	1.5214
(518.23)	h		1229.8	1270.7	1285.4	1299.4	1312.9	1325.9	1338.6	1398.6	1455.4	1511.0	1621.4	1733.2	1847.5
	s		1.4467	1.4863	1.5000	1.5129	1.5250	1.5366	1.5476	1.5972	1.6407	1.6801	1.7510	1.8146	1.8729
900	v		0.5264	0.5873	0.6089	0.6294	0.6491	0.6680	0.6863	0.7716	0.8506	0.9262	1.0714	1.2124	1.3509
(531.98)	h		1215.0	1260.1	1275.9	1290.9	1305.1	1318.8	1332.1	1393.9	1451.8	1508.1	1619.3	1731.6	1846.2
	s		1.4216	1.4653	1.4800	1.4938	1.5066	1.5187	1.5303	1.5814	1.6257	1.6656	1.7371	1.8009	1.8595
1000	v		0.4533	0.5140	0.5350	0.5546	0.5733	0.5912	0.6084	0.6878	0.7604	0.8294	0.9615	1.0893	1.2146
(544.61)	h		1196.3	1248.8	1265.9	1281.9	1297.0	1311.4	1325.3	1389.2	1448.2	1505.1	1617.3	1730.0	1845.0
	s		1.3961	1.4450	1.4610	1.4757	1.4893	1.5021	1.5141	1.5670	1.6121	1.6525	1.7245	1.7886	1.8474
1100	v			0.4532	0.4738	0.4929	0.5110	0.5281	0.5445	0.6191	0.6866	0.7503	0.8716	0.9885	1.1031
(556.31)	h			1236.7	1255.3	1272.4	1288.5	1303.7	1318.3	1384.4	1444.5	1502.2	1615.2	1728.4	1843.8
	s			1.4251	1.4425	1.4583	1.4728	1.4862	1.4989	1.5535	1.5995	1.6406	1.7130	1.7775	1.8262
1200	v			0.4016	0.4222	0.4410	0.4586	0.4752	0.4909	0.5617	0.6250	0.6843	0.7967	0.9046	1.0101
(567.22)	h			1223.5	1243.9	1262.4	1279.6	1295.7	1311.0	1379.3	1440.7	1499.2	1613.1	1726.9	1842.5
	s			1.4052	1.4243	1.4413	1.4568	1.4710	1.4843	1.5409	1.5879	1.6293	1.7025	1.7672	1.8263
	v			0.3174	0.3390	0.3580	0.3753	0.3912	0.4062	0.4714	0.5281	0.5805	0.6789	0.7727	0.8640

1400	k....	......	......	1193.0	1218.4	1240.4	1260.3	1278.5	1295.5	1369.1	1433.1	1493.2	1606.9	1723.7	1840.0
(587.10)	s....	......	......	1.3639	1.3877	1.4079	1.4258	1.4419	1.4567	1.5177	1.5666	1.6093	1.6836	1.7489	1.8063
	v....	......	......	......	0.3733	0.2936	0.3112	0.3271	0.3417	0.4034	0.4553	0.5027	0.5906	0.6728	0.7545
1600	h....	......	......	......	1187.8	1215.2	1238.7	1259.6	1278.7	1358.4	1425.3	1487.0	1604.6	1720.5	1837.5
(604.90)	s....	......	......	......	1.3489	1.3741	1.3952	1.4137	1.4303	1.4964	1.5476	1.5914	1.6669	1.7328	1.7926
	v....	......	......	......	0.2407	0.2597	0.2760	0.2907	0.3502	0.3986	0.4421	0.5218	0.5968	0.6093
1800	h....	......	......	......	1185.1	1214.0	1238.5	1260.3	1347.2	1417.4	1480.8	1600.4	1717.3	1835.0
(621.03)	s....	......	......	......	1.3377	1.3628	1.3855	1.4044	1.4765	1.5301	1.5752	1.6520	1.7185	1.7766
	v....	......	......	......	0.1936	0.2161	0.2337	0.2489	0.3074	0.3532	0.3935	0.4668	0.5252	0.6011
2000	h....	......	......	......	1145.6	1184.9	1214.8	1240.0	1336.5	1409.2	1474.5	1596.1	1714.1	1622.5
(635.82)	s....	......	......	......	1.2945	1.3300	1.3564	1.3782	1.4576	1.5139	1.5602	1.6384	1.7065	1.7886
	v....	......	......	......	......	0.1484	0.1686	0.2294	0.2710	0.3061	0.3678	0.4244	0.4784
2500	h....	......	......	......	......	1132.3	1176.8	1303.6	1387.8	1458.4	1585.3	1706.1	1826.2
(668.13)	s....	......	......	......	......	1.2687	1.3073	1.4127	1.4772	1.5273	1.6088	1.6775	1.7389
	v....	......	......	......	......	......	0.0964	0.1760	0.2159	0.2476	0.3018	0.3505	0.3966
3000	h....	......	......	......	......	......	1060.7	1267.2	1365.0	1441.8	1574.3	1698.0	1819.9
(495.36)	s....	......	......	......	......	......	1.1966	1.3690	1.4439	1.4984	1.5837	1.6540	1.7163
	v....	......	......	......	......	......	......	0.1583	0.1961	0.2288	0.2806	0.3267	0.3703
3286.2	h....	......	......	......	......	......	......	1250.5	1355.2	1434.7	1569.8	1694.6	1817.2
(706.40)	s....	......	......	......	......	......	......	1.3506	1.4309	1.4874	1.5742	1.6452	1.7080
	v....	......	......	......	......	......	0.0806	0.1364	0.1762	0.2058	0.2546	0.2977	0.3381
3500	h....	......	......	......	......	......	780.5	1224.9	1340.7	1424.5	1563.3	1689.8	1813.6
	s....	......	......	......	......	......	0.9515	1.3241	1.4127	1.4723	1.5615	1.6336	1.6968
	v....	......	......	......	......	......	0.0287	0.1062	0.1462	0.1743	0.2192	0.2581	0.2943
4000	h....	......	......	......	......	......	763.8	1174.8	1314.4	1406.8	1562.1	1681.7	1807.2
	s....	......	......	......	......	......	0.9347	1.2757	1.3827	1.4482	1.5417	1.6154	1.6795
	v....	......	......	......	......	......	0.0276	0.0798	0.1226	0.1500	0.1917	0.2273	0.2602
4500	h....	......	......	......	......	......	753.5	1113.9	1286.5	1388.4	1540.8	1673.5	1800.9
	s....	......	......	......	......	......	0.9235	1.2204	1.3529	1.4253	1.5235	1.5990	1.6640
	v....	......	......	......	......	......	0.0268	0.0593	0.1036	0.1302	0.1696	0.2027	0.2329
5000	h....	......	......	......	......	......	746.4	1047.1	1256.5	1369.5	1629.5	1665.2	1794.5
	s....	......	......	......	......	......	0.9152	1.1622	1.3231	1.4034	1.5066	1.5839	1.6499
	v....	......	......	......	......	......	0.0262	0.0463	0.0880	0.1143	0.1516	0.1825	0.2106
5500	h....	......	......	......	......	......	741.3	985.0	1224.1	1349.3	1518.2	1657.0	1788.1
	s....	......	......	......	......	......	0.9090	1.1093	1.2930	1.3821	1.4908	1.5699	1.6309

Source: Abridged from *Thermodynamic Properties of Steam*, by Joseph H. Keenan and Frederick G. Keyes, John Wiley & Sons, New York, 1937.

Table A4   Enthalpy of Compressed Water

p (t Sat.)	0				580 (467.13)				1000 (544.75)			
t	v	u	h	s	v	u	h	s	v	u	h	s
Sat.					.019748	447.70	449.53	.64904	.021591	538.39	542.38	.743
32	.016022	-.01	-.01	-.00003	.015994	.00	1.49	.00000	.015967	.03	2.99	.000
50	.016024	18.06	18.06	0.3687	.015998	18.02	19.50	.03599	.015972	17.99	20.94	.035
100	.016130	68.05	68.05	.12963	.016106	67.87	69.36	.12932	.016082	67.70	70.68	.129
150	.016343	117.95	117.95	.21504	.016318	117.66	119.17	.21457	.016293	117.38	120.40	.214
200	.016635	168.05	168.05	.29402	.016608	167.65	169.19	.29341	.016580	167.28	170.32	.292
250	.017003	218.32	218.32	.36777	.016972	217.99	219.56	.36702	.016941	217.47	220.61	.366
300	.017453	269.61	269.61	.43732	.017416	268.92	270.53	.43641	.017379	268.24	271.46	.435
350	.018000	321.59	321.59	.50359	.017954	320.71	322.37	.50249	.017909	319.83	323.15	.501
400	.018668	374.85	374.85	.56740	.018608	373.68	375.40	.56604	.018550	372.55	375.98	.564
450	.019503	429.96	429.96	.62970	.019420	428.40	430.19	.62798	.019340	426.89	430.47	.626
500	.02060	488.1	488.1	.6919	.02048	485.9	487.8	.6896	.02036	483.5	487.5	.68
510	.02087	500.3	500.3	.7046	.02073	497.9	499.8	.7021	.02060	495.6	499.4	.695
520	.02116	512.7	512.7	.7173	.02190	530.1	512.0	.7146	.02086	507.6	511.5	.712
530	.02148	525.5	525.5	.7303	.02130	522.6	524.5	.7273	.02114	519.9	523.8	.724
540	.02182	538.6	538.6	.7434	.02162	535.3	537.3	.7402	.02144	532.4	536.3	.737
550	.02221	552.1	552.1	.7569	.02198	548.4	550.5	.7532	.02177	545.1	549.2	.749
560	.02265	566.1	566.1	.7707	.02237	562.0	564.0	.7666	.02213	558.3	562.4	.763
570	.02315	580.8	580.8	.7851	.02281	576.0	578.1	.7804	.02253	571.8	576.0	.776
580					.02332	590.8	592.9	.7946	.02298	585.9	590.1	.789
590					.02392	606.4	608.6	.8096	.02349	600.6	604.9	.894
600									.02409	616.2	620.6	.818
610									.02482	632.9	637.5	.834

p (t Sat.)	1500 (596.39)				2000 (636.00)				2500 (668.31)			
t	v	u	h	s	v	u	h	s	v	u	h	s
Sat.	.023461	604.97	611.48	.80824	.025649	662.40	671.89	.86227	.028605	717.66	730.89	.9130
32	.015939	.05	4.47	.00007	.015912	.06	5.95	.00008	.015885	.08	7.43	.0000
50	.015946	17.95	22.38	.03584	.015920	17.91	23.81	.03575	.015895	17.88	25.23	.0356
100	.016058	67.53	71.99	.12870	.016034	67.37	73.30	.12839	.016010	67.20	74.61	.1280
150	.016268	117.10	121.62	.21364	.016244	116.83	122.84	.21318	.016220	116.56	124.07	.2127
200	.016554	166.87	171.46	.29221	.016527	166.49	172.60	.29162	.016501	166.11	173.75	.2910
250	.016910	216.96	221.65	.36554	.016880	216.46	222.70	.36482	.016851	215.96	223.75	.3641
300	.017343	267.58	272.39	.43463	.017308	266.93	273.33	.43376	.017274	266.29	274.28	.4329
350	.017865	318.98	323.94	.50034	.017822	318.15	324.74	.49929	.017780	317.33	325.56	.4982
400	.018493	371.45	376.59	.56343	.018439	370.38	377.21	.56216	.018386	369.34	377.84	.5609
450	.019264	425.44	430.79	.62470	.019191	424.04	431.14	.62313	.019120	422.68	431.52	.6216
500	.02024	481.8	487.4	.6853	.02014	479.8	487.3	.6832	.02004	478.0	487.3	.681?
510	.02048	493.4	499.1	.6974	.02036	491.4	498.9	.6953	.02025	489.4	498.8	.693?
520	.02072	505.3	511.0	.7096	.02060	503.1	510.7	.7073	.02048	501.0	510.4	.7051
530	.02099	517.3	523.1	.7219	.02085	514.9	522.6	.7195	.02072	512.6	522.2	.7171
540	.02127	529.6	535.5	.7343	.02112	527.0	534.8	.7317	.02098	524.5	534.2	.7292
550	.02158	542.1	548.1	.7469	.02141	539.2	547.2	.7440	.02125	536.6	546.4	.7413
560	.02191	554.9	561.0	.7596	.02172	551.8	559.8	.7565	.02154	548.9	558.8	.7536
570	.02228	568.0	574.2	.7725	.02206	564.6	572.8	.7691	.02186	561.4	571.5	.7659
580	.02269	581.6	587.9	.7857	.02243	577.8	586.1	.7820	.02221	574.3	584.5	.7785
590	.02314	595.7	602.1	.7993	.02284	591.3	599.8	.7951	.02258	587.4	597.9	.7913
600	.02366	610.4	616.9	.8134	.02330	605.4	614.0	.8066	.02300	601.0	611.6	.8043
610	.02426	625.8	432.6	.8281	.02382	620.0	628.8	.8225	.02346	615.0	625.9	.8177
620	.02498	642.5	649.4	.8437	.02443	635.4	644.5	.8371	.02399	629.6	640.7	.8315
630	.02590	660.8	668.0	.8609	.02514	651.9	661.2	.8525	.02459	644.9	656.3	.8459
640					.02603	669.8	679.4	.8691	.02530	661.2	672.9	.8610
650					.02724	690.3	700.4	.8881	.02616	678.7	690.8	.8773
660									.02729	698.4	711.0	.8954
670									.02895	722.1	735.3	.9172

Table A5 Specific Heat, Viscosity, and Thermal Conductivity of Some Common Gases at Atmospheric Pressure[a]

Temp. (°F)	Carbon dioxide C_p	μ	k	Water vapor C_p	μ	k	Nitrogen C_p	μ	k
200	.2162	.0438	.0125	.4532	.0315	.0134	.2495	.0518	.0189
400	.2369	.0544	.0177	.4663	.0411	.0197	.2530	.0608	.0219
600	.2543	.0645	.0227	.4812	.0506	.0261	.2574	.0694	.0249
800	.2688	.0749	.0274	.4975	.0597	.0326	.2624	.0776	.0279
1000	.2807	.0829	.0319	.5147	.0687	.0393	.2678	.0854	.0309
1200	.2903	.0913	.0360	.5325	.0773	.0462	.2734	.0927	.0339
1400	.2980	.0991	.0400	.5506	.0858	.0532	.2791	.0996	.0369
1600	.3041	.1064	.0435	.5684	.0939	.0604	.2846	.1061	.0399
1800	.3090	.1130	.0468	.5857	.1019	.0678	.2897	.1122	.0429
2000	.3129	.1191	.0500	.6019	.1095	.0753	.2942	.1178	.0459

Temp. (°F)	Oxygen C_p	μ	k	Sulfur dioxide C_p	μ	k	Hydrogen chloride C_p	μ	k
200	.2250	.0604	.0186	.1578	.0386	.0074	.1907	.0412	.0113
400	.2332	.0716	.0229	.1704	.0493	.0109	.1916	.0534	.0143
600	.2404	.0823	.0272	.1806	.0595	.0143	.1936	.0655	.0175
800	.2468	.0924	.0313	.1887	.0692	.0175	.1965	.0774	.0209
1000	.2523	.1021	.0352	.1950	.0784	.0205	.2002	.0892	.0245
1200	.2570	.1111	.0389	.1997	.0871	.0234	.2043	.1009	.0283
1400	.2611	.1197	.0425	.2030	.0954	.0261	.2086	.1124	.0327
1600	.2647	.1278	.0460	.2054	.1030	.0286	.2128	.1239	.0364
1800	.2678	.1353	.0492	.2069	.1103	.0310	.2168	.1351	.0407
2000	.2705	.1423	.0523	.2079	.1170	.0332	.2203	.1463	.0452

C_p = gas specific heat, Btu/lb °F; μ = viscosity, lb/ft hr; k = thermal conductivity, Btu/ft hr °F.
[a]From heat transfer considerations, the pressure effect becomes significant above 250 psig and at gas temperatures below 400°F.

Table A6 Specific Heat, Viscosity, and Thermal Conductivity of Air, Flue Gas, and Gas Turbine Exhaust Gases

Temp. (°F)	Air (dry)			Flue gas[a]			Gas turbine exhaust gases[b]		
	C_p	μ	k	C_p	μ	k	C_p	μ	k
200	.2439	.0537	.0188	.2570	.0492	.0174	0.2529	0.0517	0.0182
400	.2485	.0632	.0221	.2647	.0587	.0211	0.2584	0.0612	0.0218
600							0.2643	0.0702	0.0253
800	.2587	.0809	.0287	.2800	.0763	.0286	0.2705	0.0789	0.0287
1000							0.2767	0.0870	0.0321
1200	.2696	.0968	.0350	.2947	.0922	.0358			
1600	.2800	.1109	.0412	.3080	.1063	.0429			
2000	.2887	.1232	.0473	.3190	.1188	.0499			

[a]% vol CO_2 = 12, H_2O = 12, N_2 = 70, O_2 = 6.
[b]% vol CO_2 = 3, H_2O = 7, N_2 = 75, O_2 = 15.

Table A7   Enthalpy of Gases[a]

Temp. (°F)	A	B	C	D
200	34.98	31.85	35.52	33.74
400	86.19	78.57	87.83	83.00
600	138.70	126.57	141.79	133.42
800	192.49	175.77	197.35	184.91
1000	247.56	226.20	254.47	237.52
1400		330.15	372.93	345.77
1800		437.86	496.42	457.82

		% vol				
Gas Type		CO_2	H_2O	N_2	O_2	SO_2
A	Gas turbine exhaust	3	7	75	15	—
B	Sulfur combustion	—	—	81	10	9
C	Flue gas	12	12	70	6	—
D	Dry air	—	—	79	21	—

[a]Enthalpy in Btu/lb at 60°F.

Source: Computed with data from

Table A8 Correlations for Superheated Steam Properties

$C_1 = 80{,}870/T^2$

$C_2 = (-2641.62/T) \times 10^{C_1}$

$C_3 = 1.89 + C_2$

$C_4 = C_3(P^2/T^2)$

$C_5 = 2 + (372420/T^2)$

$C_6 = C_3C_2$

$C_7 = 1.89 + C_6$

$C_8 = 0.21878T - 126{,}970/T$

$C_9 = 2C_6C_7 - (C_3/T)(126{,}970)$

$C_{10} = 82.546 - 162{,}460/T$

$C_{11} = 2C_{10}C_7 - (C_3/T)(162{,}460)$

$v = \{[(C_8C_4C_3 + C_{10})(C_4/P) + 1]C_3 + 4.55504$
$(T/P)\}0.016018$

$H = 775.596 + 0.63296T + 0.000162467T^2 + 47.3635 \log T$
$+ 0.043557\{C_7P + 0.5C_4[C_{11} + C_3(C_{10} + C_9C_4)]\}$

$S = 1/T\{[(C_8C_3 - 2C_9)C_3C_4/2 - C_{11}]C_4/2 + (C_3 - C_7)P\}$
$\times (-0.0241983) - 0.355579 - 11.4276/T + 0.00018052T$
$- 0.253801 \log P + 0.809691 \log T$

where

P = pressure, atm

T = temperature, K

v = specific volume, ft^3/lb

H = enthalpy, Btu/lb

S = entropy, Btu/lb °F

Table A9 Coefficients to Estimate Properties of Dry, Saturated Steam with Equation[a]

$$Y = Ax + B/x + cx^{1/2} + D \ln x + Ex^2 + Fx^3 + G$$

Property	A	B	C
Temperature, °F	-0.17724	3.83986	11.48345
Liquid specific volume, ft³/lb	-5.280126×10^{-7}	2.99461×10^{-5}	1.521874×10^{-4}
Vapor specific volume, ft³/lb			
1–200 psia	-0.48799	304.717614	9.8299035
200–1,500 psia	2.662×10^{-3}	457.5802	-0.176959
Liquid enthalpy, Btu/lb	-0.15115567	3.671404	11.622558
Vaporization enthalpy, Btu/lb	0.008676153	-1.3049844	-8.2137368
Vapor enthalpy, Btu/lb	-0.14129	2.258225	3.4014802
Liquid entropy, Btu/lb °R	-1.67772×10^{-4}	4.272688×10^{-3}	0.01048048
Vaporization entropy, Btu/lb °R	3.454439×10^{-5}	-2.75287×10^{-3}	-7.33044×10^{-3}
Vapor entropy, Btu/lb °R	-1.476933×10^{-4}	1.2617946×10^{-3}	3.44201×10^{-3}
Liquid internal energy, Btu/lb	-0.1549439	3.662121	11.632628
Vapor internal energy, Btu/lb	-0.0993951	1.93961	2.428354

[a]y = property, x = pressure, psia.

Table A9

D	E	F	G
31.1311	8.762969×10^{-5}	-2.78794×10^{-8}	86.594
6.62512×10^{-5}	8.408856×10^{-10}	1.86401×10^{-14}	0.01596
-16.455274	9.474745×10^{-4}	-1.363366×10^{-8}	19.53953
0.826862	-4.601876×10^{-7}	6.3181×10^{-11}	-2.3928
30.832667	8.74117×10^{-5}	-2.62306×10^{-8}	54.55
-16.37649	-4.3043×10^{-5}	9.763×10^{-9}	1,045.81
14.438078	4.222624×10^{-5}	-1.569916×10^{-8}	1,100.5
0.05801509	9.101291×10^{-8}	-2.7592×10^{-11}	0.11801
-0.14263733	-3.49366×10^{-8}	7.433711×10^{-12}	1.85565
-0.06494128	6.89138×10^{-8}	-2.4941×10^{-11}	1.97364
30.82137	8.76248×10^{-5}	-2.646533×10^{-8}	54.56
10.9818864	2.737201×10^{-5}	-1.057475×10^{-8}	1,040.03

Bibliography

BOOKS

Fuels and Combustion

Babcock & Wilcox, *Steam, Its Generation and Use*, 38th ed., New York, 1978.

Combustion Engineering, Combustion-Fossil Power Systems, 3rd ed., Windsor, 1981.

North American, *Combustion Handbook*, 2nd ed., Cleveland, Ohio, 1978.

Spiers, H. M., Technical data on fuel, *World Energy Conference*, 6th ed., London, 1970.

Fluid Flow, Valves, Pumps

ASME, *Flow Meter Computation Handbook*, New York, 1971.

Crane Co., Flow of fluids, Technical Paper 410, New York, 1981.

Fisher Controls Co., *Control Valve Handbook*, 2nd ed., Marshalltown, Iowa, 1977.

Fisher & Porter, *Handbook of Flow Meter Orifice Sizing*, No. 10B9000, Warminster, Penna.

Ingersoll Rand, *Cameron Hydraulic Data*, 16th ed., Woodcliff Lake, N.J., 1981.

Karassik, I. J., *Centrifugal Pump Clinic*, Marcel Dekker, New York, 1981.

Marks, *Standard Handbook for Mechanical Engineers*, 7th ed., McGraw-Hill, New York, 1967.

Masoneilan, *Handbook for Control Valve Sizing*, 3rd ed., Norwood, Mass., 1971.

Miller, R. W., *Flow Measurement Engineering Handbook*, McGraw-Hill, New York, 1983.

Perry, R. H., and C. H. Chilton, *Chemical Engineers' Handbook*, 5th ed., McGraw-Hill, New York, 1974.

Heat Transfer, Boilers, Heat Exchangers

ASME, *Boiler and Pressure Vessel Code*, Secs. 1 and 8, New York, 1983.

ASME/ANSI, *Power Test Code*, steam generating units, PTC 4.1, New York, 1974.

Ganapathy, V., *Applied Heat Transfer*, PennWell Books, Tulsa, Okla., 1982.

Kern, D. Q., *Process Heat Transfer*, McGraw-Hill, New York, 1950.

Roshenow, W. M., and J. P. Hartnett, *Handbook of Heat Transfer*, McGraw-Hill, New York, 1972.

Taborek, Hewitt, and Afgan, *Heat Exchangers, Theory and Practice*, McGraw-Hill, New York, 1981.

TEMA, *Standards of Tubular Exchangers Manufacturers Association*, 6th ed., 1978.

Turner, W. C., and J. F. Malloy, *Thermal Insulation Handbook*, McGraw-Hill, New York, 1981.

Ganapathy, V., *Waste Heat Boiler Deskbook*, Fairmont Press, Atlanta, Georgia, 1991.

Journals

Chemical Engineering, McGraw-Hill, New York.

Heating Piping and Air-Conditioning, Penton/IPC, Chicago.

Heat Transfer Engineering, Hemisphere Publishing, New York.

Hydrocarbon Processing, Gulf Publishing Co., Houston, Tex.

Oil and Gas Journal, PennWell, Tulsa, Okla.

Plant Engineering, Technical Publishing, Barrington, Ill.

Power, McGraw-Hill, New York.

Power Engineering, Technical Publishing Co., Barrington, Ill.

Process Engineering, Morgan Grampian, London.

Glossary

acfh	Actual cubic feet per hour.
acfm	Actual cubic feet per minute, a term used to indicate the flow rate of gases, at any condition of temperature and pressure.
°API	A scale adopted by American Petroleum Institute to indicate the specific gravity of a liquid. Water has a API gravity of 10°API and No. 2 fuel oil, about 35 °API.
ABMA	American Boiler Manufacturers Association.
ASME	American Society of Mechanical Engineers.
ASR	Actual steam rate, a term used to indicate the actual steam consumption of steam turbines in lb/kWh.

413

BHP	Brake horsepower, a term used for power consumption or rating of turbomachinery. This does not include the efficiency of the drive.
Btu	British thermal unit, a term for measuring heat.
cP	Centipoise, a unit for measurement of absolute viscosity.
CR	Circulation ratio, a term used to indicate the ratio by weight of a mixture of steam and water to that of steam in the mixture. A CR of 4 means that 1 lb of steam–water mixture has 1/4 lb of steam and the remainder water.
dB	Decibel, a unit for measuring noise or sound pressure levels.
dBA	Decibel, scale A; a unit for measuring sound pressure levels corrected for frequency characteristics of the human ear.
fps, fpm, fph	Feet per second, minute, and hour; units for measuring the velocity of fluids.
gpm, gph	Volumetric flow rate in gallons per minute or hour.
HHV	Higher heating value or gross heating value of fuels.
HRSG	Heat recovery steam generator.
ID	Inner diameter of tube or pipe.
in. WC	A unit to measure pressure of gas streams, in inches of water column.
kW	Kilowatt, a unit of measurement of power.

LHV   Lower heating value or net heating value of a fuel.

LMTD   Log-mean temperature difference.

ln   Logarithm to base e; natural logarithm.

log   Logarithm to base 10.

M lb/hr   Thousand lb/hr

MM Btu   Million Btu

MW   Molecular weight

NO_x Oxides of nitrogen.

NPSH   Net positive suction head, a term used to indicate
               the effective head in feet of liquid column to avoid
               cavitation. Subscripts r and a stand for required and
               available.

NTU   Number of transfer units; a term used in heat ex-
               changer design.

OD   Outer diameter of tube or pipe.

oz   Ounce.

ozi   Ounces per square inch, a term for measuring fluid
               pressure.

ppm   Parts per million by weight of volume.

psia   Pounds per square inch absolute, a term for indicat-
               ing pressure.

psig   Pounds per square inch gauge, a term for measuring
               pressure.

PWL   Sound power level, a term for indicating the noise
                        generated by a source such as a fan or turbine.

RH   Relative humidity.

SBV, SBW   Steam by volume and by weight in a steam–water
                        mixture, terms used by boiler designers.

scfm, scfh   Standard cubic feet per minute or hour, a unit for
                        flow of gases at standard conditions of temperature
                        and pressure, namely at 70°F and 29.92 in. Hg, or
                        14.696 psia. Sometimes 60°F and 14.696 psia is
                        also used. The ratio of scfm at 70°F to scfm at 60°F
                        is 1.019.

SPL   Sound pressure level, a unit of measurement of
                        noise in decibels.

SSU   Seconds, Saybolt Universal, a unit of kinematic
                        viscosity of fluids.

SVP   Saturated vapor pressure, pressure of water vapor in
                        a mixture of gases.

TSR   Theoretical steam rate, a term indicating the theo-
                        retical consumption of steam to generate a kilowatt
                        of electricity in a turbine in lb/hr.

Conversion Factors

Metric to American, Metric to Metric	American to Metric, American to American

AREA

Metric to American, Metric to Metric	American to Metric, American to American
$1\ mm^2 = 0.00155\ in.^2 = 0.00001076\ ft^2$ $1\ cm^2 = 0.155\ in.^2 = 0.001076\ ft^2$ $1\ m^2 = 1550\ in.^2 = 10.76\ ft^2$	$1\ in.^2 = 645.2\ mm^2 = 6.452\ cm^2$ $\qquad = 0.0006452\ m^2$ $1\ ft^2 = 92,903\ mm^2 = 929.03\ cm^2$ $\qquad = 0.0929\ m^2$ $1\ acre = 43,560\ ft^2$ $1\ circular\ mil = 0.7854\ square\ mil$ $\qquad = 5.067 \times 10^{-10}\ m^2 = 7.854 \times 10^{-6}\ in.^2$

DENSITY and SPECIFIC GRAVITY

Metric to American, Metric to Metric	American to Metric, American to American
$1\ g/cm^3 = 0.03613\ lb/in.^3 = 62.43\ lb/ft^3$ $\qquad = 1000\ kg/m^3 = 1\ kg/liter$ $\qquad = 62.43\ lb/ft^3 = 8.345\ lb/U.S.\ gal$ $1\ \mu g/m^3 = 136\ grains/ft^3$ (for particulate pollution) $1\ kg/m^3 = 0.06243\ lb/ft^3$	$1\ lb/in.^3 = 27.68\ g/cm^3 = 27,680\ kg/m^3$ $1\ lb/ft^3 = 0.0160\ g/cm^3 = 16.02\ kg/m^3$ $\qquad = 0.0160\ kg/liter$ Specific gravity relative to water $\quad$ SGW of $1.00 = 62.43\ lb/ft^3$ at 4°C or 39.2 °F[†] Specific gravity relative to dry air $\quad$ SGA of $1.00 = 0.0765\ lb/ft^3$[‡] $\qquad = 1.225\ kg/m^3$ $1\ lb/U.S.\ gal = 7.481\ lb/ft^3 = 0.1198\ kg/liter$ $1\ g/ft^3 = 35.3 \times 10^6\ \mu g/m^3$ $1\ lb/1000\ ft^3 = 16 \times 10^6\ \mu g/m^3$

[†]$62.35\ lb/ft^3$ at 60°F, 15.6°C; 8.335 lb/U.S. gal.
[‡]$0.0763\ lb/ft^3$ for moist air.

Metric to American, Metric to Metric	American to Metric, American to American

ENERGY, HEAT, and WORK

1 cal = 0.003968 Btu

1 kcal = 3.968 Btu = 1000 cal = 4186 J
$\quad$ = 0.004186 MJ

1 J = 0.000948 Btu = 0.239 cal − 1 W sec
$\quad$ = 1 N m = 10^7 erg = 10^7 dyn cm

1 W h = 660.6 cal

3413 Btu = 1 kWh

1 Btu = 0.2929 Whr
$\quad$ = 252.0 cal = 0.252 kcal
$\quad$ = 778 ft lb
$\quad$ = 1055 J = 0.001055 MJ

1 ft lb = 0.1383 kg m = 1.356 J

1 hp hr = 1.98 × 10^6 ft lb

1 therm = 1.00 × 10^5 Btu

1 BHP (boiler horsepower) = 33,475 Btu/hr
$\qquad$ = 8439 kcal/hr = 9.81 kW

HEAT CONTENT and SPECIFIC HEAT

1 cal/g = 1.80 Btu/lb = 4187 J/kg

1 cal/cm^3 = 112.4 Btu/ft^3

1 kcal/m^3 = 0.1124 Btu/ft ^{3} = 4187 J/m^3

1 cal/g °C = 1 Btu/lb °F = 4187 J/kg K

1 Btu/lb = 0.5556 cal/g = 2326 J/kg

1 Btu/ft^3 = 0.00890 cal/cm^3
$\qquad$ = 8.899 kcal/m^3 = 0.0373 MJ/m^3

1 Btu/U.S. gal = 0.666 kcal/liter

1 Btu/lb °F = 1 cal/g °C = 4187 J/kg K

HEAT FLOW, POWER

1 N·m/s = 1 W = 1 J/s
$\quad$ = 0.001341 hp = 0.7376 ft lb/sec

1 kcal/h = 1.162 J/s = 1.162 W
$\quad$ = 3.966 Btu/hr

1 kW = 1000 J/s = 3413 Btu/hr
$\quad$ = 1.341 hp

1 hp = 33 000 ft lb/min = 550 ft lb/sec
$\quad$ = 745.7 W = 745.7 J/s
$\quad$ = 641.4 kcal/h

1 Btu/hr = 0.2522 kcal/h
$\quad$ = 0.0003931 hp
$\quad$ = 0.2931 W = 0.2931 J/s

HEAT FLUX and HEAT TRANSFER COEFFICIENT

1 cal/cm^2·s = 3.687 Btu/ft^2 sec
$\quad$ = 41.87 kW/m^2

1 cal/cm^2·h = 1.082 W/ft^2 = 11.65 W/m^2

1 kW/m^2 = 317.2 Btu/ft^2 hr

1 kW/m^2°C = 176.2 Btu/ft^2 hr °F

1 Btu/ft^2 sec = 0.2713 cal/cm^{2}s

1 Btu/ft^2 hr = 0.003 153 kW/m^2
$\qquad$ = 2.713 kcal/m^2 h

1 kW/ft^2 = 924.2 cal/cm^2 h

1 Btu/ft^2 hr °F = 4.89 kcal/m^2 h °C

LENGTH

1 mm = 0.10 cm = 0.03937 in.
$\quad$ = 0.003281 ft

1 m = 100 cm = 1000 mm = 39.37 in.
$\quad$ = 3.281 ft

1 km = 0.6214 mile

1 in. = 25.4 mm = 2.54 cm = 0.0254 m

1 ft = 304.8 mm = 30.48 cm = 0.3048 m

1 mile = 5280 ft

1 μm = 10^{-6} m

1 Angstrom unit = 1 Å = 10^{-10} m

PRESSURE

1 N · m^2 = 0.001 kPa = 1.00 Pa

1 mm H$_2$O = 0.0098 kPa

1 mm Hg = 0.1333 kPa = 13.60 mm H$_2$O
$\quad$ = 1 torr = 0.01933 psi

1 kg/cm^2 = 98.07 kPa = 10,000 kg/m^2

1 in. H$_2$O = 0.2488 kPa = 25.40 mm H$_2$O
$\quad$ = 1.866 mm Hg
$\quad$ = 0.00254 kg/cm^2 = 2.54 g/cm^2

1 in. Hg = 3.386 kPa = 25.40 mm Hg
$\quad$ = 345.3 mm H$_2$O = 13.61 in. H$_2$O

Metric to American, Metric to Metric	American to Metric, American to American

Metric to American, Metric to Metric:

= 10,000 mm H_2O = 394.1 in. H_2O
= 735.6 mm Hg = 28.96 in. Hg
= 227.6 oz/in.^{2} = 14.22 psi
= 0.9807 bar
1 bar = 100.0 kPa = 1.020 kg/cm^2
= 10,200 mm H_2O = 401.9 in. H_2O
= 750.1 mm Hg = 29.53 in. Hg
= 232.1 oz/in.^{2} = 14.50 psi
= 100,000 N/m^2
1 g/cm^2 = 0.014 22 psi
= 0.2276 oz/in.^{2}
= 0.3937 in. H_2O
(For rough calculations,
1 bar = 1 atm = 1 kg/cm^2
= 10 m H_2O = 100 kPa)

American to Metric, American to American:

1 psi
= 7.858 oz/in.^{2} = 0.491 psi
= 25.4 torr
= 6.895 kPa = 6895 N/m^2
= 703.1 mm H_2O = 27.71 in. H_2O
= 51.72 mm Hg = 2.036 in. Hg
= 16.00 oz/in.^{2}
= 0.0703 kg/cm^2 = 70.31 g/cm^2
= 0.068 97 bar

1 oz./in.^{2}
= 0.4309 kPa
= 43.94 mm H_2O = 1.732 in. H_2O
= 3.232 mm Hg
= 0.004 39 kg/cm^2 = 4.394 g/cm^2

1 atm†
= 101.3 kPa = 101,325 N/m^2
= 10,330 mm H_2O = 407.3 in. H_2O
= 760.0 mm Hg = 29.92 in. Hg
= 235.1 oz/in.^{2} = 14.70 psi
= 1.033 kg/cm^2
= 1.013 bar

TEMPERATURE

$°C = 5/9 \, (°F - 32)$
$°F = (9/5°C) + 32$
$K = °C + 273.15$
$°R = °F + 459.67$

THERMAL CONDUCTIVITY

1 W/m K = 0.5778 Btu ft/ft^2 hr °F
= 6.934 Btu in./ft^2 hr °F
1 cal cm/cm^2·s·°C = 241.9 Btu ft/ft^2 hr °F
= 2903 Btu in./ft^2 hr °F
= 418.7 W/m K

1 Btu ft/ft^2 hr °F = 1.730 W/m K
= 1.488 kcal/m h K
1 Btu in./ft^2 hr °F = 0.1442 W/m K
1 Btu ft/ft^2 hr °F = 0.004139 cal cm/cm^2 s °C
1 Btu in./ft^2 hr °F = 0.0003445 cal cm/cm^2 s °C

THERMAL DIFFUSIVITY

1 m^2/s = 38 760 ft^2/hr
1 m^2/h = 10.77 ft^2/hr

1 ft^2/hr = 0.0000258 m^2/s = 0.0929 m^2/h

VELOCITY

1 cm/s = 0.3937 in./sec = 0.032 81 ft/sec
= 10.00 mm/s = 1.969 ft/min

1 in./sec = 25.4 mm/s = 0.0254 m/s
= 0.0568 mph

†Normal atmosphere = 760 torr (mm Hg at 0°C)—not a "technical atmosphere," which is 736 torr or 1 kg/cm^2. Subtract about 0.5 psi for each 1000 ft above sea level.

Metric to American, Metric to Metric	American to Metric, American to American

VELOCITY (*continued*)

1 m/s $= 39.37$ in./sec $= 3.281$ ft/sec $= 196.9$ ft/min $= 2.237$ mph $= 3.600$ km/h $= 1.944$ knot	1 ft/sec $= 304.8$ mm/s $= 0.3048$ m/s $= 0.6818$ mph 1 ft/min $= 5.08$ mm/s $= 0.00508$ m/s $= 0.0183$ km/h 1 mph $= 0.4470$ m/s $= 1.609$ km/h $= 1.467$ ft/sec 1 knot $= 0.5144$ m/s 1 rpm $= 0.1047$ radian/sec

VISCOSITY, absolute, μ

0.1 Pa $\cdot$ s $= 1$ dyne s/cm^2 $= 360$ kg/h m $= 1$ poise $= 100$ centipoise (cP) $= 242.1$ lb$_m$/hr ft $= 0.002089$ lb/$_f$ sec/ft^2 1 kg/h $\cdot$ m $= 0.672$ lb/hr/ ft $= 0.00278$ g/s cm $= 0.00000581$ lb$_f$ sec/ft^2 μ of water[†] $= 1.124$ cP $= 2.72$ lb$_m$/hr ft $= 2.349 \times 10^{-5}$ lb sec/ft^2	1 lb$_m$/hr $\cdot$ ft $= 0.000008634$ lb$_f$ sec/ft^2 $= 0.413$ cP $= 0.000413$ Pa $\cdot$ s 1 lb$_f$ sec/ft^2 $= 115,800$ lb$_m$/hr ft $= 47,880$ cP $= 47.88$ Pa s 1 reyn $= 1$ lb$_f$ sec/in.^{2} $= 6.890 \times 10^6$ cP μ of air[†] $= 0.0180$ cP $= 0.0436$ lb/hr ft $= 3.763 \times 10^{-7}$ lb sec/ft

VISCOSITY, kinematic, ν

1 cm^2/s $= 0.0001$ m^2/s $= 1$ stokes $= 100$ centistokes (cS) $= 0.001076$ ft^2/sec $= 3.874$ ft^2/hr 1 m^2/s $= 3600$ m^2/h $= 38,736$ ft^2/hr $= 10.76$ ft^2/sec ν of water[†] $= 1.130$ centistokes $= 32$ SSU $= 1.216 \times 10^{-5}$ ft^2/sec	1 ft^2/sec $= 3600$ ft^2/hr $= 92,900$ cS $= 0.0929$ m^2/s 1 ft^2/h $= 0.000278$ ft^2/sec $= 25.8$ cS $= 0.0000258$ m^2/s ν of air[†] $= 14.69$ cS $= 1.581 \times 10^{-4}$ ft^2/sec

VOLUME

1 cm^3 (cc) $= 0.000\ 001\ 00$ m^3 $= 0.0610$ in.^{3} $= 0.0338$ U.S. fluid oz. 1 liter (dm^3) $= 0.0010$ m^3 $- 1000$ cm^3 $= 61.02$ in.^{3} $= 0.03531$ ft^3 $= 0.2642$ U.S. gal	1 in.^{3} $= 16.39$ cm^3 $= 0.0001639$ m^3 $= 0.01639$ liter 1 ft^3 $= 1728$ in.^{3} $= 7.481$ U.S. gal $= 6.229$ Br gal $= 28,320$ cm^3 $= 0.02832$ m^3 $= 28.32$ liters $= 62.427$ lb of $39.4°$F ($4°$C) water $= 62.344$ lb of $60°$F ($15.6°$C) water

[†]Viscosity at STP.

Metric to American, Metric to Metric	American to Metric, American to American

$1\ m^3$ = 1000 liter = $1 \times 10^6\ cm^3$
         = 61,020 in.^{3} = 35.31 ft^3
         = 220.0 Br gal
         = 6.290 bbl
         = 264.2 U.S. gal
         = 1.308 yd^3

1 U.S. gal = 3785 cm^3 = 0.003785 m^3
            = 3.785 liters = 231.0 in.^{3}
            = 0.8327 Br gal = 0.1337 ft^3
            = sp gr $\times$ 8.335 lb
            = 8.335 lb of water
            = 1/42 barrel (oil)
1 Br gal = 277.4 in.^{3}
          = 0.004 546 m^3 = 4.546 liters
          = 1.201 U.S. gal
1 bbl, oil = 9702 in.^{3} = 5.615 ft^3
            = 0.1590 m^3 = 159.0 liters
            = 42.00 U.S. gal
            = 34.97 Br gal

VOLUME FLOW RATE

$1\ cm^3/s$ = $1 \times 10^{-6}\ m^3/8$
$1\ liter/s$ = $1 \times 10^{-3}\ m^3/s$
$1\ m^3/h$ = 4.403 U.S. gpm (gal/min)
         = 0.5887 ft^3/min

1 gpm (gal/min) = 60.0 gph (gal/hr)
                 = 0.01667 gps (gal/sec)
                 = 0.00223 cfs (ft^3/sec)
                 = 0.1337 cfm (ft^3/min)
                 = 0.8326 Br gpm
                 = 0.227 m^3/h
                 = 1.429 bbl/hr
                 = 34.29 bbl/day
1 gph (gal/hr) = 0.00105 liter/s
               = 0.000037 1 cfs (ft^3/sec)
1 cfm (ft^3/min) = 6.18 Br gpm
                  = 0.000471 m^3/s
1 cfs (ft^3/sec) = 448.8 gpm
                  = 22,250 Br gph

WEIGHT, FORCE, MASS

1 g = 0.03527 oz avdp mass
1 kg mass = 1000 g mass
          = 35.27 oz. avdp mass
          = 2.205 lb avdp mass
1 kg force = 1000 g force = 9.807 N
           = 2.205 lb avdp force
1 metric ton = 1000 kg = 2205 lb

1 oz avdp mass = 28.35 g = 0.02835 kg
1 lb avdp mass = 453.6 g = 0.4536 kg
              = 4.536 $\times$ 10$^6\mu$g
1 lb avdp force = 0.4536 kg force
               = 4.448 N
1 lb = 7000 grains
1 short ton = 2000 lb = 907.2 kg
1 long ton = 2240 lb = 1015.9 kg

Index

Printed in the United States
by Baker & Taylor Publisher Services

Printed in the United States
by Baker & Taylor Publisher Services